# Deltas of the World

**Series Editor**

Abdelazim Negm, Zagazig, Egypt

Worldwide, over 600 million people live in deltas. Extensive human activities in deltas, their supplying rivers, and related maritime currents increase the pressure on deltas' natural resources and habitat equilibrium, leading to degradation of the available resources and jeopardizing inhabitants' livelihoods – and in some cases even their lives.

This book series is intended to promote a sustainable future for deltas around the globe and the communities that live in them. Accordingly, we publish studies focusing on various aspects of deltas, including hydrogeology, soil pollution, water pollution, environmental hydrology, environmental hydraulics, groundwater engineering and management, morphology, anthropology, socioeconomics, wastewater management, water resources engineering and management, agriculture, greenhouse agriculture, irrigation and drainage engineering, hydraulic structures, geophysics and geology, sedimentology, mineralogy, remote sensing and GIS studies, climate change variabilities and impacts, marine-delta interactions and sustainability, and solid waste management. The findings presented here support the SDGs and Agenda 2030 with regard to helping countries with deltas and related environments to preserve their natural resources for future generations.

The editors invite scientists, researchers and scientific communities to contribute to the series by submitting delta atlases, edited books, monographs, and high-quality conference proceedings focused on deltas around the world.

Mahmoud Nasr · Balasubramani Ravindran
Editors

# Solid Waste Management in Delta Region for SDGs Fulfillment

## Delta Sustainability by Waste Management

Springer

*Editors*
Mahmoud Nasr
Department of Environmental Engineering
Egypt-Japan University of Science
and Technology (E-JUST)
Alexandria, Egypt

Balasubramani Ravindran
Department of Environmental Energy
and Engineering
Kyonggi University
Suwon-si, Korea (Republic of)

ISSN 2731-832X ISSN 2731-8338 (electronic)
Deltas of the World
ISBN 978-3-031-58252-3 ISBN 978-3-031-58253-0 (eBook)
https://doi.org/10.1007/978-3-031-58253-0



This Springer imprint is published by the registered company Springer Nature Switzerland AG
The registered company address is: Gewerbestrasse 11, 6330 Cham, Switzerland

# Contents

1 **Solid Waste Management and Sustainability in Deltas: Introduction and Updates** . . . . . 1
Mahmoud Nasr

2 **Towards Sustainable Delta Ecosystems: Pollution Mitigation for Achieving SDGs in Indian Delta Region** . . . . . 27
Anita Gupta, Saroj Kanta Barik, and Puneet Singh Chauhan

3 **Sustainable Solid Waste Disposal to Mitigate Water Pollution Problem and Its Social Environmental Impact** . . . . . 47
Shyamal Datta, Mouli Mitra, and Subhasis Roy

4 **Integrated Waste Management System for Environment Protection in Deltas** . . . . . 67
Leila Soudani, Meriem Chafaa, Moulkheir Selmani, Mohamed Islem Bouacha, Omar Safa, Mhamed Maatoug, and Hermann Heilmeier

5 **Eco-utilization and Sustainable Management of Solid Wastes of Steel Industries** . . . . . 101
Trinath Biswal and Pratap Kumar Swain

6 **Valorization of Deep Eutectic Solvent Pretreated Lignocellulosic Biomass for Improved Biogas Production** . . . . . 127
Akinola David Olugbemide, Ikhazuagbe Hilary Ifijen, Chandan Mahata, Filipa A. Vicente, and Blaž Likozar

7 **Development of Novel Sustainable Waste-Based Adsorbents for Water Treatments** . . . . . 155
Oscar E. Medina, Yurany Villada Villada, Farid B. Corteś, and Camilo A. Franco

**8 Integrated Waste Management System for Environmental Protection and Sustainability of Resources** ........ 187
Yahya Jani and Monica Odlare

**9 Evaluation of the Water Quality of the Jucu/ES (Brazil) River, Through Physical–Chemical and Biological Parameters in Favor of the Sustainability** ........ 215
Taina Turial da Silva, Geisamanda Pedrini Brandão Athayde, Maria Tereza Weitzel Dias Carneiro, and Josimar Ribeiro

**10 Solid Waste Management and Sustainability in Deltas: Concluding Remarks and Outlook** ........ 245
Mahmoud Nasr

# Solid Waste Management and Sustainability in Deltas: Introduction and Updates 1

Mahmoud Nasr

**Abstract**

Deltas, comprising approximately 5% of the Earth's land surface, occupy a high population density globally. The distribution of population density depends on the geographical location and the countries or regions of Deltas under investigation. Each Delta is unique and influenced by specific hydrological, geological, and environmental conditions. Deltas face the pressing need for effective waste management strategies to mitigate environmental degradation and health risks. This chapter is an introduction to the critical nexus of solid waste management (SWM) and sustainability in Deltas, shedding light on the corresponding challenges and barriers. The chapter's main findings would strengthen the contribution of stakeholders, policy-makers, and private and public sectors in achieving sustainable SWM strategies.

## 1.1 Introduction

A river delta refers to the area of land formed at the mouth of a river where the river splits into several channels (i.e., a landform shaped like a triangle) (Elsadek et al. 2024). They are characterized by a network of distributaries, which are channels that branch off from the main river and distribute its water and sediment into the surrounding area. Deltas

M. Nasr (✉)
Environmental Engineering Department, Egypt-Japan University of Science and Technology (E-JUST), Alexandria, Egypt
e-mail: mahmmoudsaid@gmail.com; mahmoud-nasr@alexu.edu.eg; mahmoud.nasr@ejust.edu.eg

Sanitary Engineering Department, Faculty of Engineering, Alexandria University, Alexandria, Egypt

 
M. Nasr and B. Ravindran (eds.), *Solid Waste Management in Delta Region for SDGs Fulfillment*, Deltas of the World, https://doi.org/10.1007/978-3-031-58253-0_1

occupy urban areas, industries, infrastructure, farmlands, and aquaculture ponds (Youssef et al. 2024). The compacted sediment layers of Deltas can be used to extract hydrocarbon (natural gas, oil, and coal) and harvest salts, providing crucial economic significance to the country. Table 1.1 lists the major Deltas worldwide.

During the past decade, most of the world's Deltas experienced severe pollution from various anthropogenic practices associated with illegal solid waste disposal (Deng and Liu 2024). There is a great interest in understanding the interaction of physical, ecological, and social variations within delta systems in relation to human waste disposal. Hence, this chapter discusses the commonly performed reliable waste management practices in Deltas, including (i) implementing separation-at-source initiatives that sort recyclable items into their respective recycling streams, (ii) providing beneficial solutions to the disposal of organic waste materials, such as garden waste, food waste, wood waste, and other manufactured biodegradables, (iii) reducing the percentage of waste that is sent to landfill

**Table 1.1** Largest Deltas worldwide and their features

| Delta name | Location | Notable features |
|---|---|---|
| Ganges–Brahmaputra Delta | India | The largest delta, highly fertile, supports the highest population density |
| Mekong Delta | Southeastern Asia | Known as the Nine Dragon River Delta, rich in biodiversity, feeds the Vietnamese population |
| Lena Delta | Russia | Frozen tundra turned lush wetland, supports trade and breeding of fish |
| Huang He Delta | China | Delta's location changes, engineering to protect oil and gas wells |
| Mississippi Delta | United States | Cultural significance, shrinking due to rising sea level |
| Indus Delta | Pakistan | Suffering from loss of vegetation, shrinking mangrove forests |
| Volga Delta | Russia | Europe's largest delta, vital for migratory birds and fish |
| Orinoco Delta | Venezuela, South America | Uninhabited, national park with diverse wildlife |
| Niger Delta | Nigeria, Western Africa | Rich in marine life, historical importance, supports agriculture and fishery |
| Tigris-Euphrates Delta | Southeastern Asia | Drains rivers across 5 countries, birthplace of civilizations |

(a bottom liner and leachate collection system are required), and (iv) promoting a green approach to business and delivering maximum value to our clients.

## 1.2 Bibliometric Analysis of Solid Waste Sustainability

The bibliometric study analyzes data on the published articles about "Waste", "Sustainability", and "Deltas" searched within the Article title, Abstract, and Keywords. The results were analyzed according to the number of articles per year, journal titles, subject areas, affiliation, country, and citation analysis. The SCOPUS search engine was used in this bibliometric study. Table 1.2 lists the essential parameters and results of this search strategy in the 2010–2023 period. The published articles address the main waste management and disposal, pollution reduction from waste, and the associated toxicological features. The categories of these articles include:

- strategy and management,
- general environmental science, and
- renewable energy, sustainability, and the environment

Subject areas include, but are not limited to:

- sources and occurrences of pollutants related to solid waste generation in the Delta region.
- synergism of contaminant exposure and biological, ecological, and human health impacts.
- contaminants of emerging concerns, such as antibiotic-resistant microorganisms or genes, microplastics/nanoplastics, and electronic wastes (Demarema et al. 2024)
- remediation/mitigation of environmental contaminants via innovative techniques (Dogu et al. 2021)
- modeling of pollution processes, patterns, or trends, demonstrating environmental occurrences, transport, and behavior.
- synthesis/fabrication of novel materials used for remediation and/or mitigation of pollution
- waste management policy, education, economic assessment (Life cycle cost, LCC)), and environmental assessments (life cycle assessment, LCA and social-LCA). For instance, LCA is a method for the environmental assessment of products and services, covering their life cycle from raw material extraction to waste treatment.
- technical, political, and environmental concerns associated with material cycles and waste management.
- studies on eutrophication and secondary pollution by eutrophication caused by the disposal of solid wastes into the aquatic environment

**Table 1.2** Bibliometric analysis data on solid waste sustainability in Deltas

| Parameter | Data/observation |
|---|---|
| Database | SCOPUS |
| Query string | TITLE-ABS-KEY (waste AND sustainability AND deltas) |
| Period time | Year of publication 2010–2023 |
| Language | English |
| Number of documents | 66 document results |
| Document type | Article (71.2%)<br>Conference review (10.6%)<br>Conference Paper (9.1%)<br>Review (4.6%)<br>Book Chapter (4.5%) |
| Top 10 subject area | Environmental Science (34)<br>Engineering (20)<br>Energy (10)<br>Earth and Planetary Sciences (8)<br>Agricultural and Biological Sciences (6)<br>Business, Management and Accounting (4)<br>Computer Science (4)<br>Materials Science (4) |
| Top 10 journals | IOP Conference Series Earth and Environmental Science (3)<br>Science of the Total Environment (3)<br>Buildings (2)<br>Construction and Building Materials (2)<br>Environmental Pollution (2)<br>Journal of Cleaner Production (2)<br>Water Environment Research (2)<br>Advanced Materials Research (1)<br>Advances in Global Change Research (1)<br>African Journal of Hospitality Tourism and Leisure (1) |
| Top 5 affiliations | ARENA Research for Sustainable Resources (2)<br>Ministry of Education of the People's Republic of China (2)<br>Nagasaki University (2)<br>RUDN University (2)<br>Chinese Academy of Sciences (2) |

(continued)

**Table 1.2** (continued)

| Parameter | Data/observation |
|---|---|
| Top 5 countries | China (15)<br>Nigeria (9)<br>United States (8)<br>Egypt (7)<br>Viet Nam (5) |
| Top 5 funding sponsors | National Natural Science Foundation of China (8)<br>Fundamental Research Funds for the Central Universities (2)<br>Ministry of Higher Education, Malaysia (2)<br>Natural Science Foundation of Guangdong Province (2)<br>Anhui Provincial Department of Science and Technology (1) |

The main objectives and aims of these articles are to investigate:

- socio-economic benefits and policy implications of generating sustainable energy from municipal solid waste (MSW).
- tracking the carbon flows in municipal waste management: Paving the Way for Sustainable Development
- resource utilization of drinking water treatment aluminum sludge in green cementing material production.
- landfill site selection for sustainable solid waste management (SWM)
- carbon reduction trade-off between bioenergy generation and that spent on collection, transport, and treatment.
- environmental and economic trade-off-based approaches towards biogas recovery from solid waste.
- the avoidance of the detrimental effects of toxic chemicals and materials on the receiving environmental dimensions (soil, water, and atmosphere)
- sustainability assessment of biofuel and value-added product from organic portion of solid waste (Emmanouilidou et al. 2023)
- valorization of cattle manure via a thermo-chemical process to produce environmentally friendly products (Mannaa et al. 2024)
- sustainable energy recovery from mixed agri-food waste as a feedstock for bio-circular-green economy (Blanchard et al. 2023)
- application of waste biomass for the production of biofuels and catalysts
- life cycle environmental benefits, and technology development and challenges of recycling waste liquor and chemicals in the recovery of bioethanol
- zero-waste approaches for sustainable production and biorefineries of microalgal biomass

**Table 1.3** Characterization methods of solid wastes

| Characterization method and/or equipment | Role in defining the biomass features |
|---|---|
| Scanning electron microscope | Scanning electron microscope (SEM) is one of the common methods for imaging the microstructure and morphology of the materials |
| Energy-dispersive X-ray spectroscopy | It is a method of elemental detection via the collection of characteristic X-rays induced by electron beam radiation |
| Transmission electron microscopy | It is a technique of imaging the internal structure of solids (etc., morphology, and size) using a beam of high-energy electrons transmitted through the solid |
| Thermogravimetric analysis (TGA) | Is a powerful technique for the measurement of thermal stability of materials including solid wastes |
| Fourier transform infrared spectroscopy | It is used to characterize the different functional groups such as lipids, proteins, nucleic acids, and polysaccharides present on the solid waste surface |
| X-ray photoelectron spectroscopy (XPS) | It is a quantitative technique used to analyze the solid waste surface chemistry, e.g., determining the surface composition (elemental identification) |

The chapters' outputs are important for a multidisciplinary and diverse audience of scientists, policymakers, the broad public, and the environment-caring stakeholders.

## 1.3 Solid Waste Characterization Methods

The proper characterization techniques of this biomass comprise pore-size distribution, ash fusion and deformation temperatures, calorific values, and thermogravimetric analysis (Table 1.3) (Palansooriya et al. 2023). These characterization methods would support the choice of specific biomass for the most appropriate implementation ($H_2$, $CH_4$, or ethanol production) (Srimalanon and Kachapongkun 2023).

## 1.4 Physical, Thermochemical and Biological Waste-to-Energy Conversion Technologies

The unrestricted disposal of solid wastes severely influences the growth and activities of aquaculture and deteriorates the relationships between zooplankton, phytoplankton, and fish (Lopez et al. 2018). The thermo-chemical transformation methods (e.g., pyrolysis, gasification, and combustion) are used to generate syngas, charcoal, and biofuel (Ahmad et al. 2023). These methods are summarized in Table 1.4.

**Table 1.4** Typical operating conditions for thermochemical and biological waste-to-energy conversion technologies

| Technology | Temperature (°C) | Heating rate (°C/s) | Residence time | Typical product distribution | Note |
|---|---|---|---|---|---|
| Incineration | 800–1200 | Variable | Variable | Bottom ash > Fly ash > Slag > Gas | It is the primary and most conventional thermal waste-to-energy conversion technology (He et al. 2023)<br>Incineration of fossil fuels is considered one of the main sources of GHG emissions into the atmosphere |
| Flash pyrolysis | 800–1000 | 1000 | 0.5 s | Bio-oil > Gas > Biochar | Pyrolysis (devolatilization) is the thermal degradation of an organic substance in the absence of air to produce char, pyrolysis oil, and synthesis gas, e.g., the conversion of wood to charcoal<br>Fast pyrolysis of biomass is an endothermic process, where heat is supplied to the biomass for initiating depolymerization at the reaction temperature |
| Slow pyrolysis | 300–700 | 0.1–1 | 10–100 min | Biochar > Bio-oil > Gas | Slow pyrolysis is characterized by low temperatures, slow heating rates, and long residence times<br>Slow pyrolysis takes place in a temperature range of 300 °C–550 °C and heating rates of 0.01 °C/min–80 °C/min and has a vapor residence time that can last minutes, hours, or even days |
| Intermediate pyrolysis | 500–600 | 2–10 | 10–20 s | 10–20 s | In this process, heating rates are significantly lower than in fast pyrolysis and solids residence times are much longer (few minutes) |
| Liquefaction | 250–450 | 0.05–5 | 1–40 min | Bio-oil > Tar | For example, coal liquefaction is the process of making a liquid fuel from coal |

(continued)

**Table 1.4** (continued)

| Technology | Temperature (°C) | Heating rate (°C/s) | Residence time | Typical product distribution | Note |
|---|---|---|---|---|---|
| Gasification | 700–1000 | Variable | 30–90 min | Syngas > Biochar | It is a high-temperature process in which carbonaceous fuels and organic solid matters are transformed into a gaseous mixture of hydrogen ($H_2$) + carbon monoxide (CO), called synthesis gas (or syngas) further used as a fuel gas |
| Anaerobic digestion | 35–37 | None | Days | Methane > Other gases > Bio-sludge | It is an anaerobic biological process that converts organic matter into biogas, including ≈ 50–70% $CH_4$, 30–50% $CO_2$, and trace amounts of other gases, e.g., $H_2S$, and $NH_3$ |
| Composting | 25–55 | – | 1–2 months | Humus > Nutrients > Minerals | It is a treatment process in which the organic constituents of biosolids are biologically decomposed under controlled aerobic conditions (Isibika et al. 2023) |

### 1.4.1 Physical-Based Conversion Using Cavitation

Cavitation-based technologies can be employed to improve the anaerobic digestion (AD) efficiency, maintaining sustainable management of solid wastes (e.g., food wastes and agricultural residues) (Sonu et al. 2023). In this process, the liquid state is rapidly transformed into the gaseous state as an action of growth and collapse of bubbles. The creation of cavitation bubbles in the liquid phase results in the generation of intense shock waves with tremendous destructive forces. The efficiency of the cavitation process performance is influenced by vapor pressure, viscosity, temperature, surface tension, the existence of dissolved gases, and the presence of other components, such as surfactants. Some physical (mechanical) factors, e.g., shock waves, microjets, and shear stress, also affect the completion of the cavitation process (e.g., lignin depolymerization). Cavitation is classified into hydrodynamic cavitation, vaporous cavitation, gaseous cavitation, and vibrational cavitation, so-called acoustic cavitation according to the conditions of formation and development of cavitation bubbles. The collapse energy of cavitation bubbles destroys material structure, further affecting material characteristics (e.g., particle size, specific surface area, and porosity). The modification in these features facilitates lignocellulosic biomass solubilization as well as biogas productivity. The success of particles' disintegration by cavitation depends on the efficient breakdown of the hydrogen bonds, Van der Waal forces, and covalent bonds of the complex waste structure, further generating simpler byproducts (high-value aromatic monomers) soluble in the medium. The extremely high temperatures and pressures accompanied by cavitation can create a series of chemical reactions, yielding multiple reactive species.

## 1.4.2 Biological-Based Conversion

### 1.4.2.1 Anaerobic Digestion

Anaerobic digestion (AD) is a sustainable management technique employed to convert the organic fraction of solid wastes and agricultural feedstock (e.g., crop residue and cattle manure) into biogas and digestate (Wang et al. 2023). Biogas can be used to generate fuel for vehicles, electricity, and heat, whereas digestate can be employed for producing fertilizers, and composts (Midden et al. 2023). This step is crucial to reduce the landfilling of these organic wastes that are responsible for greenhouse gas (GHG) emissions, and can reduce the value of the surrounding areas and cause soil and water contamination (Heidari-Maleni et al. 2023). In particular, integrating AD into government policies to treat organic wastes would assist developing countries in providing direct financial returns and maintaining environmental conservation. The advantages of this system include the treatment of wastes with lower energy consumption, and less sludge production compared to aerobic-based bioprocesses. However, the disadvantages include the requirement of further treatment to meet discharge standards, insufficient N and P removals, and a higher sensitivity to environmental conditions (temperature and medium pH) fluctuations. However, the absence of expertise in AD technology to design, construct, and operate the technology may delay the biogas production project.

### 1.4.2.2 Composting

Microorganisms perform multiple bio-reactions to facilitate the biological decomposition of biodegradable solid waste under an aerobic condition, describing the composting process (Mengqi et al. 2023). This process yields a compost material rich in nutrients, minerals, and beneficial microorganisms called humus, which can be used to replace chemical fertilizers for crop production (Blanchard et al. 2023). In this process, black soldier fly larvae (*Hermetia illucens*) can sufficiently utilize various types of organic wastes (animal meat and seafood wastes), maintaining a cradle-to-cradle system, where the waste is effectively managed and converted into useful resources (Syarifinnur et al. 2023). However, composting is a slow process because not all organics are readily biodegradable and suitable for composting. The composite material should be free of damaging components, metal ions, and emerging contaminants to avoid the spreading of diseases (Md et al. 2023).

#### 1.4.2.3 Vermicomposting

Vermicomposting is a process that describes the degradation of organic wastes by the action of earthworms together with microorganisms, causing recovery of a humus-like material called vermicompost. The main advantages of vermicomposting include the reduction of the waste volume and stabilization of the organic solid residues (Chowdhury and Sarkar 2023). However, vermicomposting requires an extended period (a treatment period of over six months) to convert the organic matter into usable forms. Moreover, vermicomposting is not efficient in killing pathogens and it releases a very foul and unpleasant odour (Zhang et al. 2023).

### 1.4.3 Thermal-Based Conversion

#### 1.4.3.1 Combustion

The combustion process expresses the burning of biomass in the presence of an air supply, generating heat energy along with $CO_2$ and water (Yadav et al. 2023). Direct combustion utilizes the heat energy of combustion reaction to produce the steam to run the turbine compressor or pump; i.e., a mechanism recognized as heat-to-mechanical energy. However, this process might suffer from the creation of some environmental concerns associated with the release of $CO_2$, $NO_X$, $SO_X$, and $H_2S$ emissions (Tawfik et al. 2024). As such, the operating factors of the combustion process should be optimized using precise numerical models and simulations to avoid the creation of the GHG effect problems (Mohod and Bagal 2023).

#### 1.4.3.2 Incineration

Incineration is used to convert the waste materials into incinerator bottom ash, clinker, fly ash, and particulates, further reducing waste volume (Chen et al. 2023). The released ash can be used as a partial replacement for cement in concrete mixes (e.g., a supplementary cementitious material) because it contains some quantities of typical cement minerals. The incineration process is also used to eliminate pathogenic organisms and harmful toxic substances (Arteaga et al. 2023). The generated emissions ($SO_X$, $NO_X$, and $CO_2$) should be controlled to avoid air pollution (Halba et al. 2023).

#### 1.4.3.3 Gasification

Gasification describes a thermochemical process used to convert solid waste rich in carbonaceous matter into fuel gases or syngas under a controlled amount of oxygen and/or steam (Chang et al. 2023). Water and tar are also generated as unwanted products from the gasification process. The produced syngas, containing a mixture of carbon monoxide (CO), hydrogen ($H_2$), and traces of methane ($CH_4$), can be utilized directly as fuels for engines and turbines. The variation of the generated byproduct amounts depends on the temperature applied, ranging between 700 and 1300 °C (Ahmad et al. 2023). The accomplishment of solid waste gasification also relies on the type of gasifying agents, such as air, oxygen, steam, or supercritical water inside a gasifier (Chu et al. 2023).

#### 1.4.3.4 Pyrolysis

Pyrolysis is a thermal treatment method, where the solid waste is heated under an oxygen-deprived condition at temperatures ranging from 500 to 800 °C (Hasan et al. 2023). This process tends to generate liquid, solid, and gaseous substances based on the pyrolysis temperature applied, e.g., slow pyrolysis (300 °C ~ 700 °C;<1 °C/s), fast pyrolysis (450 °C ~ 800 °C; 300 °C/min), and flash pyrolysis (>600 °C; 1000 °C/s). However, most studies used the pyrolysis process to convert solid wastes into biochar rich in carbon (Ezz et al. 2023), where this material can be employed to restore contaminated soils, enhance crop production, and mitigate the risk of GHG emissions (e.g., $CO_2$, $CH_4$, and $N_2O$) in farmland (Ganesan et al. 2023). Rotary kilns, microwave reactors, and screw kilns are robust and affordable units used to perform the pyrolysis of tires, complex plastic, electronic and electric materials, and wood waste (Carregosa et al. 2023).

## 1.5 Advantages and Challenges of Biochemical Conversion Processes

Waste-to-energy (WtE) is one of the most important technology considerations for the sustainable management of waste. Biochemical conversion technology defines the microbial degradation and breakdown of the organic portions in solid wastes, including (e.g., food/kitchen residues, agricultural crops, and cattle manure). The main advantages and drawbacks of this technique are summarized in Table 1.5. Anaerobic digestion utilizes microorganisms to convert biodegradable and moist waste (biomass) such as food waste and livestock sludge into biogas (Eraky et al. 2023). The biogas (methane) produced in the anaerobic biological process is a mixed gas, and its main contents are $CH_4$, $CO_2$, $H_2S$, and water. This process can be used for biogas recovery and generation of nutrient-rich digestate (Midden et al. 2023).

Composting denotes an aerobic method of decomposing organic solid waste. It is a major type of recycling organic waste to produce useful fertilizers. In this process, biological decomposition of organic materials occurs under aerobic conditions with part of the process conducted under thermophilic (>50 °C) conditions (Nanda et al. 2023). Microorganisms play a significant role in this process by bio-transforming contaminants into less hazardous components and immobilizing pollutants.

Phycoremediation (microalgae-based wastewater treatment technology) is used for removing nutrients from wastewater while generating biomass that can be a substitute energy source replacing conventional feedstock. Algal biomass can also be processed to prepare a wide range of chemical and bio-medical products.

**Table 1.5** Advantages and disadvantages of biochemical conversion processes

| Biochemical processes | Advantages | Disadvantages |
|---|---|---|
| Composting | • Low cost, and consumes less resources (e.g., energy and water)<br>• Used to make fertilizers (e.g., composts)<br>• Protects soil against erosion<br>• Increases nutrients in soil | • Extended time of microbial adaptation<br>• Requires more land area<br>• May need nutrient supplementation before compost utilization<br>• Possibility of contamination by heavy metals |
| Vermicomposting | • Improves water holding capacity<br>• Supply various micro and macronutrients to agricultural lands<br>• Protects soil from erosion | • Proper growth conditions for worms should be maintained |
| Anaerobic digestion | • Simple operation<br>• Less maintenance<br>• Reduce biochemical oxygen demand (BOD) concentrations using a low-cost method<br>• Biogas production | • Energy requirement to maintain mesophilic methanogenic process<br>• Acidification may occur due to volatile fatty acids accumulation at higher organic loads |
| Landfilling | • Easy to deal with waste<br>• Cheap way to get rid of the waste when land is available<br>• Prevents illegal solid waste dumping | • May pollute the soil (e.g., leas to deforestation)<br>• Contributes to groundwater pollution<br>• Leads to ecological imbalance |

## 1.6 Meeting SDGs from "Waste-to-Energy" Strategy

The "waste-to-energy" strategies aim at transforming non-recyclable waste materials into valuable energy forms (e.g., heat, electricity, and fuel) (Aboughaly and Fattah 2023), achieving a sustainable waste management approach (see Table 1.5) (Santulli et al. 2023). These techniques include incineration, gasification, decarbonization, combustion, pyrolysis, AD, and landfill gas recovery. The implementation of these emerging SWM methods would assist in fulfilling the targets of the sustainable development goals (SDGs) (Shea and Thornton 2019).

### 1.6.1 SDG1: No Poverty

Energy generated by waste management is consumed by individuals and society worldwide for adopting several human-related activities (e.g., transportation, manufacturing, lighting, heating, cooling, standby power) (Manjunath et al. 2023). For example, almost every institution in society, e.g., schools, workplaces, offices, universities, and colleges, requires electricity. Nations and communities (e.g., connection between researchers and policymakers) employ the waste management practice to fulfill their energy requirements to perform their duties (Target 1.1) (Banik et al. 2023). Several job opportunities can be created by establishing waste-to-energy conversion projects (Target 1.4), further reducing the unemployment rate trend (Piske et al. 2023). In Malaysia, the design, establishment, and operation of energy recovery facilities using wastes as an input feedstock have created jobs for the local community (Noor et al. 2013). Stakeholders can work towards reducing people living in poverty by considering economic profit, ecological costs, and social benefits from sustainable waste management (Target 1.A).

### 1.6.2 SDG2: Zero Hunger

Several issues worldwide, such as pandemics, wars, and inflations, have recently participated in reducing the quantities of available natural resources (e.g., natural resources include a wide array of categories, e.g., agricultural, conservational, forestry, oceanic, water, energy, and mineral resources). This situation adversely impacts the food security status of developing and poor countries (Heiba et al. 2023a). The management of organic wastes contributes to food production via several routes, including the implementation of the pyrolysis process to produce biochar used to enhance the soil's physical properties (Target 2.3). Moreover, the composting of organic waste feedstock has been employed to produce organic fertilizer used for increasing crop yields (Czekała et al. 2023). This management method also minimizes the spreading of illegal waste dumping and landfilling sites (Akhtar et al. 2023). Biochar can also be used as an active component in food processing and packaging applications (Target 2.5), further addressing various emerging ecological challenges (Podder et al. 2589). For instance, biochar is required for producing alcoholic beverages (Target 2.B), and decolorizing fruit juice and sugarcane syrup (Fang et al. 2023). Biochar was also successfully employed to improve the quality of sheep manure composts and reduce $N_2O$, $CH_4$, and $NH_3$ emissions by 68.64%, 65.23%, and 42.05%, respectively (Wang et al. 2023a).

### 1.6.3 SDG3: Good Health and Wellbeing

Inappropriate management of solid wastes, especially wastes from pharmaceutical industries, is responsible for secondary pollution, such as air and soil pollution (Sapkota and Pariatamby 2023). The unmanaged and uncontrolled solid waste disposal sites disturb public health and negatively affect the quality of life of all individuals. Some of the diseases and drawbacks associated with illegal solid waste disposal include anemia, mental disorders, nausea/vomiting, inhalation injuries, and leukaemia (Awino and Apitz 2023).

The management of electrical and electronic waste is essential to avoid the endangering of human health (e.g., a rise in blood pressure and irritation of the eyes, skin, and lungs), causing thousands of deaths annually (Target 3.2). Most of these concerns accompanied by solid waste pollution can be avoided using sustainable, eco-friendly (Target 3.3), and cost-efficient biological and thermo-chemical waste management processes (Dar et al. 2023). Establishing well-designed landfill sites (these sites should not be placed within surface water or water resources protection areas), as well as implementing waste recycling plants and infrastructure, is crucial in reducing the negative impacts of waste on both the environment and human health (Target 3.5).

### 1.6.4 SDG4: Quality Education

Ensuring high-quality education is an essential strategy of SWM, providing equitable education for boys and girls (Target 4.1). Developing waste management programs can support the awareness and behavior in low-income countries that suffer from the health impact of solid waste disposal (e.g., waste dumping in unauthorized locations) (Heiba et al. 2023b). High-quality teaching can be ensured by involving trainees (men and women) in the learning processes for waste sorting and recycling (Target 4.3). Knowledge transfer (sharing) is essentially required to overcome the limitations accompanied by insufficient resources, grants, or facilities used to employ waste recycling projects (Uba et al. 2023).

### 1.6.5 SDG5: Gender Equality

Energy is one of the main human needs, influencing the types of activities performed by men and women (Wang et al. 2023b). For instance, the waste recycling habit assists women who face great challenges in transferring household and domestic wastes to disposal sites. Moreover, waste management facilitates the women's work to pick recyclable waste from bins, followed by the sorting and processing wastes at plastic and organic waste recycling facilities. Agricultural waste management plays a vital role in the economic stability of smallholder women farmers by reducing reliance on external markets, lowering expenses, and generating additional income through the sale of crop residues in local markets (Target 5.A). It's important to ensure women's active participation and equal opportunities in establishing effective waste management projects (Ansari et al. 2023). Women should be involved in higher-income and decision-making positions in the sustainable waste sector, supporting the gender-waste management nexus (Target 5.5).

### 1.6.6 SDG6: Clean Water and Sanitation

Dumping large quantities of solid wastes, e.g., hospital, medical, and healthcare wastes, into the aquatic environment negatively affects the surface water (streams, rivers, and lakes) quality (Palansooriya et al. 2023). Waste management is crucial for avoiding the deterioration of water resources because the recent population growth (Peng et al. 2023), urbanization, and socioeconomic development have increased the industrial and domestic water demand (Target 6.1). Moreover, this waste management pattern tends to save water needed for food production, and reduce groundwater contamination from landfills and industrial reservoirs (Target 6.6) (Dadebo et al. 2023a). To achieve sustainable freshwater production (Saqr et al. 2024), SWM and soil and wastewater remediation should be considerably improved (Target 6.3) (Aboughaly and Fattah 2023).

### 1.6.7 SDG7: Affordable and Clean Energy

Different forms of energy can be produced from the thermochemical treatment of solid wastes. These treatment technologies include (i) gasification to produce syngas and ash/slag (mineral residues), (ii) pyrolysis to generate liquid fuel (bio-oil), and (iii) combustion to obtain power and heat (Target 7.1) (Hasan et al. 2023). Biogas recovery from the AD of organic wastes can produce electricity and heat or as an energy source for cooking (Target 7.2) (Subbarao et al. 2023). Waste management offsets are a specific type of carbon footprint and used to eliminate and avoid $CO_2$ emissions because the projects (thermal treatment of solid waste) either capture emissions from waste or reduce the total amount of waste (Target 7.3) (Aboughaly and Fattah 2023). As such, carbon credits represent the difference between how much GHG is created from making the product from raw materials and that obtained when using recycled materials.

### 1.6.8 SDG8: Decent Work and Economic Growth

Environmental-friendly projects are adopted to raise the growth in employment and income, supporting the green economy concept. New green job opportunities (e.g., environmental engineer, waste collector, and waste management engineer, project manager, and consultant) are generated through implementing the waste recycling facilities (Target 8.2). These job opportunities have recently increased because waste management companies are expanding their services to include recycling, composting, and other environmentally friendly practices (Target 8.3). These practices provide beneficial products, such as bioenergy, bioplastic, organic fertilizer/biofertilizer, and biochar (Sorout et al.

2023). Increasing the number of biogas-producing plants that receive biogenic feedstock has reduced environmental concerns caused by uncontrolled dumping, open-air incinerators, and landfills, promoting sustainable economic growth (Target 8.4).

### 1.6.9 SDG9: Industry, Innovation and Infrastructure

An integrated dark- and photo-fermentation of organic waste feedstock is used to produce hydrogen gas. This gas is an innovative energy carrier that has several applications in fuel cells to produce electricity, petroleum refining, vehicular transportation, and aerospace (Eraky et al. 2023). Moreover, ethanol production through the fermentation of agricultural and fruit wastes (sugar transformation into carbon dioxide and alcohol typically by yeasts under an anaerobic condition) is progressively attracting investments and inventiveness. Industrial waste management includes segregation, land application (composting), landfill, and recycling of several types of wastes, such as paper, glass, organic, metal, plastics, wood, and grass (Target 9.1). Stakeholders have recently focused on overcoming the techno-economic challenges that can face the success of waste management technologies, including collection and disposal infrastructure, and ineffective recycling or composting (Target 9.2). Other challenges include financial constraints, inadequate technological advances, and a lack of clear regulations and support from localities (Target 9.3). Overcoming these limitations would govern the climate-energy nexus (Bhatia et al. 2023).

### 1.6.10 SDG10: Reduced Inequality

Equitable resource distribution is one of the main social aims that can be achieved by reducing cities' carbon footprints (Karthik et al. 2023). The international trade of waste between countries is used for further treatment, disposal, or recycling (Target 10.a). Toxic or hazardous wastes are often transferred (exported) from developed countries to developing countries, and they are often disposed of in open landfills or burned in incinerators. People in some developing countries may rationally accept increased exposure to hazardous pollutants in exchange for opportunities to increase their productivity (Ngulube et al. 2024), and, hence, their income. As such, the hazardous waste trade should be regulated (Target 10.3), specifically to prevent the dumping of hazardous waste in the lands of less developed countries. Moreover, the waste recycling projects support the countries that have no many natural resources (e.g., minerals, metals, and forests), or those subjected to natural resource depletion (Target 10.6) (Chowdhury and Sarkar 2023).

### 1.6.11 SDG11: Sustainable Cities and Communities

People prefer living in urban cities that have proper SWM systems, e.g., collection and transportation, reuse or recycling, and treatment or safe disposal. As such, poorly designed and planned urbanization tends to cause multiple barriers, such as air and water pollution, land degradation, emissions of methane and hazardous leachate, and climate change (Target 11.2). The uncontrolled dumping and accumulation of solid wastes in cities increase land and atmosphere pollution, plant death, loss of habitats, and infection and transmit diseases (Ducasse et al. 2022). The Internet of Things (IoT) technique has been recently employed to maintain a smart garbage monitoring system, showing the garbage status through an IoT application. This approach attempts to optimize resource allocation, reduce running expenses, and improve the sustainability of waste services (Target 11.6). Moreover, using IoT in waste management attempts to provide separate areas, e.g., green and public spaces, and commercial areas, further improving the traffic flow (Target 11.7) (Karthik et al. 2023). Local communities in smart sustainable cities should be engaged and empowered to participate in these waste management activities. This step should be supported by acquiring sufficient information on the city corporations' and municipalities' average annual waste production.

### 1.6.12 SDG12: Responsible Consumption and Production

The manufacturing of domestic materials (e.g., food, shelter, clothing) is increasing exponentially to satisfy the economic requirements because all human wants to have some basic common characteristics. Some solid wastes can be used as a resource needed for the creation of a good or service, further achieving economic growth (Target 12.4). The quantities and types of solid wastes recycled to recover valuable items (e.g., computers, chemicals, nutrients, furniture, phones, and bags) vary according to industrialization and elevated population expansion (Target 12.5) (Mohod and Bagal 2023). Renewable natural resources (air, solar, plant, soil, and water) should be utilized efficiently (Target 12.2), and a large amount of generated waste should be reused and recycled to avoid inadequate sequential production/consumption schemes (Aboughaly and Fattah 2023). This approach attempts to eliminate the barriers in order to transform the linear economy into the circular economy (Subbarao et al. 2023), needing awareness and education programs (Target 12.1).

### 1.6.13 SDG13: Climate Change

The world is facing climate crises and catastrophes, including sea levels rising, Arctic ice melting, coral reefs dying, ocean acidification, and forest burning. Proper SWM can mitigate the increase in the anthropogenic emissions caused by various human activities (Target 13.1), e.g., domestic and industrial waste dumping, fossil fuel combustion, transportation, and unplanned landfilling (Themelis 2023). Waste transformation reduces $CH_4$ emissions from cattle digestion (enteric emissions) and stored manure, and $CO_2$ released road transport, and offshore drilling, contributing to climate change (Syarifinnur et al. 2023). The implementation of AD to treat biogenic organic wastes addresses the targets of SDG13 by capturing $CH_4$, which is then used beneficially. Raising awareness of climate change and health (Target 13.3) supports developing countries in securing grant funding and access to key experts under the United Nations Framework Convention on Climate Change (UNFCCC) framework (Zhang et al. 2023). This framework promotes effective climate change-related plans and mechanisms to avoid human influence on the climate system (Target 13.b) (Shea and Thornton 2019).

### 1.6.14 SDG14: Life Below Water

Several strategies are proposed to avoid the dumping of solid wastes into aquatic ecosystems, which are used to protect the quality of oceans and seas (the planet's largest ecosystem) (Target 14.2) (Zidan et al. 2023). This approach is highly linked to marine species' growth and health and maritime transport services (boat, ship, sailboat, or barge) (Dadebo et al. 2023b). Moreover, it avoids the escape of various fish types and many planktonic organisms (Target 14.1), making the seafood supply chain more sustainable (Midden et al. 2023).

### 1.6.15 SDG15: Life on Land

Pollution (e.g., leachate pollution) can transfer from solid wastes to terrestrial ecosystems, causing significant environmental problems especially if the waste contains toxic substances. For instance, deforestation and desertification caused by human activities and climate change deteriorate biological diversity and healthy ecosystems. An efficient stabilization of this waste before reaching the soil matrix, obtaining less soluble, mobile, or toxic forms of waste, is of significant importance (Target 15.1) (Mohod and Bagal 2023). Reducing the amount of waste from the source is required to maintain sustainable management of land for agriculture and tourism (Podder et al. 2589). Education program on ecosystems is essential to avoid biodiversity loss (Target 15.9).

### 1.6.16 SDG16: Peace, Justice and Strong Institutions

Waste management projects aim at providing advanced solutions to waste and pollution, supporting a global shift toward environmental justice (Target 16.3). The purpose of institutions, along with their function and philosophy, is to allow the organizational requirements to initiate and establish waste management projects. SDG 16 ensures that other goals act synergistically together to reduce all causes of environmental depletion, such as the accumulation of solid wastes in streets, landscapes, and urban areas (Target 16.7). The four primary sociological institutions, e.g., family, religion, education, and government, should be involved in maintaining a healthy environment through the recovery of green material and renewable and clean energy from solid waste recycling (Target 16.8). As such, proper environmental management and governance are important for long-term peace, stability, and security in disaster-prone countries, mainly, in communities that depend on natural resources for their survival (Chiaramonti and Panoutsou 2018).

### 1.6.17 SDG17: Partnership for Goals

Several developing countries are facing considerable challenges in achieving the "waste-to-energy" strategy. These barriers include (i) operational challenges (e.g., site selection, and waste segregation, collection, and transportation), (ii) technical challenges (e.g., advanced and modern technologies of recycling, personnel's knowledge and awareness, lower calorific value for high-moisture wastes, and atmospheric damage), (iii) economic challenges (e.g., project budget, installation, and maintenance), and (iv) managerial challenges (e.g., training and continual support, and communication gap between researchers and the government, collaboration from stakeholders) (Shea and Thornton 2019). International collaboration between developed and developing countries should include technical assistance and capacity building to strengthen the solid waste recycling approach.

## 1.7 Suggestions for Maintaining Efficient Utilization of Solid Wastes

- The waste management systems and infrastructure should be upgraded to receive the large amounts of solid waste.
- Organic wastes should not be contaminated with potentially toxic bio-contaminants, e.g., antibiotics and heavy metals, before entering the digesters.
- This biogas can be used in electric power generation, reducing the dependence on fossil fuels (e.g., coal, petroleum, natural gas) (Themelis 2023).

## 1.8 Conclusions

Deltaic systems are among the most dynamic and productive environments on Earth and many have a high population density. Deltas play a central role in food and water security, and riverine and coastal vulnerability of urban and rural areas. This chapter generally addresses the main subjects provided by the book titled "*Solid Waste Management in Delta Region for SDGs Fulfillment: Delta Sustainability by Waste Management*". These topics cover the recently applied techniques of solid waste management to meet the sustainable strategies of waste recycling and material recovery. As such, this book focuses on the fulfilment of the three pillars of sustainable development through maintaining reliable biological and thermochemical techniques of solid waste conversion. The given information can be utilized by waste management professionals in academia, government, industry, engineering, management, planning, and public health.

**Acknowledgements** The author acknowledges Nasr Academy for Sustainable Environment (NASE).

**Contributions**
MN: conceptualization, and writing—original draft.

**Ethics declarations**
**Conflict of Interest**

The author has no relevant financial or non-financial interests to disclose.

**Ethical Approval**
Not applicable.

**Consent to Participate**
Not applicable.

**Consent for Publication**
The author consents for publication.

## References

Aboughaly M, Fattah I (2023) Environmental analysis, monitoring, and process control strategy for reduction of greenhouse gaseous emissions in thermochemical reactions. Atmosphere 14(4):655. https://doi.org/10.3390/atmos14040655

Ahmad, Ahmad M, Md Ali U, Ken K (2023) Gasification char residues management: assessing the characteristics for adsorption application. Arab J Chem 16(9):104993. https://doi.org/10.1016/j.arabjc.2023.104993

Akhtar S, Hollaender H, Yuan Q (2023) Impact of heat and contaminants transfer from landfills to permafrost subgrade in arctic climate: a review. Cold Reg Sci Technol 206:103737. https://doi.org/10.1016/j.coldregions.2022.103737

Ansari, Nasr M, Rawat , Bux F (2023) Meeting sustainable development goals (SDGs) through progression of pilot-scale algal system to commercial raceway pond (300,000 L). Biomass Convers Biorefinery 13(9):7663–7676. https://doi.org/10.1007/s13399-021-01661-0

Arteaga, Silva J, Yarasca-Aybar C (2023) Solid waste management and urban environmental quality of public space in Chiclayo, Peru. City Environ Interact 20:100112. https://doi.org/10.1016/j.cacint.2023.100112

Awino Apitz S (2023) Solid waste management in the context of the waste hierarchy and circular economy frameworks: an international critical review. Integr Environ Assess Manag. https://doi.org/10.1002/ieam.4774

Banik, Chakraborty D, Rizwan M, Shaik A, Chandan M (2023) Review on disposal, recycling and management of waste polyurethane foams: a way ahead. Waste Manag Res 41(6):1063–1080. https://doi.org/10.1177/0734242X221146082

Bhatia L, Jha H, Sarkar T, Sarangi P (2023) Food waste utilization for reducing carbon footprints towards sustainable and cleaner environment: a review. Int J Environ Res Public Health 20(3):2318. https://doi.org/10.3390/ijerph20032318

Blanchard C, Harris P, Pocock C, McCabe B (2023) Food and garden organic waste management in AUSTRALIA: co-benefits for regional communities and local government. Sustainability (Switzerland) 15(13):9901. https://doi.org/10.3390/su15139901

Carregosa J, Carregosa, W. Silva, T. Santos A. Wisniewski Jr (2023) Thermochemical conversion of aquatic weed biomass in a rotary kiln reactor for production of bio-based derivatives. J Anal Appl Pyrolysis 173:106048. https://doi.org/10.1016/j.jaap.2023.106048

Chang Y-J, Chang J-S, Lee D-J (2023) Gasification of biomass for syngas production: Research update and stoichiometry diagram presentation. Bioresour Technol 387:129535. https://doi.org/10.1016/j.biortech.2023.129535

Chen, Perumal P, Illikainen M, Ye G (2023) A review on the utilization of municipal solid waste incineration (MSWI) bottom ash as a mineral resource for construction materials. J Build Eng 71:106386. https://doi.org/10.1016/j.jobe.2023.106386

Chowdhury Sarkar A (2023) Vermicomposting—the sustainable solid waste management. Waste Manage Resour Recycl Dev World 701–719. https://doi.org/10.1016/B978-0-323-90463-6.00013-0

Chu Z, Li Y, Zhang C, Fang Y, Zhao J (2023) A review on resource utilization of oil sludge based on pyrolysis and gasification. J Environ Chem Eng 11(3):109692. https://doi.org/10.1016/j.jece.2023.109692

Czekała W, Nowak M, Piechota G (2023) Sustainable management and recycling of anaerobic digestate solid fraction by composting: a review. Bioresour Technol 375:128813. https://doi.org/10.1016/j.biortech.2023.128813

Chiaramonti D, Panoutsou C (2018) Low-ILUC biofuel production in marginal areas: Can existing EU policies support biochar deployment in EU MED arid lands under desertification? Chem Eng Trans 65:841–846. https://doi.org/10.3303/CET1865141

Dadebo, Atukunda A, Ibrahim M, Nasr M (2023) Integrating chemical coagulation with fixed-bed column adsorption using rice husk-derived biochar for shipboard bilgewater treatment: scale-up design and cost estimation. Chem Eng J Adv 16:100520. https://doi.org/10.1016/j.ceja.2023.100520

Dadebo, Ibrahim M, Fujii M, Nasr M (2023) Sequential treatment of surfactant-laden wastewater using low-cost rice husk ash coagulant and activated carbon: modeling, optimization, characterization, and techno-economic analysis. Bioresour Technol Rep 22:101464. https://doi.org/10.1016/j.biteb.2023.101464

Dar S, Sharma P, Shukla B (2023) Enhanced landfill mining waste valorization techniques: a methodological review. Lect Notes Civil Eng 281:109–125. https://doi.org/10.1007/978-981-19-4731-5_10

Demarema S, Nasr M, Ookawara S, Abdelhaleem A (2024) Enhanced synergistic system for the persulfate activation under visible light using novel N–ZnO photocatalyst supported on Lantana camara-based biochar. Chemosphere 349:140840. https://doi.org/10.1016/j.chemosphere.2023.140840

Deng M, Liu H (2024) Driving factors and potential offshore pollution of plastic mulch residue in farmland in the Yellow River Delta, China. Agric Ecosyst Environ 364:108902. https://doi.org/10.1016/j.agee.2024.108902

Dogu O, Pelucchi M, Van de Vijver R, Van Steenberge P, D'hooge D, Cuoci A, Mehl M, Frassoldati A, Faravelli T, Van Geem K (2021) The chemistry of chemical recycling of solid plastic waste via pyrolysis and gasification: state-of-the-art, challenges, and future directions. Prog Energy Combust Sci 84:100901. https://doi.org/10.1016/j.pecs.2020.100901

Ducasse V, Capowiez V, Peigné J (2022) Vermicomposting of municipal solid waste as a possible lever for the development of sustainable agriculture. A review. Agron Sustain Dev 42(5):89. https://doi.org/10.1007/s13593-022-00819-y

Elsadek E, Zhang K, Hamoud Y, Mousa A, Awad A, Abdallah M, Shaghaleh H, Hamad A, Jamil M, Elbeltagi A (2024) Impacts of climate change on rice yields in the Nile River Delta of Egypt: a large-scale projection analysis based on CMIP6. Agric Water Manag 292:108673. https://doi.org/10.1016/j.agwat.2024.108673

Emmanouilidou E, Mitkidou S, Agapiou A, Kokkinos N (2023) Solid waste biomass as a potential feedstock for producing sustainable aviation fuel: a systematic review. Renew Energy 206:897–907. https://doi.org/10.1016/j.renene.2023.02.113

Eraky M, Nasr M, Elsayed M, Ai P, Tawfik A (2023) Synergistic interaction of tween 20 and magnesium@ functionalized graphene oxide nano-composite for dual productivity of biohydrogen and biochar from onion peel waste. Renew Energy 216:119082. https://doi.org/10.1016/j.renene.2023.119082

Ezz H, Ibrahim M, Fujii M, Nasr M (2023) Dual biogas and biochar production from rice straw biomass: a techno-economic and sustainable development approach. Biomass Convers Biorefinery 13:10807–10821. https://doi.org/10.1007/s13399-021-01879-y

Fang, Huang T, Lu H, Wu X-L, Chen Z, Yang H, Wang S, Tang Z, Li Z, Hu B, Wang X (2023) Biochar-based materials in environmental pollutant elimination, H2 production and CO2 capture applications. Biochar 5(1):42. https://doi.org/10.1007/s42773-023-00237-7

Ganesan S, Padmapriya G, Omoikhoje I, Tharudini J, De Zoysa S (2023) Biochar: its characteristics application and utilization of on environment. Basic Sci Sustain Dev 2:115–124. https://doi.org/10.1515/9783111071206-009

Halba, Thengane S, Arora P (2023) A critical outlook on lignocellulosic biomass and plastics co-gasification: a mini-review. Energy Fuels 37(1):19–35. https://doi.org/10.1021/acs.energyfuels.2c02907

Hasan M, Anzar N, Sharma P, Malode S, Shetti N, Narang J, Kakarla R (2023) Converting biowaste into sustainable bioenergy through various processes. Bioresour Technol Rep 23:101542. https://doi.org/10.1016/j.biteb.2023.101542

He, Hu H, Jiao F, Zuo W, Liu C, Xie H, Dong L, Wang X (2023) Thermal separation of heavy metals from municipal solid waste incineration fly ash: a review. Chem Eng J 467:143344. https://doi.org/10.1016/j.cej.2023.143344

Heiba Y, Ibrahim M, Mohamed A, Fujii M, Nasr M (2023) Developing smart sustainable irrigation matrix (SIM)-based model for selection of best irrigation techniques: a framework to achieve SDGs. J Clean Prod 420:138404. https://doi.org/10.1016/j.jclepro.2023.138404

Heiba Y, Nasr M, Fujii M, Mohamed A, Ibrahim M (2023) Improving irrigation schemes using sustainable development goals (SDGs)-related indicators: a case study of tomato production in pot-scale experimentation. Environ Dev Sustain. https://doi.org/10.1007/s10668-023-03357-z

Heidari-Maleni, Taheri-Garavand A, Rezaei M, Jahanbakhshi A (2023) Biogas production and electrical power potential, challenges and barriers from municipal solid waste (MSW) for developing countries: a review study in Iran. J Agric Food Res 13:100668. https://doi.org/10.1016/j.jafr.2023.100668

Isibika, Simha P, Vinnerås B, Zurbrügg C, Kibazohi O, Lalander C (2023) Food industry waste-an opportunity for black soldier fly larvae protein production in Tanzania. Sci Total Environ 858:159985. https://doi.org/10.1016/j.scitotenv.2022.159985

Karthik, Sreevidya L, Nithya Devi R, Hemalatha G, Yamini R (2023) An efficient waste management technique with IoT based smart garbage system. Mater Today Proceed 80:3140–3143. https://doi.org/10.1016/j.matpr.2021.07.179

Lopez G, Artetxe M, Amutio M, Alvarez J, Bilbao J, Olazar M (2018) Recent advances in the gasification of waste plastics. A critical overview. Renew Sustain Energy Rev 82:576–596. https://doi.org/10.1016/j.rser.2017.09.032

Manjunath, Di Mare M, Ouellet-Plamondon C, Bhojaraju C (2023) Exploring the potential use of incinerated biomedical waste ash as an eco-friendly solution in concrete composites: a review. Constr Build Mater 387:131595. https://doi.org/10.1016/j.conbuildmat.2023.131595

Mannaa M, Mansour A, Park I, Lee D-W, Seo Y-S (2024) Insect-based agri-food waste valorization: agricultural applications and roles of insect gut microbiota. Environ Sci Ecotechnol 17:100287. https://doi.org/10.1016/j.ese.2023.100287

Md A-A-M, Hossain N, Hossain M, Sultana R (2023) Conversion of leather industry solid waste to organic fertilizer by vermicomposting: use for plant growth. Text Leather Rev 6:37–56. https://doi.org/10.31881/TLR.2022.82

Mengqi Z, Shi A, Ajmal M, Ye L, Awais M (2023) Comprehensive review on agricultural waste utilization and high-temperature fermentation and composting. Biomass Convers Biorefinery 13(7):5445–5468. https://doi.org/10.1007/s13399-021-01438-5

Mohod M. Bagal (2023) Technological developments in the energy generation from municipal solid waste (landfill gas capture, combustion, pyrolysis and gasification). 360-Degree Waste Management, Volume 1: Fundamentals, Agricultural and Domestic Waste, and Remediation, pp. 139–157. https://doi.org/10.1016/B978-0-323-90760-6.00005-9

Nanda S, Sarker T, Kang K, Li D, Dalai A (2023) Perspectives on thermochemical recycling of end-of-life plastic wastes to alternative fuels. Materials 16(13):4563

Ngulube K, Abdelhaleem A, Fujii M, Nasr M (2024) Synergism of artificial intelligence and techno-economic for sustainable treatment of methylene blue dye-containing wastewater by photocatalysis. Sustainability 16:529. https://doi.org/10.3390/su16020529

Noor Z, Yusuf R, Abba A, Abu Hassan M, Mohd Din M (2013) An overview for energy recovery from municipal solid wastes (MSW) in Malaysia scenario. Renew Sustain Energy Rev 20:378–384. https://doi.org/10.1016/j.rser.2012.11.050

Palansooriya K, Dissanayake P, Igalavithana A, Tang R, Cai Y, Chang S (2023) Converting food waste into soil amendments for improving soil sustainability and crop productivity: a review. Sci Total Environ 881:163311. https://doi.org/10.1016/j.scitotenv.2023.163311

Peng, Li Y, Li Q, Liang C, Nasr M, Xu Y, Liu Y, Zhou Y (2023) The effect of free ammonia on ammonium removal and N2O production in a consortium of microalgae and partial nitritation cultures. Chem Eng J 474:145572. https://doi.org/10.1016/j.cej.2023.145572

Piske B, Lopes F, Utzig J, Wiggers V (2023) Small-scale waste heat recovery through incineration–a brief review. Renew Energy Power Qual J 21:154–159. https://doi.org/10.24084/repqj21.255

Podder, Patra B, Pattnaik F, Nanda S, Dalai A (2023) A review of carbon capture and valorization technologies. Energies 16(6):2589. https://doi.org/10.3390/en16062589

Sapkota B, Pariatamby A (2023) Pharmaceutical waste management system–are the current techniques sustainable, eco-friendly and circular? A review. Waste Manag 168:83–97. https://doi.org/10.1016/j.wasman.2023.05.052

Santulli, Fragassa C, Pavlovic A, Nikolic D (2023) Use of sea waste to enhance sustainability in composite materials: a review. J Mar Sci Eng 11(4):855. https://doi.org/10.3390/jmse11040855

Saqr, Nasr M, Fujii M, Yoshimura C, Ibrahim M (2024) Delineating suitable zones for solar-based groundwater exploitation using multi-criteria analysis: a techno-economic assessment for meeting sustainable development goals (SDGs). Groundw Sustain Dev 25:101087. https://doi.org/10.1016/j.gsd.2024.101087

Shea M, Thornton T (2019) Tracing country commitment to Indigenous peoples in the UN framework convention on climate change. Glob Environ Change 58:101973. https://doi.org/10.1016/j.gloenvcha.2019.101973

Sonu, Rani G, Pathania D, Abhimanyu, Umapathi R, Rustagi S, Huh Y, Gupta V, Kaushik A, Chaudhary V (2023) Agro-waste to sustainable energy: a green strategy of converting agricultural waste to nano-enabled energy applications. Sci Total Environ 875:2023.https://doi.org/10.1016/j.scitotenv.2023.162667

Sorout J, Raj S, Kaur D, Lamba P (2023) Waste-based bricks: evaluation of strength behaviour of bricks containing different waste materials as an additive. Water Air Soil Pollut 234(7):424. https://doi.org/10.1007/s11270-023-06438-x

Srimalanon en, Kachapongkun P (2023) Synthesis gas production with gasification technology from municipal solid waste. Appl Sci Eng Prog 16(4):6633. https://doi.org/10.14416/j.asep.2023.01.002

Subbarao P, Silva TD, Adlak K, Kumar S, Chandra R, Vijay V (2023) Anaerobic digestion as a sustainable technology for efficiently utilizing biomass in the context of carbon neutrality and circular economy. Environ Res 234:116286. https://doi.org/10.1016/j.envres.2023.116286

Syarifinnur S, Nuraini Y, Prasetya B, Handayanto E (2023) Comparing compost and vermicompost produced from market organic waste. Int J Recycl Org Waste Agric 12(3):279–289. https://doi.org/10.30486/ijrowa.2022.1944251.1368

Tawfik, Nasr M, Ni S-Q (2024) Paper mill sludge rich enzymes and microbial community promotes the hydrogenesis of black liquor containing furfural. Int J Hydrog Energy 52(Part D):58–73. https://doi.org/10.1016/j.ijhydene.2023.01.242

Themelis (2023) Energy and materials recovery from post-recycling wastes: WTE. Waste disposal and sustainable energy. https://doi.org/10.1007/s42768-023-00138-2

Uba U, Efut E, Bassey Obeten U, Asuquo E, Uba J (2023) Sociodemographic factors and environmental workers' knowledge of the impact of awareness creation on sustainable disposal of solid wastes. Heliyon 9(7):e18122. https://doi.org/10.1016/j.heliyon.2023.e18122

van Midden C, Harris J, Shaw L, Sizmur T, Pawlett M (2023) The impact of anaerobic digestate on soil life: a review. Appl Soil Ecol 191:105066. https://doi.org/10.1016/j.apsoil.2023.105066

Wang W, Chang J-S, Lee D-J (2023) Anaerobic digestate valorization beyond agricultural application: current status and prospects. Bioresour Technol 373:128742. https://doi.org/10.1016/j.biortech.2023.128742

Wang Z, Xu Y, Yang T, Liu Y, Zheng T, Zheng C (2023) Effects of biochar carried microbial agent on compost quality, greenhouse gas emission and bacterial community during sheep manure composting. Biochar 5:3. https://doi.org/10.1007/s42773-022-00202-w

Wang B, Gupta R, Bei L, Wan Q, Sun L (2023) A review on gasification of municipal solid waste (MSW): syngas production, tar formation, mineral transformation and industrial challenges. Int J Hydrog Energy 48(69):26676–26706. https://doi.org/10.1016/j.ijhydene.2023.03.086

Yadav S, Katiyar P, Mesfer M, Danish M (2023) Syngas production from thermochemical conversion of mixed food waste: a review. Wiley Interdiscip Rev Energy Environ 12(3):e468. https://doi.org/10.1002/wene.468

Youssef Y, Gemail K, Atia H, Mahdy M (2024) Insight into land cover dynamics and water challenges under anthropogenic and climatic changes in the eastern Nile Delta: inference from remote sensing and GIS data. Sci Total Environ 913:169690. https://doi.org/10.1016/j.scitotenv.2023.169690

Zhang Y, Wang X, Zhu W, Zhao Y, Wang N, Gao M, Wang Q (2023) Anaerobic fermentation of organic solid waste: Recent updates in substrates, products, and the process with multiple products co-production. Environ Res 233:116444. https://doi.org/10.1016/j.envres.2023.116444

Zidan, Nasr M, Fujii M, Ibrahim M (2023) Environmental and economic evaluation of downflow hanging sponge reactors for treating high-strength organic wastewater. Sustainability (Switzerland) 15(7):6038. https://doi.org/10.3390/su15076038

# 2 Towards Sustainable Delta Ecosystems: Pollution Mitigation for Achieving SDGs in Indian Delta Region

Anita Gupta, Saroj Kanta Barik, and Puneet Singh Chauhan

**Abstract**

Being exceptionally fertile, the Indian delta regions contribute significantly to India's food supply and serve as source of surface and ground water for millions of people. These areas contribute immensely to the livelihoods of the people through fishery, tourism, transportation, and commerce, improving country's economy and people's wellbeing besides supporting rich biodiversity. Although these ecosystems have great ecological and economic importance, they face serious threat from climate change and human activities such as deforestation, industrial and domestic sewage discharge, increasing urbanization, and solid and agricultural waste disposal. Anthropogenic pollutants, primarily heavy metals such as As, Co, Cr, Pb, Ni, Mn, Fe, and Zn, macro- and micro-plastics, pesticides, polycyclic aromatic hydrocarbons, and polychlorinated biphenyl are being released in great quantity threatening the overall health of the ecosystems, and existence of several plant and animal species. Excessive groundwater removal has resulted in salt water intrusion into groundwater aquifers, rendering the groundwater unfit for human consumption and cultivation. The conservation of these areas needs effective pollution abatement measures taking a multidisciplinary approach. This article aims to examine anthropogenic pollutants' ecological and economic impact on Indian delta regions and propose multidisciplinary mitigation measures that include sewage discharge, urbanization, garbage management, and pesticide

A. Gupta · P. S. Chauhan (✉)
CSIR-National Botanical Research Institute (CSIR-NBRI), Rana Pratap Marg, Lucknow 226001, India
e-mail: puneet@nbri.res.in

Academy of Scientific and Innovative Research (AcSIR), Ghaziabad 201002, India

S. K. Barik
Department of Botany, North-Eastern Hill University, Shillong, India

 
M. Nasr and B. Ravindran (eds.), *Solid Waste Management in Delta Region for SDGs Fulfillment*, Deltas of the World, https://doi.org/10.1007/978-3-031-58253-0_2

usage. The goal is to integrate pollution-reduction efforts with SDGs, such as protecting livelihoods, boosting the economy, and maintaining biodiversity in the face of climate change and human activity.

## 2.1 Introduction

On a global scale, the world's oceans serve as the mother for more than 3 billion people, providing crucial marine and coastal biodiversity that sustains their livelihoods (Chen et al. 2020). However, over the years, marine spaces and their deltas have been taken for granted and often treated as a universal dumping ground, leading to environmental degradation (Guzzetti et al. 2018). This disregard for the health of the oceans is particularly concerning given their vital role in supporting diverse ecosystems and human well-being. Meanwhile, India is undergoing a notable transformation marked by a surge in population growth, increased economic activity, and rapid industrialization, all unfolding within the context of accelerated urbanization (Franco et al. 2017). This evolution has resulted in a heightened demand for water resources to support the expanding agricultural, industrial, and domestic sectors (Huang et al. 2021). Rivers and the deltas play a crucial role in India's hydrological system, serving as essential sources to meet the growing water needs (Loucks 2019). Paradoxically, these very sources, critical for sustaining the nation's water supply, are also becoming conduits for runoff from urban and agricultural areas, industrial effluents, and untreated or inadequately treated sewage (Singh et al. 2022). The intricate interplay between rising water demand and the prevalent existence of diverse pollutants underscores the difficulties associated with sustaining water systems in India (Akhtar et al. 2021) restricting the progress towards achieving SDGs. As human activities continue to impact these vital water sources, it becomes imperative to address the interconnected issues of global marine degradation and local water resource management to ensure the sustainable use of water for current and future generations. Therefore, this article focuses on the following objectives:

i. Evaluation of the ecological and economic impact of anthropogenic pollutants, including heavy metals, plastics, pesticides, and other chemicals, on the Indian delta regions' ecosystems, biodiversity, and the health of the people living in these areas,
ii. Provide multidisciplinary mitigation measures to address the threat of pollution, focusing on reducing industrial and domestic sewage discharge, controlling urbanization, managing solid and agricultural waste disposal, and regulating the use of agrochemical pollutants to ensure the conservation of water resources and the sustenance of flora and fauna, and
iii. Propose and promote pollution abatement measures that align with various SDGs, aiming to safeguard the livelihoods of the people, improve the economy, and preserve the rich biodiversity of the delta regions while mitigating the adverse effects of climate change and human activities.

## 2.2 The Delta Regions of India

A delta region is delineated as a wetland that takes shape when a river in the course of its flow towards a distinct aquatic body, discharges the transported water and the sedimentary materials into an alternate body of water encompassing oceans, lakes, or other rivers (Nyanga et al. 2021). Deltas are of critical ecological and economic importance across the world because they are linked to the world's greatest coastal marine fisheries and include the bulk of the planet's coastal wetlands (Day et al. 2016). They are often centering of intense economic activities and sustain the greatest level of ecosystem goods and services (Day et al. 2019).

The deltaic landforms of India are mainly found along its eastern coast, starting from the Ganges river delta in the northern end to the Cauvery river delta in the southern reach (Fig. 2.1). Other significant deltas include the Mahanadi, Godavari, and Krishna deltas, with the Godavari delta being India's third largest after the Ganges and Mahanadi deltas (Kubo et al. 2018). The Gangetic delta encompasses the littoral regions of India and Bangladesh with an area of 105,000 km$^2$ extending from 22°53″ N to 24°11″ N, and 88°09″ E to 88°48″ E. Within the geographical expanse of India, the Ganges courses through the territories of Uttarakhand, Uttar Pradesh, Bihar, Jharkhand, and West Bengal, ultimately discharging its water into the Bay of Bengal (Helmer and Hespanhol 1997; Raha et al. 2021). The Mahanadi delta with a total area of 9063 km$^2$, is located within the state of Odisha, in close proximity to the confluence point of the Mahanadi river and the Bay of Bengal. This delta in the eastern coast exhibits a characteristic arcuate morphology and is geographically positioned between the longitudes of 85°25′E and 87°0′E, and latitudes of 19°30′N and 20°40′N (Behera, et al. 2004). The Godavari deltaic region is positioned within the East Godavari District of Andhra Pradesh, situated along the eastern coast of India. Geographically, the Godavari delta extends between the longitudes of 81°30′ and 82°35′E and the latitudes of 16°18′ and 17°05′N, spread over 5,100 km$^2$ (Sarma et al. 2001).

## 2.3 Possible Threats to the Indian Delta Regions

Approximately 25% of India's population resides within 50 km distance from the coastline (Abarna et al. 2023). These are dynamic and productive landforms at the meeting point of terrestrial and marine ecosystems. According to Kuenzer and Renaud (2012), deltas tend to be heavily inhabited due to the opportunities they offer for income generation, primarily through agriculture and aquaculture production, and associated trade. In addition to natural hazards, increased economic activities and other human-induced processes, pose serious threat to many river delta's worldwide (Syvitski et al. 2009; Tessler et al. 2015; Szabo et al. 2015). Newly emerging coastal hazards such as climate change-related sea-level rise, increased salinity, flooding, and cyclones, are jeopardizing the socio-ecological systems

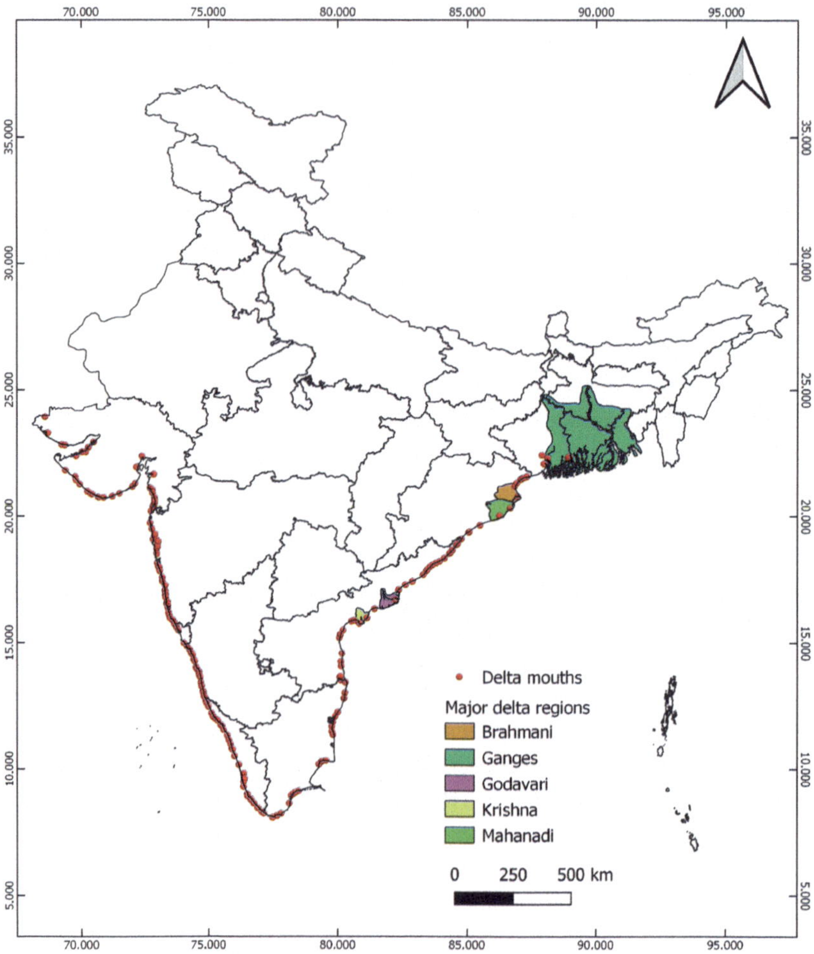


**Fig. 2.1** The map of the main delta regions of India with Delta mouths

of the deltas (Renaud et al. 2013; Renaud et al. 2016). While it is important to acknowledge that deltas and their associated river basins are natural dynamic systems, human interventions, both in-situ (e.g., urbanization, industrialization, deforestation, groundwater and hydrocarbon extraction, agricultural intensification, and channel regulation) and ex-situ (e.g., increased upstream water withdrawals, diversion, and sediment trapping), pose significant threats to these regions (Renaud et al. 2013). Consequently, these areas like any other water-based environments have been consistently undervalued and used as a common repository for waste disposal. Figure 2.2 illustrates some of the major natural and anthropogenic activities posing threat to these regions. Collectively, these processes exacerbate the loss of livelihoods and exacerbate existing disparities, leading to population migration and the emergence of new pockets of poverty, especially in metropolitan areas within the delta regions (Martin et al. 2013). A graphical representation illustrating the evolution of author's keywords pertaining to the Indian Delta Regions over the years is depicted in Fig. 2.3. This article presents an analysis of the growing problem of pollutants in the deltas of India, with a focus on possible mitigation measures in order to achieve SDGs.

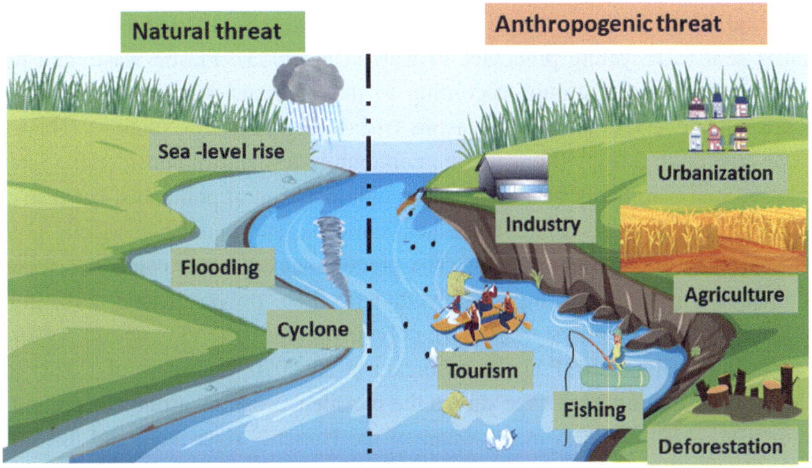


**Fig. 2.2** The major natural and anthropogenic threats to Indian Delta Region

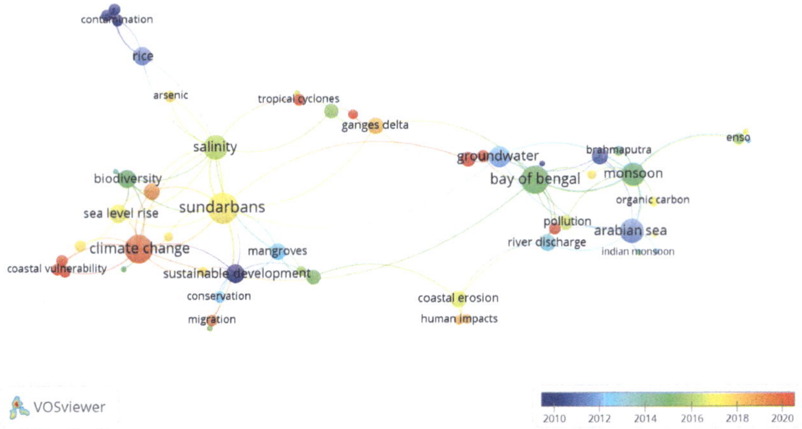


**Fig. 2.3** Visualization of author's keywords on Indian Delta Regions over the years

## 2.4 Pollution in Indian Deltas

### 2.4.1 Plastic Waste Pollution

Plastic contaminants originating mostly from terrestrial sources have emerged as a significant menace to the marine ecosystem. Various leisure pursuits, such as engaging in picnics, boating, swimming, and fishing in marine environments, have the potential to produce plastic waste. This waste is either directly disposed of in the sea or it finds its way into the marine ecosystems through water channels. Annually, India produces a total of 15 million metric tons (Mt) of plastic trash and only 25% of this material is sent to recycling processes (Tiwari et al. 2023). Plastic waste can be categorized into two discrete groups according to their dimensions: macroplastics (with a size of 25 mm or larger) and microplastics (measuring 5 mm or smaller) (Naidu et al. 2021; Oswalda et al. 2021). The scientific community has shown considerable interest in microplastics and commonly describes it as conglomeration of pollutants, because of their capacity to adsorb a range of additives, heavy metals, and persistent organic pollutants (POPs) ubiquitous in the surroundings. The predominant polymer categories recognized as microplastics in India encompass polyethylene (PE), polypropylene (PP), polyethylene terephthalate (PET), polystyrene (PS), polyvinyl chloride (PVC), nylon, polyurethane (PU), as well as polyamide (PA) (Veerasingam et al. 2020; Srihari et al. 2022; Ranjani et al. 2022). Research findings have shown the prevalence, spatial arrangement, and origin of microplastic (MP) particles in the water of the Hooghly river delta and the silt along its riverbanks inside the Sundarbans Biosphere Reserve (Neelavannan et al. 2023). The findings of the research indicate that the concentrations of MPs in water were $718 \pm 244$ items/$m^3$ (average $\pm$ standard deviation, $n = 10$), while in sediments, the concentrations were $428 \pm 266$ items/kg dw (mean $\pm$ standard deviation, $n = 17$) (Neelavannan et al. 2023). These values were generally higher in comparison to other river systems. Napper et al. (2021) estimated that a significant amount of microplastics, ranging from 1 to 3 billion pieces is released into the Bay of Bengal daily where the concentration of microplastics exhibits an upward trend from the point of origin to the open sea. Another serious problem with plastic pollution is the use of mercury as a catalyst in its production. Mercury was found in the Gangetic Bay of Bengal region and may occur in several forms, including inorganic as well as organic and elemental forms (Bhandari and Nakamura 2016). Methyl mercury, a neurotoxic substance, has the ability to readily traverse the blood–brain barrier and placenta, hence has been posing a potential threat to the developing brain. Additionally, it has been shown that this condition may contribute to the development of Alzheimer's disease, allergies, depression, renal failure, speech and visual impairments, as well as hypospermia (Bhandari and Nakamura 2016). Table 2.1 presents data on the presence of microplastic pollution in several deltas in India, including information on the shape, color, quantity, and polymer type of the microplastics.

**Table 2.1** Microplastic concentration at major Indian Delta rivers

| Study area | Shape | Color | Abundance (d. w.) | Polymer type[a] | References |
|---|---|---|---|---|---|
| Ganga River | Fibers, filaments, foams/beads, films, as well as fragmented plastics | – | ~99 to 410 items per kg | PET, PE, PP, PS | Sarkar et al. (2019) |
| | Fragments, foams, films, and filaments | White, transparent, blue, black, orange, red, yellow, green, and brown | 380 to 684 items/1000 $m^3$ | PE, PP, PS | Singh et al. (2021) |
| | Lines, spheres, fibers, beads and fragments | – | 153 to 2069 pieces/$m^3$ | – | Ghosh et al. (2021) |
| | Fiber, fragment, film, foam, and pellet | White, red, blue, green, blank, and yellow | 93 to 1200 items/kg dw, 450 to 1200 items/$m^3$ | HDPE, POM, PPS, PAM, PP, PTFE, PBD | Neelavannan et al. (2023) |
| Cauvery River | Films, fibers, fragments, foams | Orange, white, green, saffron red and more | ~386 to 85 items per kg | PA, PP, PVC, PS, PET, PE, PEG | Maheswaran et al. (2022) |
| | Fragment, fibrous, pellets | White, blue, green, blank, yellow, red | 8 to 119 particles/$m^2$ | HDPE, LDPE, PE, PA, PVC, PS | Neelavannan et al. (2022) |
| Krishna-Godavari | Filament, fragments, films | Black, transparent, yellow, white, green, blue, red | 1.7–13 items/ 50 g DW | PP, PE, PS | Dhineka et al. (2022) |

[a] HDPE—High-density polyethylene; POM—polyoxymethylene or polyacetal; PPS—polyphenylene sulfide; PAM—polyacrylamide; PP—polypropylene; PTFE—polytetrafluoroethylene; PBD—polybutadiene; PET—polyethylene terephthalate; PVC—polyvinyl chloride; PA—polyvinyl alcohol; PP—polypropylene; PE—polyethylene; PS—polystyrene; and PA—polyamide

As on date, the smallest microplastic particles are often overlooked due to technical limitations. With improvements in equipment and methods such as Fourier Transform Infrared microscopy, and Raman microscopy, the smaller size fractions must be investigated, and their impacts on aquatic life from ecological and toxicological perspectives must be evaluated. This is important as the smaller-sized micro plastic particles are preferentially ingested by zooplankton and other lower trophic level aquatic organisms and may translocate across a cell membrane (Bermúdez and Swarzensk 2021).

It is clear that managing plastic waste is linked to many other SDGs, including addressing climate change (SDG 13), protecting life below water (SDG 14), encouraging responsible consumption and production (SDG 12), supporting sustainable cities and communities (SDG 11), encouraging good health and well-being (SDG 3), and conserving life on land (SDG 15).

### 2.4.2 Aquaculture Waste Pollution

The proliferation of aquaculture infrastructure in coastal regions has resulted in unprecedented damage to the mangrove ecosystems. Waste from aquaculture ponds and marine aquaculture is mainly produced as aquaculture sludge consisting of settled food remnants, fish or shrimp excrement, and wastes stemming from the processing of fish or shrimp (Nagaraju et al. 2021). From 2019 to 2021, Nagaraju et al. (2021) collected sludge samples from aquaculture ponds at 35 sites around the delta of Andhra Pradesh during the pre-monsoon and post-monsoon seasons. The sample collection sites were divided into zones depending on the farming intensity as well as proximity to potentially polluted regions. They found that ammonia and nitrate concentrations are greater during the pre-monsoon season, leading to eutrophication. In addition to that, intense aquaculture activities have resulted in addition of organic effluents with high level of alkalinity, salinity, total $NH_3$, $NO_3^-$, Ca, K, Fe, and $Cl^-$ contents. The pollutants reduced biological oxygen demand (BOD) of the aquatic ecosystems significantly affecting the aquatic life. Further, several negative socioeconomic impacts were recorded including deterioration of critical coastal wetlands, increased salinity of groundwater and farmlands, and the loss of productivity. The aquaculture operations in the delta region of Andhra Pradesh were of high intensity and expanded in northeastward direction, as revealed from the geospatial information and on-site survey results. The period from 2016 to 2020 witnessed a substantial growth of 6.08% in the area under aquaculture (Nagaraju et al. 2022). In the western delta area of Andhra Pradesh, it was observed that at least 78% of the water samples were of low quality and were unsuitable for consumption and household use. The mean $NH_3$ concentration in the aquaculture sample was 0.15 mg/l (Nagaraju et al. 2023). It was observed that 78% of the water samples were beyond the permissible limit notified by WHO. The findings indicate that $NH_3$ concentrations are above the acceptable thresholds, posing serious concerns in aquatic farming environments owing to its hazardous properties (Nagaraju et al. 2023). From 1973 to 2006, the Mahanadi delta witnessed a reduction of 26.06 $km^2$ in mangrove cover, and a concurrent expansion of 3657 ha in aquaculture (Pattanaik and Prasad 2011). Between 1999 and 2009, the expansion of aquaculture in the Indian Sundarbans region resulted in the conversion of 10,536.67 ha land (3.71%) into agricultural land, 730.40 ha (about 3.87%) of mudflats, and 623.23 ha (approximately 0.28%) of mangrove forests. Furthermore, from 2009 to 2019, the expansion of aquaculture resulted in the decline of 13,471.50 ha (equivalent to 6.02%) of agricultural land, 1,583.64 ha (approximately 9.98%) of mudflat, and 382.35 ha (about 0.18%) of mangrove forests. The practice of aquaculture in coastal regions, particularly when it comes at the cost of agricultural activities and conservation of mangrove forests, presents a significant obstacle to the achievement of Sustainable Development Goals 14 and 15 (Giri et al. 2022).

### 2.4.3 Agricultural Waste Pollution

The landholding in deltaic landscape is characterized by a large number of small-scale and subsistence farmers, who typically cultivate rice and have little agricultural drainage infrastructure (Chattopadhya et al. 2019; Nath et al. 2021). In the event of inadequate drainage facilities, agricultural residues are transported into the freshwater ecosystems of the delta through surface runoff. In the aquatic ecosystems of the Sundarbans, Sarkar and Bhattacharya (2010) noted significant rise in nitrate and phosphate concentrations during the July–October monsoon season which was ascribed to the assumed inflow of excess nitrate and phosphate compounds from runoff of the nearby agricultural areas. Similar occurrences in the Sundarbans region were documented by Mitra (2020). Moreover, organochlorine pesticide residues were found in sediment cores inside the Indian Sundarbans (Bhattacharya et al. 2003). Chakraborty et al. (2015) also reported the presence of agricultural pollutants, including those linked to phosphate-based herbicides and pesticides, which had a notable effect on bacterial diversity in different locations of Sundarbans delta. The surface water quality in the upper delta areas of the Cauvery River was compared with WHO guidelines (WHO 2004) and drinking water standards of Bureau of Indian Standards by Solaraj et al. (2010). The study revealed contamination of water due to total dissolved solids, COD, and phosphate, which were originated from the sewage and industrial effluents as well as agricultural runoff. Similarly, wells along the Cauvery Delta's shoreline showed higher amounts of Na, K, Mg, $Cl^-$, $SO_4{}^{2-}$, and $Br^-$, indicating seawater intrusion. This phenomenon was related to various pollution sources, including irrigation return flow and the use of synthetic fertilizers, as well as agricultural waste, all of which had a notable influence on the groundwater quality (Vetrimurugan et al. 2013). The analysis of sediments from the Mahanadi delta revealed anthropogenic contamination, mostly originating from agricultural practices, particularly the intensive use of pesticides in the southeast sector of the delta. The contamination was evidenced by the presence of cobalt, lead and zinc (Samal et al. 2023). To address agricultural waste pollution in the Indian delta areas, adoption of sustainable agriculture practices, better waste management, and protection of water quality must be promoted. These activities would ensure to achieve SDG 2, SDG 6, SDG 11, SDG 13, and SDG 15, and will assist India in meeting the 2030 Agenda for Sustainable Development.

### 2.4.4 Heavy Metals Pollution from Industrial and Agricultural Runoff

Heavy metal pollution is a significant cause for worry because of their long-lasting presence in the environment, their ability to be recycled via biogeochemical processes, and the potential concerns they provide in terms of toxicity. The coastal environment has seen a substantial rise in the accumulation of heavy metals in the past few decades, primarily because of human activity including industrial emissions and agro-pesticides. Sediments serve as an effective mechanism for the sequestration of heavy metals, primarily via formation of stable complexes with organic matter, carbonates, and iron-manganese oxides present in the sediment (Peng et al. 2009). Geogenic processes, such as volcanic eruptions, leaching, and weathering of rocks, have also contributed to an increased concentration of heavy metals in the natural environment. Furthermore, heavy metals such as copper, cadmium, nickel and zinc have polluted the river, contaminating many fish species and potentially endangering human health through the process of biomagnification. Samal et al. (2023) analyzed the mean concentrations of Co and Pb in water of the deltaic areas of the Mahanadi River and found to exceed the global average values. Specifically, in the Kujang region, the mean Co concentration was 47.25 ppm, while the average Pb concentration was 44.70 ppm. Likewise, in Subala region, the average Co concentration was 37.03 ppm, and the average Pb concentration was 61.27 ppm (Samal et al. 2023). Similarly, Fe, Mn, Cr, Zn, Cu, Ni, and Pb were the most abundant metals in the sediments of three reservoirs from the Delta area of Cauvery river. *Catla catla* and *Etroplus suratensis* had the highest concentration of metals among the fish species, indicating a concern for human consumption (Dhanakumar et al. 2013). Drinking groundwater samples from Cauvery river contained Pb and Ni much above the permissible level of 0.05 ppm and, 0.1 ppm, respectively (WHO 2004; Vetrimurugan et al. 2017). The gill samples of *C. catla* and *Cirrhinus reba*, obtained from the Ganga basin, exhibited Zn concentrations of $32.41 \pm 2.55$ mg/kg and $4.77 \pm 0.34$ mg/kg, respectively. Among the heavy metals analyzed, Zn had the highest concentration, followed by Pb, Cu, Cd, and Cr, in the chosen fish tissues (Maurya et al. 2019). The findings of Mitra et al. (2010) revealed a similar trend in the accumulation pattern of Zn, Cu, and Pb in the muscles of juvenile fish from the Gangetic delta area in the northern Bay of Bengal. Water contaminated with heavy metals is used for irrigation, resulting in the subsequent buildup of these metals in crops as well. The current contamination of waterbodies and fish stock in the river with heavy metals calls for urgent and imperative actions from all pertinent parties. It is imperative to take part in these activities since they contribute significantly towards protection of the health of the local communities.

### 2.4.5 Pesticide Pollution

The use of pesticides in contemporary agricultural practices has resulted in nonpoint contamination of aquatic ecosystems impacting drinking water sources and non-targetted organisms (Stehle et al. 2023). In the Indian context, the presence of these pesticides has been observed in several major river basins. These compounds have shown the ability to undergo bio-magnification, and accumulate inside animal tissues, human blood, adipose tissue, and breast milk (Rano and Singh 2021). For instance, the analysis for chlorpyriphos, ethion, and dicofol in the river water of Deomoni from the Terai region of West Bengal yielded average pesticide residue concentrations of $0.0091 \pm 0.0020$ mg/l, $0.0180 \pm 0.0071$ mg/l, and $0.0892 \pm 0.0375$ mg/l, respectively. In sediments, the corresponding concentrations were $0.0513 \pm 0.0085$ mg/l, $0.0414 \pm 0.0045$ mg/l, and $0.1271 \pm 0.0122$ mg/l. In the muscle tissues of *Puntius* sp. fish, the average concentrations were $5.0371 \pm 1.4236$ ppm, $3.7700 \pm 0.6391$ ppm, and $2.9599 \pm 0.4027$ ppm for chlorpyriphos, dicofol, and ethion, respectively (Singh et al. 2015). The concentrations of Hexachlorocyclohexanes, Dichlorodiphenyltrichloroethanes, and endosulfan residues were determined in water samples from the Cauvery river, and the concentrations were as high as 2,300 ng/l, 3,600 ng/l, and 15,400 ng/l, respectively. In the sediment, the highest concentrations of HCHs and DDTs were reported i.e. 158 ppb and 9.15 ppb of dry weights, respectively. The samples of biota, namely fish and prawns, that were obtained from the river exhibited notable concentrations of HCHs (228 ppb) and DDTs (2,805 ppb) (Patil et al. 2015). Chakraborty et al. (2016) quantified the geometric mean concentration of organochlorine pesticides ($\Sigma$OCPs) in the river Hoogly, which was 53 ppb, and polychlorinated biphenyls ($\Sigma$19PCBs) which was 108 ppb. These values were greater in the river Brahmaputra with the corresponding values of 24 and 77 ppb. Within the category of OCPs, it was observed that $\gamma$-HCH exhibited the highest frequency of detection in both the rivers, indicating the continued use of lindane. Residues of DDT and endosulfan were detected in specified geographical regions where historical or current sources were present. Higher levels of heavier congeners (namely, penta-hepta) were detected in locations along the Hooghly river area where there was a significant presence of port and industrial activity, including informal operations involved in the processing of electronic waste scrap. Moreover, in the suburban industrial areas of Howrah district, elevated levels of PCB-126 were observed along river Hooghly. Polychlorinated biphenyls (PCBs) were widely dispersed throughout the river Brahmaputra.

## 2.5 Mitigation Strategies for Combating Pollution in Delta Regions of India for Achieving Sustainable Development Goals

The concept of sustainable development, first introduced in 1987, emerged as a prominent focal point at the UNECD in 1992. Over the past 20 years, worldwide efforts for sustainable development have been greatly influenced by the text known as "Agenda 21" (Karlsson-Vinkhuyzen et al. 2012). A total of 193 countries agreed in September 2015 to work towards a set of 17 Sustainable Development Goals (SDGs) and 169 related subgoals that must be achieved by 2030 (Wang et al. 2023; Chien et al. 2021; Bilan et al. 2019). The SDGs are a set of international directives that need adherence from the global community in order to effectively tackle global concerns. The purpose of their design is to guarantee the preservation of the natural environment and the sustenance of life on the planet. This will enable the future generations to live in a state of complete dignity, peace, and prosperity (Labadi et al. 2021; Guerra et al. 2021). The objectives include three key domains, viz., social development (zero hunger, no poverty, gender equality, peace and justice and strong institutions), environmental development (climate action, life below water, life on land), and economic growth (reduced inequalities, decent work and economic growth), while considering the well-being of individuals, the planet, the promotion of peace, prosperity, and collaborative efforts (Kutan et al. 2018). Clean water and sanitation (SDG 6) is one of the 17 SDGs which addresses a number of issues, including drinking water, hygiene, wastewater treatment and reuse, and ecosystem health, and is also linked to other goals such as Zero Hunger (SDG 2), Good Health and Wellbeing (SGD 3), Sustainable Cities (SDG 8), and Life Below Water (SGD 14) (Weerasooriya et al. 2021). Water resources are essential for many aspects of development, including sustaining a healthy ecosystem, ensuring food security, promoting health, and reducing poverty (UN SGD-6 Report 2018). SDG 6 is particularly relevant in delta areas such as the Indian delta region, where proper water management is critical for the lives of millions of people who rely on these fertile fields. However, these regions are highly susceptible to pollution, which threatens the quality of water and the overall ecosystem health. Therefore, addressing pollution in delta regions is of paramount importance addressing SGD 6. Consequently, it becomes imperative to enact efficient measures for pollution control and foster a culture of responsible environmental management within these geographical areas. The Indian Delta region's progress towards various SDGs and its possible effects on pollution are shown in the Table 2.2. The accomplishments and endeavors in the area offer a promising direction in tackling numerous types of pollution under the SDGs.

**Table 2.2** The achievement of the SDGs in the Indian Delta Region and its impact on pollution

| SDGs | Achievement in Indian Delta Region | Impact on pollution |
|---|---|---|
| SDG 1: No poverty | Progress is being made towards social and economic goals, notably in southern states such as Kerala and Tamil Nadu (Bajpai and Biberman 2020) | Reduced pollution is connected to economic development and poverty alleviation (Sebesvari et al. 2016) |
| SDG 6: Clean water and sanitation | The Swachh Bharat initiative has achieved 100% compliance in rural regions of states such as Rajasthan, Punjab, Maharashtra, Kerala, and Gujarat in terms of establishing home toilets. Despite continued obstacles, Gujarat, Haryana, and Madhya Pradesh have made significant progress in sanitation and availability to safe drinking water, with about 100% of the population having access to safe water (Bajpai and Biberman 2020) | Addressing water pollution and improving sanitation facilities |
| SDG 11: Sustainable cities and communities | Currently, 73.6% of urban wards nationwide offer door-to-door waste collection services, with the goal of reaching 100% by 2030 in alignment with SDG 11.6. Leading in waste collection are Andhra Pradesh (95.8%), Madhya Pradesh (96.7%), Rajasthan (94.3%), and Delhi (86.4%) (Bajpai and Biberman 2020) | Mitigating urban pollution and promoting sustainable practices in communities |
| SDG 13: Climate action | Contribution potential through programmes such as the IAAS Delta Project, which attempts to improve water management in deltas | Addressing pollution linked to climate change and promoting sustainable environmental practices |

## 2.6 Policy Framework for Mitigating Pollution at Indian Deltas

The Indian delta area, renowned for its distinctive ecological and cultural importance, confronts urgent issues stemming from pollution, notably water pollution. These pollutants pose significant risks to both human welfare and environmental sustainability. Nevertheless, there have been endeavors to address these concerns and adhere to the objectives outlined in the SDGs. India has a rich historical background concerning the implementation of environmental conservation measures, which can be traced back to the 1970s. Notably, the 42nd Amendment Act, enacted in 1976, was a significant milestone as it expressly integrated environmental protection into the Indian Constitution (Ramakrishna 1984). A large number of policies/acts/rules have been implemented by Indian government to safeguard our environment. For example, the National Water Policy-2012 (MoJS, http://mowr.gov.in), the National Environment Policy-2006 (MoEF&CC-http://moef.gov.in), the National Plan for Conservation of Aquatic Ecosystem-2013 (NPCA; MoEF&CC, http://moef.gov.in), the National Action Plan on Climate Change-2008 (NAPCC; MoJS, http://mowr.gov.in), the Atal Mission for Rejuvenation and Urban Transformation (Ministry of Housing and Urban Affairs-MoHUA-(http://mohua.gov.in), and The Mahatma Gandhi National Rural Employment Guarantee Act-2005 (MGNREGA; Ministry of Rural Development-MoRD, https://rural.nic.in) are notable policies and initiatives executed by the Indian government. The proliferation of regulations and programmes implemented across many ministries exacerbates the complexities associated with the protection, restoration, and management of waterbodies (Yadav and Goyal 2022). The implementation of the Integrated Coastal Zone Management (ICZM) of Sundarbans is one such initiative by the Government of India aiming to facilitate sustainable development (Pitchaikani 2020).

The Plastic Waste Management Rules (2016) (Samuel and Ahlawat, 2021) were adopted by the Government of India in accordance with provisions outlined in sections 3, 6 and 25 of the Environment (Protection) Act (1986). These rules were issued by the MoEFCC, the designated Ministry of the Government of India. There is a pressing need to implement strategies to mitigate plastic use in everyday life. Rule 4 establishes specific requirements for the authorized production, distribution, and use of carry bags, plastic sheets, coverings made of plastic sheets, and multi-layered packaging. The aim is to promote environmentally sustainable local plastic waste management practices. According to Rule 17(3), the Pollution Control Boards at the national, and state levels are authorized to compile the report about the execution of these regulations. Furthermore, it grants these agencies the ability to seek aid from the District Magistrates or the Deputy Commissioners to enforce these regulations inside their respective districts.

Additionally, the monitoring and management of water bodies is of utmost importance depending on the local circumstances. It is essential to maintain the quality of water from all sources, particularly the groundwater and surface water used for drinking, residential, and irrigation purposes free from various contaminants. The urgency of preventing more contamination necessitates implementing precise management strategies, including the treatment of industrial waste and adherence to regulatory guidelines for wastewater disposal. It is essential to implement stringent pollution control measures by rigorously enforcing regulations to guarantee the effective functioning and upkeep of wastewater treatment facilities. It is also recommended that more government initiatives be designed to ensure the availability of clean and drinkable water. Moreover, it is suggested that all stakeholders adopt prompt and enduring collaborative measures to mitigate the pollution levels, given the significant public health risks associated with this matter. By persistently implementing and enhancing these policies, promoting cooperation among various stakeholders, and enhancing public awareness, the Indian Delta region can advance towards a more sustainable and prosperous future. This progress will be achieved while maintaining a harmonious relationship with its distinctive natural environment and cultural legacy and making substantial contributions to global endeavors to achieve the Sustainable Development Goals (Table 2.2).

## 2.7 Conclusion

In conclusion, India's delta regions, known for their natural diversity and cultural significance, are under threat from many forms of pollution, particularly water pollution. This study investigated the ecological and economic effects of anthropogenic contaminants on these critical ecosystems, which included heavy metals, plastics, and pesticides. Climate change, industrial activity, urbanization, and agricultural practices have all posed several challenges to delta regions, jeopardizing both biodiversity and the well-being of millions of people who rely on them. Aligning these efforts with the SDGs, particularly SDG 6 (Clean Water and Sanitation), is critical. A comprehensive plan incorporating various SDGs is required, including Zero Hunger (SDG 2), Good Health and Wellbeing (SDG 3), Sustainable Cities (SDG 11), and Life Below Water (SDG 14). SDG 12 (Responsible Consumption and Production) is equally important for fostering sustainable waste management. While commendable initiatives like the Plastic Waste Management Rules and Integrated Coastal Zone Management have been set in motion by the government, sustaining these positive trajectories requires continuous efforts. Fostering heightened public awareness and promoting collaborative endeavors across sectors are imperative to uphold the ecological integrity and socio-economic resilience of these vital delta regions in the face of escalating environmental changes.

**Acknowledgement** The authors are grateful to the Director, CSIR- NBRI for encouragement and providing the necessary facilities.

**Conflict of Interest** The authors declare that they have no conflict of interest.

## References


Abarna R, George SL, Balasubramani K, Yuvaraj S, Shekhar S, Gnanappazham L, Prasad KA (2023) Estimating built-up risk from multi-natural hazards: a case study of Northern coastal plains of Tamil Nadu. Nat Hazards Res 3(1):49–65

Akhtar N, Syakir Ishak MI, Bhawani SA, Umar K (2021) Various natural and anthropogenic factors responsible for water quality degradation: a review. Water 13(19):2660

Bajpai N, Biberman J (2020) India and the SDGs (No. 22). ICT India Working Paper

Behera L, Sain K, Reddy PR (2004) Evidence of underplating from seismic and gravity studies in the Mahanadi delta of eastern India and its tectonic significance. J Geophys Res Solid Earth 109(B12)

Bermúdez JR, Swarzenski PW (2021) A microplastic size classification scheme aligned with universal plankton survey methods. MethodsX 8:101516

Bhandari BB, Nakamura R (2016) The Bay of Bengal: a forgotten sea. In The proceeding of the symposium on the Bay of Bengal, Yangoon, Myanmar, vol 4

Bhattacharya B, Sarkar SK, Mukherjee N (2003) Organochlorine pesticide residues in sediments of a tropical mangrove estuary, India: implications for monitoring. Environ Int 29(5):587–592

Bilan Y, Streimikiene D, Vasylieva T, Lyulyov O, Pimonenko T, Pavlyk A (2019) Linking between renewable energy, CO2 emissions, and economic growth: challenges for candidates and potential candidates for the EU membership. Sustainability 11(6):1528

Chakraborty A, Bera A, Mukherjee A, Basak P, Khan I, Mondal A, Bhattacharyya M (2015) Changing bacterial profile of Sundarbans, the world heritage mangrove: impact of anthropogenic interventions. World J Microbiol Biotechnol 31:593–610

Chakraborty P, Khuman SN, Selvaraj S, Sampath S, Devi NL, Bang JJ, Katsoyiannis A (2016) Polychlorinated biphenyls and organochlorine pesticides in River Brahmaputra from the outer Himalayan Range and River Hooghly emptying into the Bay of Bengal: occurrence, sources and ecotoxicological risk assessment. Environ Pollut 219:998–1006

Chattopadhyay K, Gayan S, Mondal I, Mishra SK, Mukherjee AK, Reddy JN, Marndi BC, Sarkar RK (2019) Stress tolerant rice and on-farm seed production ensure food security and livelihood to small and marginal farmers of Sundarbans (Indian site). SAARC J Agric 17(2):127–139

Chen S, De Bruyne C, Bollempalli M (2020) Blue economy: community case studies addressing the poverty–environment nexus in ocean and coastal management. Sustainability 12(11):4654

Chien F, Anwar A, Hsu CC, Sharif A, Razzaq A, Sinha A (2021) The role of information and communication technology in encountering environmental degradation: proposing an SDG framework for the BRICS countries. Technol Soc 65:101587

Census Data (2011) Census of India Website: Office of the Registrar General & Census Commissioner. http://www.censusindia.gov.in/2011census/population_enumeration.html. Accessed 28 Nov 18

Day JW, Agboola J, Chen Z, D'Elia C, Forbes DL, Giosan L, Yanez-Arancibia A (2016) Approaches to defining deltaic sustainability in the 21st century. Estuar Coast Shelf Sci 183:275–291

Day JW, Ramachandran R, Giosan L, Syvitski J, Kemp GP (2019) Delta winners and losers in the Anthropocene. In Coasts and estuaries, pp 149–165. Elsevier

Dhanakumar S, Murthy KR, Solaraj G, Mohanraj R (2013) Heavy-metal fractionation in surface sediments of the Cauvery River Estuarine Region, Southeastern Coast of India. Arch Environ Contam Toxicol 65:14–23

Dhineka K, Kaviarasan T, Sambandam M, Mishra P, Murthy MR (2022) Comparison of microplastic abundance in varying depths of deep-sea sediments, Bay of Bengal. In OCEANS 2022-Chennai, pp 1–4. IEEE

Franco S, Mandla VR, Ram Mohan Rao K (2017) Trajectory of urban growth and its socioeconomic impact on a rapidly emerging megacity. J Urban Plann Dev 143(3):04017002
Ghosh S, Das R, Bakshi M, Mahanty S, Chaudhuri P (2021) Potentially toxic element and microplastic contamination in the river Hooghly: implications to better water quality management. J Earth Syst Sci 130:1–16
Giri S, Samanta S, Mondal PP, Basu O, Khorat S, Chanda A, Hazra S (2022) A geospatial assessment of growth pattern of aquaculture in the Indian Sundarbans Biosphere Reserve. Environ Dev Sustain 24(3):4203–4225
Government of India. Solid Waste Management Rules (2016)
Guerra JBSOA, Hoffmann M, Bianchet RT, Medeiros P, Provin AP, Iunskovski R (2021) Sustainable development goals and ethics: building "the future we want. Environ Dev Sustain 1–22
Guzzetti E, Sureda A, Tejada S, Faggio C (2018) Microplastic in marine organism: environmental and toxicological effects. Environ Toxicol Pharmacol 64:164–171
Helmer R, Hespanhol I (1997) Water pollution control: a guide to the use of water quality management principles. CRC Press
Huang Z, Yuan X, Liu X (2021) The key drivers for the changes in global water scarcity: water withdrawal versus water availability. J Hydrol 601:126658
Karlsson-Vinkhuyzen SI (2012) From Rio to Rio via Johannesburg: Integrating institutions across governance levels in sustainable development deliberations. In Natural resources forum, vol 36, no 1, pp 3–15. Blackwell Publishing Ltd, Oxford, UK
Krishna KS (2005) Science plan for coastal hazard preparedness. Curr Sci 1339–1347
Kubo S, Kumar KCVN, Demudu G, Malini BH, Rao KN, Agrawal R, Ramakrishnan R, Rajawat AS (2018) Monitoring of eco-restoration of mangrove wetlands through time series satellite images: a study on Krishna-Godavari Delta region, east coast of India. Geogr Rev Japan Ser B 90(2):66–75
Kuenzer C, Renaud FG (2012) Climate and environmental change in river deltas globally: expected impacts, resilience, and adaptation. The Mekong delta system: interdisciplinary analyses of a river delta. Springer, Netherlands, Dordrecht, pp 7–46
Kumar D (2017) River Ganges-Historical, cultural and socioeconomic attributes River GangesHistorical, cultural and socioeconomic attributes. Aquat Ecosyst Heal Manag 20:8–20. https://doi.org/10.1080/14634988.2017.1304129
Kutan AM, Paramati SR, Ummalla M, Zakari A (2018) Financing renewable energy projects in major emerging market economies: evidence in the perspective of sustainable economic development. Emerg Mark Financ Trade 54(8):1761–1777
Labadi S, Giliberto F, Rosetti I, Shetabi L, Yildirim E (2021) Heritage and the sustainable development goals: policy guidance for heritage and development actors. Int J Herit Stud
Loucks DP (2019) Developed river deltas: are they sustainable? Environ Res Lett 14(11):113004
Maheswaran B, Karmegam N, Al-Ansari M, Subbaiya R, Al-Humaid L, Raj JS, Govarthanan M (2022) Assessment, characterization, and quantification of microplastics from river sediments. Chemosphere 298:134268
Martin M, Kang YH, Billah M, Siddiqui T, Black R, Kniveton D (2013) Policy analysis: climate change and migration Bangladesh. Dhaka, Bangladesh: Refugee and Migratory Movements Research Unit (RMMRU)
Maurya PK, Malik DS, Yadav KK, Kumar A, Kumar S, Kamyab H (2019) Bioaccumulation and potential sources of heavy metal contamination in fish species in River Ganga basin: possible human health risks evaluation. Toxicol Rep 6:472–481
Mitra A, Mondal K, Banerjee K (2010) Concentration of heavy metals in fish juveniles of Gangetic delta of West Bengal, India. Res J Fish Hydrobiol 5(1):21–26

Mitra A, Mitra A (2020) Mangroves: a nutrient retention box. Mangrove For India Explor Ecosyst Serv 87–114

Nagaraju TV, Malegole SB, Chaudhary B, Ravindran G (2022) Assessment of environmental impact of aquaculture ponds in the western delta region of Andhra Pradesh. Sustainability 14(20):13035

Nagaraju TV, Sunil BM, Chaudhary B, Prasad CD, Gobinath R (2023) Prediction of ammonia contaminants in the aquaculture ponds using soft computing coupled with wavelet analysis. Environ Pollut 331:121924

Nagaraju TV, Sunil BM, Chaudhary B (2021) Impact of aquaculture solid waste on environment in the delta region of Andhra Pradesh: a case study. In Indian geotechnical conference, pp 369–374. Springer Nature Singapore, Singapore

Naidu SA, Mawii L, Rao VR, Anitha G, Mishra P, Narayanaswamy BE, Kumar VA, Murthy MR, Gupta GVM (2021) Characterization of plastic debris from surface waters of the eastern Arabian Sea-Indian Ocean. Mar Pollut Bull 169:112468

Napper IE, Baroth A, Barrett AC, Bhola S, Chowdhury GW, Davies BF, Koldewey H (2021) The abundance and characteristics of microplastics in surface water in the transboundary Ganges River. Environ Pollut 274:116348

Nath A, Samanta S, Banerjee S, Danda AA, Hazra S (2021) Threat of arsenic contamination, salinity and water pollution in agricultural practices of Sundarban Delta, India, and mitigation strategies. SN Appl Sci 3:1–15

Nations U (2018) Sustainable development goal 6 synthesis report on water and sanitation. United Nations, New York

Neelavannan K, Achyuthan H, Sen IS, Krishnakumar S, Gopinath K, Dhanalakshmi R, Rajalakshmi PR, Sajeev R (2022) Distribution and characterization of plastic debris pollution along the Poompuhar Beach, Tamil Nadu, Southern India. Mar Pollut Bull 175:113337

Neelavannan K, Sen IS, Sinha N, Thakur AK, Misra S (2023) Microplastics in the Ganga-Brahmaputra delta: sources and pathways to the sundarbans biosphere reserve-an UNESCO World Heritage Centre. Environ Adv 11:100350

Nyanga C, Obegi BN, Bich LTT (2021) Anthropogenic activities as a source of stress on species diversity in the Mekong river delta. River Deltas Research-Recent Advances

Oswalda SB, Schoorb MM, Buschmanc F, Collasa FP (2021) Macro-and mesoplastic abundance and composition in the water column of the river Waal. Rivers Uncertain Future 38

Patil NN, Selvaraj KK, Krishnamoorthy V, Elaiyaraja A, Ramaswamy BR (2015) Organochlorine pesticide contamination in the Kaveri (Cauvery) river, India: a review on distribution profile, status, and trends. Water Chall Solut Glob Scale 115–128

Pattanaik C, Prasad SN (2011) Assessment of aquaculture impact on mangroves of Mahanadi delta (Orissa), East coast of India using remote sensing and GIS. Ocean Coast Manag 54(11):789–795

Peng JF, Song YH, Yuan P, Cui XY, Qiu GL (2009) The remediation of heavy metals contaminated sediment. J Hazard Mater 161(2–3):633–640

Pitchaikani JS (2020) Integrated coastal zone management practices for Sundarbans, India

Raha UK, Kumar BR, Sarkar SK (2021) Policy framework for mitigating land-based marine plastic pollution in the Gangetic Delta region of Bay of Bengal-a review. J Clean Prod 278:123409

Ramakrishna K (1984) The emergence of environmental law in the developing countries: a case study of India. Ecology LQ 12:907

Ranjani M, Veerasingam S, Venkatachalapathy R, Jinoj TPS, Guganathan L, Mugilarasan M, Vethamony P (2022) Seasonal variation, polymer hazard risk and controlling factors of microplastics in beach sediments along the southeast coast of India. Environ Pollut 305:119315

Rano S, Singh M (2021) Strategy for the inspection of pesticide residues in food and agriculture. Sustain Agric Rev 47 Pestic Occur Anal Remediat Vol. 1 Biol Syst 309–333

Renaud FG, Syvitski JP, Sebesvari Z, Werners SE, Kremer H, Kuenzer C, Ramesh R, Jeuken AD, Friedrich J (2013) Tipping from the holocene to the anthropocene: how threatened are major world deltas? Curr Opin Environ Sustain 5(6):644–654

Renaud FG, Szabo S, Matthews Z (2016) Sustainable deltas: livelihoods, ecosystem services, and policy implications. Sustain Sci 11:519–523

Samal P, Singarasubramanian SR, Manoj MC, Srivastava J, Dsouza N, Balakrishna K, Ali S et al (2023) Heavy metal contamination assessment and its associated human health risk evaluation in the Mahanadi River sediments, India. Int J Environ Sci Technol 20(10):10673–10694

Samuel M, Ahlawat S (2021) EPR in plastic waste management: policy tool yet to be realised. Asian J Manage 12(4):394–400

Sarkar SK, Bhattacharya BD (2010) Water quality analysis of the coastal regions of Sundarban mangrove wetland, India using multivariate statistical techniques. Environ Manag 24:159–172

Sarkar DJ, Sarkar SD, Das BK, Manna RK, Behera BK, Samanta S (2019) Spatial distribution of meso and microplastics in the sediments of river Ganga at eastern India. Sci Total Environ 694:133712

Sarma VVLN, Krishna GM, Malini BH, Rao KN (2001) Landuse/Landcover change detection through remote sensing and its climatic implications in the godavari delta region. J Indian Soc Remote Sens 29:85–91

Sastry JS, Vethamony P, Swamy GN (1991) Morphological changes at Godavari delta region due to waves, currents and the associated physical processes

Sebesvari Z, Renaud FG, Haas S, Tessler Z, Hagenlocher M, Kloos J, Kuenzer C et al (2016) A review of vulnerability indicators for deltaic social-ecological systems. Sustain Sci 11:575–590

Singh S, Bhutia D, Sarkar S, Rai BK, Pal J, Bhattacharjee S, Bahadur M (2015) Analyses of pesticide residues in water, sediment and fish tissue from river Deomoni flowing through the tea gardens of Terai Region of West Bengal, India. Int J Fish Aquat Stud 3(2):17–23

Singh N, Mondal A, Bagri A, Tiwari E, Khandelwal N, Monikh FA, Darbha GK (2021) Characteristics and spatial distribution of microplastics in the lower Ganga River water and sediment. Mar Pollut Bull 163:111960

Singh N, Poonia T, Siwal SS, Srivastav AL, Sharma HK, Mittal SK (2022) Challenges of water contamination in urban areas. In Current directions in water scarcity research, vol 6, pp 173–202. Elsevier

Solaraj G, Dhanakumar S, Rutharvel Murthy K, Mohanraj R (2010) Water quality in select regions of Cauvery Delta River basin, southern India, with emphasis on monsoonal variation. Environ Monit Assess 166:435–444

Srihari S, Subramani T, Prapanchan VN, Li P (2022) Human health risk perspective study on characterization, quantification and spatial distribution of microplastics in surface water, groundwater and coastal sediments of thickly populated Chennai coast of South India. Hum Ecol Risk Assess 29:1–12

Stehle S, Ovcharova V, Wolfram J, Bub S, Herrmann LZ, Petschick LL, Schulz R (2023) Neonicotinoid insecticides in global agricultural surface waters–exposure, risks and regulatory challenges. Sci Total Environ 867:161383

Syvitski JPM, Kettner AJ, Overeem I, Hutton EWH, Hannon MT, Brakenridge GR, Day J, Vörösmarty C, Saito Y, Giosan L, Nicholls RJ (2009) Sinking deltas due to human activities. Nat Geosci 2:681–686

Szabo S, Renaud FG, Hossain S, Sebesvari Z, Matthews Z, Foufoula-Georgiou E, Nicholls RJ (2015) Sustainable development goals offer new opportunities for tropical delta regions. Environ Sci Policy Sustain Dev 57:16–23

Tessler ZD, Vörösmarty CJ, Grossberg M, Gladkova I, Aizenman H, Syvitski JPM, Foufoula-Georgiou R (2015) Profiling risk and sustainability in coastal deltas of the world. Science 349:638–643

Tiwari R, Azad N, Dutta D, Yadav BR, Kumar S (2023) A critical review and future perspective of plastic waste recycling. Sci Total Environ 163433

Veerasingam S, Ranjani M, Venkatachalapathy R, Bagaev A, Mukhanov V, Litvinyuk D, Verzhevskaia L, Guganathan L, Vethamony P (2020) Microplastics in different environmental compartments in India: analytical methods, distribution, associated contaminants and research needs. TrAC Trends Anal Chem 133:116071

Vetrimurugan E, Elango L, Rajmohan N (2013) Sources of contaminants and groundwater quality in the coastal part of a river delta. Int J Environ Sci Technol 10:473–486

Vetrimurugan E, Brindha K, Elango L, Ndwandwe OM (2017) Human exposure risk to heavy metals through groundwater used for drinking in an intensively irrigated river delta. Appl Water Sci 7:3267–3280

Wang D, Dong L, Di S (2023) Data-driven comparison of urban sustainability towards sustainable urban development under sustainable development goals (SDGs): a review based on bibliometric analysis. Front Energy Res 11:1168126

Weerasooriya RR, Liyanage LPK, Rathnappriya RHK, Bandara WBMAC, Perera TANT, Gunarathna MHJP, Jayasinghe GY (2021) Industrial water conservation by water footprint and sustainable development goals: a review. Environ Dev Sustain 1–49

World Health Organization (2004) Guidelines for drinking-water quality, vol 1. World Health Organization

Yadav S, Goyal VC (2022) Current status of ponds in India: a framework for restoration, policies and circular economy. Wetlands 42(8):107

# Sustainable Solid Waste Disposal to Mitigate Water Pollution Problem and Its Social Environmental Impact

3

Shyamal Datta, Mouli Mitra, and Subhasis Roy

**Abstract**

Solid waste disposal is a widespread problem worldwide, more precisely in urban areas. Solid waste disposal involves several complex technical challenges not limited to environmental pollution but also harm the water ecology. Solid waste can create unsanitary conditions and damage the ground surface water, leading to water pollution and adversely affecting the human population worldwide. The government authorities, stakeholders, and policymakers are responsible for not taking proper measures to mitigate this serious issue that may cause worldwide water scarcity. Therefore, this chapter discusses recent innovations in sustainable solid waste disposal to mitigate the water pollution problem and its social and environmental impact with several steps. Utilization of surfactants is prioritized, which attributes to eco-friendly and cost-optimization processes for wastewater treatment. The application of surfactants and biosurfactants in water treatment shows a promising future due to its increased bioavailability, low cost, easy applicability, and affordability for poor countries. Various challenges regarding the implementation of water treatment to stop water pollution are thoroughly reviewed, and relevant data are analyzed to develop a sustainable water treatment plan using biosurfactants. Different types of waste, categorizing it, and developing approaches for scalable and sustainable methods to deal with water pollution are discussed. Moreover, proper monitoring of the situation and selection of regions based on the urgency and selection of particular sustainable methods in a scientific way can help in the

All authors made equal contributions

S. Datta · M. Mitra · S. Roy (✉)
Department of Chemical Engineering, University of Calcutta, 92 A. P. C. Road, Kolkata 700009, India
e-mail: srchemengg@caluniv.ac.in

 
M. Nasr and B. Ravindran (eds.), *Solid Waste Management in Delta Region for SDGs Fulfillment*, Deltas of the World, https://doi.org/10.1007/978-3-031-58253-0_3

implementation process to improve the quality of freshwater, the water ecosystem, and the environment. The sustainable waste management method also aims to minimize the health-related risk associated with water pollution with a promising approach to mitigate water scarcity worldwide.

## 3.1 Introduction

Rapid development in the twentieth century followed by population growth, sequentially with unpredictable climate change abruptly mounting pressure on the limited water resources on our planet. According to the report, 62 million tonnes of garbage is collected yearly from a country like India. 70% of the total waste is collected, 31 million of which is deposited at the landfill site (Mastakar et al. 2019). Several relevant data indicate that due to the limited resources of fresh water, it is crucial to figure out a balanced approach between human needs and the development of technologies to minimize environmental damage (Mastakar et al. 2019; Shrivastava 1995). Solid wastes are the consequences of the day-to-day activities of the human population as well as the activities of industrial or technological innovation. In the era of technological development, solid waste management to mitigate water pollution has become a critical approach worldwide, mainly in underdeveloped countries (Abdel-Shafy and Mansour 2018). Poor solid waste management leads to more wastewater, which pollutes our environment, freshwater resources, and water ecosystems. Therefore, it harms agriculture, soil quality, and human health, leading to a severe global water crisis (Abdel-Shafy and Mansour 2018; Jury and Vaux 2007). The data shows that drinking water resources are adversely polluted due to 70% of water pollution incidents, which cause severe direct harm to human health (Geissen et al. 2015). Delta sustainability, therefore, aims to strengthen the measure of waste management procedures. It can thus help implement wastewater recycling and increase the reuse and utilization of water resources in an environment-friendly and cost-optimized process.

The generation of factory waste and its discharge becomes the burden of the municipalities and government stakeholders due to a lack of infrastructure and knowledgeable and skilled manpower to deal with a huge amount of waste under scientific treatment (Abdel-Shafy and Mansour 2018; Guerrero et al. 2013). The disposal of solid waste, including sewage and non-sewage waste, without proper treatment, accumulates in the environment, which causes water pollution in both surface and groundwater, damaging the balance in the water ecosystem and affecting the water lives in the aquatic habitats at disposal sites (Sasakova et al. 2018). According to the consumption of unsafe drinking water and poor sanitation, almost 829,000 deaths were reported in the year 2021, by UNESCO (Lin et al. 2022). Less involvement of the government and authorities and lack of awareness among them left the whole waste management and water treatment system either broken or not running smoothly. Problems associated with poor waste management and water pollution demand an implementation in the whole process.

Considering the situation regarding water pollution, the discussion on different sources of water pollution and possible strategies for mitigating the problems are the main concerns of this chapter. Economy and sustainability are vital in any long-term project execution (Zaman et al. 2022). Therefore, water pollution's social and economic impacts and different possible outcomes are vividly discussed. The possible reasons and challenges regarding water pollution control are considered important factors to analyze the condition to propose the potential prospect of minimizing the challenges and implementing the mitigation procedure. Relevant data are collected on a global scale from various scholarly articles, and the graphical representation of the data lights on the proper picture of the problem of solid waste and how it is related to water pollution (Shrivastava 1995; Zanoni and Fungaroli 1973). This chapter, therefore, focuses on understanding the different possible ways of using chemicals and biosurfactants in water treatment as it helps in cost optimization throughout the wastewater treatment process. The use of surfactants can be a well-improved and well-planned scientific procedure that includes water recycling and reusing wastewater in a more promising way through an area-crisis-based model.

The aims of this chapter are

i. Understanding the areas of implementation in the waste management procedure which is affecting the environment leading to water pollution
ii. To incorporate more scientific and sustainable methods of waste management and wastewater treatment procedures to mitigate water pollution and water scarcity problems at the same time
iii. Understanding the importance of delta sustainability in mitigating waste management and water pollution to develop low-cost, continual, and feasible procedures.

## 3.2 The Need for Wastewater Management

Wastes are regarded as no longer in use and affect disposal sites such as soil, environment, and freshwater resources if left untreated. When the wastes are discharged into the river basin areas, the water ecology is adversely affected, and the agricultural fields are exposed to massive sewage sludge this way (Correction to Lancet Respir Med 2021). The impact of untreated waste on the environment is considered one of the leading causes of water scarcity worldwide. The solid wastes deposited in the water sites are toxic, flammable, and contain reactive substances. These are hazardous and harmful agents responsible for causing waterborne diseases and affecting human health (Fuller et al. 2022). Waterborne diseases such as cholera, dysentery, hepatitis-A, typhoid fever, etc. are transmitted due to the poor management of waste, poor sanitation practices, and untreated industrial wastes accumulation and thus affect human health mainly in rural villages or slums areas, leading to the occurrence of epidemic or an endemic situation (Haseena et al. 2017). The situation worsens in the rainy season, specifically in the urban area, after heavy rain causes floods due to the lack of a proper water drainage system. The need for wastewater management can be assumed to protect or maintain the water quality in terms of physical, chemical, and biological properties. Various organic and inorganic substances present in water can influence water quality; more precisely, nitrogenous and phosphorous compounds, arsenic, and microplastics beyond the limit can reduce the quality of the water ecosystem dramatically (Pathak and Singh 2023). Lack of awareness and less participation of the government stakeholders in the urban or rural area causes unplanned accommodation and disposal of wastes. The less involvement of the local authorities complicates the situation regarding waste management. Political interference plays a huge role in this case by weakening the system and preventing the leaders or government stakeholders from actively participating in waste management treatment procedures, thus worsening water pollution (Anuardo et al. 2022). Due to human activities and climate change, the impact on water resources should be analyzed thoroughly to predict future scarcities and to help in future innovation. To ensure pure and safe water for future generations, it is necessary to study with data that helps find out the gap in mitigating future water scarcity and the problem associated with water pollution in advance.

## 3.3 Classification of Wastes and Their Characterization

Waste management or waste treatment is based on the reuse or recycling of waste materials or solid wastes to minimize the burden and reduce its harmful effects on the environment. Untreated waste is one of the leading causes of environmental and water pollution. Waste classification is generally difficult due to the different types of resources, properties of their discharge, and types of disposals. Wastes are generated by the day-to-day activities of humans kinds to meet their daily requirements (Mastakar et al. 2019; De and Debnath 2016). These wastes can be solid, liquid, or organic based on their physical state and properties. When the wastes are discharged into the water or at the disposal sites without proper scientific treatment, it pollutes both the ground and surface water directly or indirectly. It harms the water ecology, leading to huge water scarcity (Picó and Barceló 2019). In this case, the wastewater can be classified as sewage or non-sewage. The sewage water contains urine, feces, and other contaminated biodegradable small solid particles and pollutants. The non-sewage contains stormwater, flood water, and wastewater from industry or garages (Haseena et al. 2017). There are several types of waste that directly or indirectly pollute water, including most commonly domestic waste, animal waste, biomedical waste, commercial waste, industrial waste, and factory waste.

Therefore the wastes can be classified into some major groups according to their sources, such as industrial solid waste, agricultural solid waste, municipal solid waste, and hazardous solid waste. Industrial waste contains electronic waste, solid garbage, radioactive waste, and some recyclable waste materials (Gutberlet and Uddin 2017).

Agricultural solid wastes are not generally hazardous; they are non-hazardous and contain crop residues, forest wastes, sawdust, animal product wastes, medical on-farm solid wastes, and horticultural production wastewaters. Most of the waste generated in developing countries' urban areas falls in the municipal solid waste (MSW) category, which mainly contains residential, commercial, construction, and other demolition wastes (Buenrostro et al. 2001). Based on the instrument and legislation, wastes can also be classified as hazardous and non-hazardous, which may be biodegradable or non-biodegradable. The activities of microbes such as vegetable, fruit, or wood wastes generally decompose biodegradable wastes. The non-biodegradable wastes are not decomposed by the bioactivities of microbes such as plastics, glass, toxic metals, or chemical wastes (Rastogi et al. 2020). Classification and the waste's characterization help manage and categorize the wastewater treatment procedures cost-effectively according to the requirements (Fig. 3.1).

Table 3.1 shows that the microplastic concentration affects the water level in different ways, such as through the entrance source of stormwater, stormwater drainage system, and mainly through the ground surface and road runoff. In this case, microplastic of the larger size range (125–5000 μm) is found on the ground surface, and catchment area of a developed. Iran has the lowest range of a microplastic size (<5000 μm) on the ground surface (Wang et al. 2022). Table 3.1 also shows the microplastic concentration in particles per liter is higher in developed countries than in underdeveloped countries (Wang et al. 2022; Alkhadra et al. 2022).

Table 3.2 defines the scenario of heavy metal contamination and the ecological risk due to heavy metal contamination (Roy et al. 2022a; Ritchie et al. 2023). In this case, the contamination is measured according to the degree of contamination suggested by Hakanson in 1980 (Roy et al. 2022b). According to the Duong and Lee. 2011, the degree of contamination, denoted by $C_{\text{deg}}$ is classified as (Roy et al. 2022a, b):

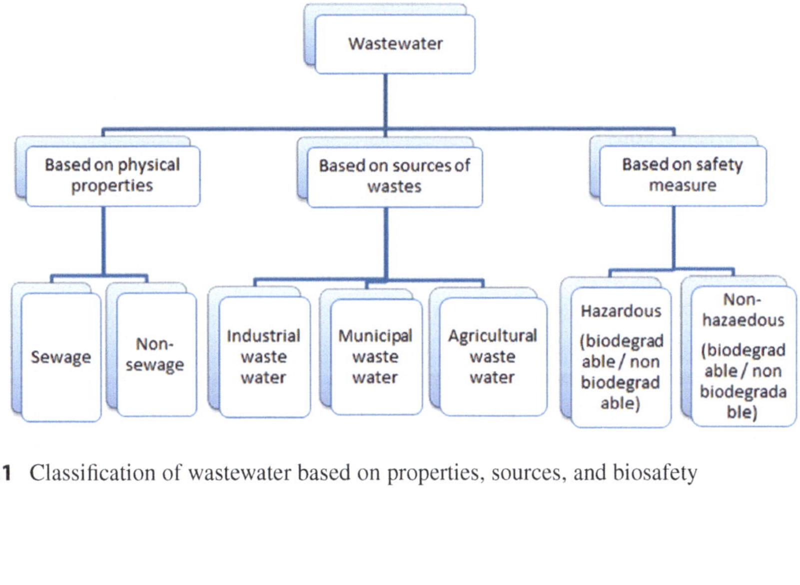


**Fig. 3.1** Classification of wastewater based on properties, sources, and biosafety

**Table 3.1** Comparative data on microplastic (MP) contamination in water in developed and underdeveloped countries (Wang et al. 2022)

| Country | Underdeveloped country | Developed country |
|---|---|---|
| Microplasticconcentration Particles/L | 1.2–22 | 1.1–292 |
| Size range (μm) | 37–5000 | 125–5000 |

**Table 3.2** Comparative measurement of metal contamination in water in different continents (Roy et al. 2022a)

| Continents | Asia | Europe | America | Australia | Africa |
|---|---|---|---|---|---|
| Heavy metal Degree of contamination | Very high | Very high | Considerable | Considerable | Considerable |
| Ecological risk index | Very high risk | Very high risk | Considerable | Considerable | High |

very high—($20 \leq C_{deg}$)
Considerable—($10 \leq C_{deg} < 20$)
Moderate—($5 \leq C_{deg} < 10$)
Low—($C_{deg} < 5$).

The continent regions that exceed the $C_{deg}$ values are on the highest priority list to take action to prevent contamination immediately (Li et al. 2023).

According to Hakanson (1980), the potential Ecological Risk (PER) can define the toxic response of heavy metals to the environment (Ritchie et al. 2023; Li et al. 2023). In this case, according to Islam et al. (2015), the Ecological risk index ($Er^i$) can be explained according to the level of risk, such as (Roy et al. 2022b; Li et al. 2023):

$Er^i \geq 320$—very high risk
$160 \leq Er^i < 320$—high risk
$80 \leq Er^i < 160$—considerable risk
$40 \leq Er^i < 80$—moderate risk.
$Er^i < 40$—low risk.

According to this range, the Ecological risk index is defined in the different continent zone. In this case, the Table 3.3 shows that Asia and Europe are at the highest environmental risk compared to Australia and Africa. America is at a moderate ecological risk index.

Table 3.3 shows the comparative rate of death between the years 1990 and 2019, which offers a picture of the effect of water pollution in developing (underdeveloped) and developed countries. According to the data in the Table 3.3, the death rate due to water contamination has decreased in 29 years in underdeveloped countries. In the case

**Table 3.3** Death rate due to water pollution in underdeveloped and developed countries in the year 1990 and 2019 (Ritchie et al. 2023)

| Year | Underdeveloped countries (%) | Developed countries (%) |
|---|---|---|
| 1990 | 5.046 | 0.053 |
| 2019 | 2.126 | 0.036 |

of some developed countries, there is a lower death rate in these years due to water pollution. Therefore it can be said that underdeveloped countries have a higher rate of death due to the consumption of polluted water. Water pollution is an excellent problem for countries where the healthcare system is also not developed.

## 3.4 Implementation of Waste Management to Mitigate Water Pollution

Demand for developing mitigation strategies for water pollution created a huge concern regarding solid waste accumulation and wastewater treatment procedures. Water pollution, causing water scarcity worldwide due to improper handling of waste management and water treatment technologies, is now a threat to the global environmental crisis (Massoud et al. 2009). To implement mitigation strategies for the risks and challenges associated with solid waste mishandling and water pollution are focused on developing sustainable waste management along with water treatment procedures. The factors related to solid waste management and water pollution incidents are the economic condition of a country, the technological strategies of an industrial region, geographical structures, political interferences, and the internal infrastructure (Fig. 3.2).

The water treatment procedure is carried out through several techniques depending on the type of pollutants, such as physical, biological, and chemical water treatment (Saravanan et al. 2021). Water recycling and wastewater reuse can solve the problem through the help of sustainable solid management in the disposal sites. In this case, to increase the number of freshwater resources and minimize water pollution in the environment, the most promising water treatment is to remove the pollutants from the wastewater by using surfactants according to the types of pollutants (Saleh et al. 2022). To get maximum efficiency in the mitigation of water pollution, the removal of pollutants by application of surfactant on the water bodies should be carried out according to sources of the wastes based on the economic condition, geographical region, and internal infrastructure of a region in a country. These surfactants are structurally a class of molecules known as amphiphilic molecules which can be classified depending on their activities on different types of waste or pollutants which also helps in wastewater recycling (Saleh et al. 2022; Sarubbo et al. 2015).

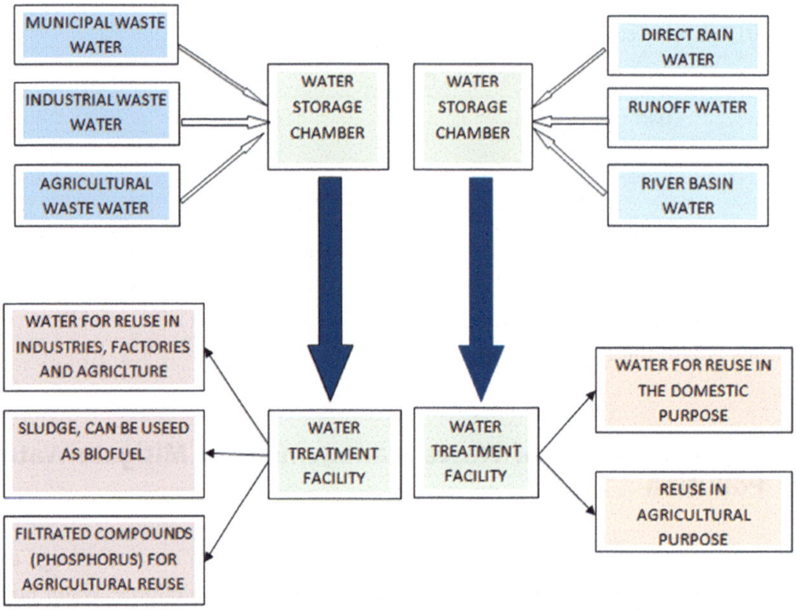


**Fig. 3.2** Future approach of wastewater treatment to minimize the effective stress on freshwater resources

### 3.4.1 Petroleum Hydrocarbon Removal

Level of affiliation with components of crude oil, aliphatic hydrocarbons are efficiently removed by sodium dodecyl sulfate (SDS) compound. SDS are low-cost and easily available. In contrast, aromatic hydrocarbons are removed by the interaction of saponin compounds (Rasheed et al. 2020). These surfactants act according to their affiliation level with the pollutants to be removed from the water bodies. SDS (chemical formula: $CH_3$ $(CH_2)11SO_4Na$) is mainly an anionic surfactant that consists of an alkyl chain at $C^{12}$. This alkyl chain can penetrate the oil particles in the water. SDS also possesses amphiphilic properties, with the change in the concentration it shows self-assemblies, and above a certain concentration that is "critical micelle concentration" (CMC) it can form micelles (Rasheed et al. 2020).

### 3.4.2 Pesticide Removal

Surfactants like alkyl polyoxyethylene or the same compound containing a hydrophobic chain as a structural agent are also helpful. These polyoxyethylene surfactants are anionic components containing an alkyl chain and formed by self-clustering in aqueous media.

In the application of heat and agitation, it forms a bilayer structure. They act through the remediation process in pesticide removal from the soil particles in agricultural wastewater (Ugorji et al. 2022).

### 3.4.3 Toxic Heavy Metal Removal

Soil-washing, desorption, and phytoremediation surfactants or bio-surfactants can eliminate heavy metal. Surfactants also can be used in the ultra-filtration method that works by increasing the membrane pre-treatment and removing metal ions. The bio-surfactant are biologically active compounds consisting of microbial products which possess a strong affinity for the heavy metal in the wastewater. The bioavailability of these biosurfactants is good along with the ability to reduce the surface tension and form micelles (Malik et al. 2022).

### 3.4.4 Commercial and Healthcare Waste Removal

Surfactants integrated into an emulsion-liquid membrane (ELM) work through the membrane-based selective permeability of solutes. In this case, the simultaneous extraction of selective particles is involved in one simple and cost-effective way. The mass transfer rate of the pollutants can be enhanced by the higher interfacial area of ELM (Kumar et al. 2019).

### 3.4.5 Volatile Organic Compounds and Dye Removal

Mesoporous silica with a nanoparticle cap removes volatile organic compounds utilizing the absorbing mechanism. The presence of toxic dye in water is very harmful, which also can be removed by the absorption method. These nanoparticles are chemically stable and also resistant to different environmental conditions which make it easy to perform photocatalyst degradation of several types of dye present in the wastewater (Urbina-Suarez and (2022).

Reusing and recycling wastewater by applying surfactants and bio-surfactants can improve the condition sustainably and effectively as these are lower cost, affordable, and have good bioavailability. In this case, a step-by-step procedure can be followed as mentioned bellow

i. Determination of the types of wastes, discharge, their resources, and the identification of disposal sites (Stefania et al. 2019).

ii. Determination of the microplastic concentration of an area where the water quality is severely damaged and taking proper steps to control the use of the plastic.
iii. Identifying the freshwater resources and measuring the DO (Dissolve Oxygen) and BOD (Biological Oxygen Demand) to take water pollution prevention measures (Stefania et al. 2019).
iv. Reuse of the waste material such as in agriculture as a fertilizer and then treatment of final waste through the scientific method developed by the expertise.
v. Holding campaigns and public conferences to spread concerns to increase awareness among the population about solid waste and water scarcity problems to limit water overuse (Manika et al. 2023).
vi. To measure the geographical structure of a particular area or region that includes transportation, economic, and energy utilization condition to select the most effective model for dealing with the problem by the expertise which can give a cost-effective plan according to the condition of the region (Chen et al. 2017).

## 3.5 Socio-economic Benefits

According to UNESCO, around 829,000 death were reported in 2021 due to diarrhea caused by the consumption of unsafe drinking water (Bijekar et al. 2022). Pure and safe drinking water is still a big challenge for some countries. Countries with the most effective areas with untreated solid waste must invest in solid waste treatment and water-recycling infrastructure to avoid future water scarcity and environmental crises. As water can be considered a part of national security and breaching, the economic condition of a country is also affected. In this case, the main purpose of the program should be the eco-friendly cost optimization procedure that is affordable for underdeveloped countries also (He et al. 2021).

### 3.5.1 Social Impact

Socio-economic adaptations need financial inclusions not limited to only government policies but also depend on factors like financial flow and future investment strategies. Continuous investment and rapid implementations in the management procedure are required to maintain a proper method of waste management, find a sustainable route for accessing safe drinking water, and keep an environmental balance on water ecology (Chocat et al. 2007). According to the UN global water analysis and assessment, there a particular gap exists in policy-making due to insufficient funds. Around 80% of the country faces a financial crunch for continuing operations and periodic maintenance, which can be described as the critical factor that stops providing proper service to the general population (c).

Government initiatives lead and control the development and progress activities through various models of training, activities, campaign, workshops, and investment in higher education, which helps sequentially to promote the area of interest among skilled workers and mitigate the knowledge gaps (Akhmouch and Clavreul 2016).

Financial support for human capital by government initiatives and programs influences more people to participate in the green revolution for solid waste management, which leads to the mitigation of water pollution and water scarcity problems (Marshall and Farahbakhsh 2013). Therefore government policies play a more significant role toward the society in economic and technological development by implementing new policies and strategies to reduce environmental damage.

### 3.5.2 Environmental Quality Improvements

The continuous uprising of loads of untreated solid wastes that lead to water pollution threatens the global environmental condition. Hence improving water quality and recycling wastewater for various uses can help overcome the environmental crisis (Massoud et al. 2009). In this case, the surfactant based wastewater treatment gives an eco-friendly approach that should be incorporated under the governmental policies. Although the local governments should take initiatives to spread awareness among the general people and the staff in every workplace. Therefore some environment-friendly workplace guidelines can be developed to avoid causing harm to the environment (Wu et al. 2017). In this case, the government investigates appropriately. It applies the environmental policies while approving any industrial, technological, or municipal project so that the effects of the harmful and hazardous wastes cannot damage the freshwater resources to secure public health and public access to safe drinking water for the people of a region or people of a country.

## 3.6 Finance and Sustainability

To invest in developing waste management and water-treatment procedures to secure public health and keep environmental balance, government initiative, and private–public partnership is crucial to overcome the challenges and barriers toward a pollution-free healthy world. Water pollution-related project sustainability risks might face challenges due to economic instability, political unrest, poor public finances, and vulnerable geopolitical tensions which might hinder future investment in water security (Dominique and Money 2022). Public–private partnerships and investments can also provide several benefits not limited to solving the water scarcity problem but also to helping the potential investors financially and the young generation by creating work opportunities (Surachman et al. 2022). The economic growth of a country can be dampened by water scarcity. The solid waste leading problem can worsen the environmental condition along with the breakdown of the healthcare infrastructure in a country due to waterborne diseases outbreak. A sustainable solution to this problem can help grow an environmentally friendly infrastructure and thus involve more people in developing new work opportunities related to waste-management and water-treatment projects. Furthermore, economic stability will provide a strengthened technological exchange, greater information systems, and access to more scientific instruments that help mitigate problems in a cost-effective and fewer times consuming manner (Hao et al. 2022).

## 3.7 Sustainable Development Goal Achieved

Ensuring water security, wastewater recycling, proper scientific sanitation, environmental, and socioeconomic improvement is the basic necessity of a modern technology-driven economy that can be achieved through sustainable development goals (SDG) reforms. The first and foremost priority of SDG strengthening the system to make it better accessible for the population such as providing safe drinking water to the common people and which can be achieved by installing proper water treatment projects. Access to purified and safe drinking water reduces the vulnerability caused by a waterborne disease outbreak. Medium-size water projects attract investment as well as create job opportunities in the local community. Poor management and disposal of different types of waste affect the drinking water resources as well as compromise hygiene, which causes various types of diseases. So proper sanitation model developed and approved by the local government authorities helps to minimize surface water pollution through the sustainable management of waste and wastewater which can be sequentially achieved by the SDG.

There are several sources of waste that are polluting the water and freshwater resources. Industrial waste disposal adversely damages the fresh water ecosystem. In this case, the sustainable development goal approaches the proper cost-effective scientific technique to treat the waste before disposal at the waterside. Agricultural waste or the inorganic fertilizer used in agriculture may affect surface water resources. SDG encourages the use of organic fertilizer as well as recycle of agricultural waste into organic fertilizer to minimize water pollution, which is primarily achieved by the SDG. Dumping solid wastes directly into the ground causes contamination of the surface water resources with heavy metals. The use of surfactants in wastewater treatment can be considered a sustainable approach for the management of wastes and wastewater as it is a lower-cost, eco-friendly procedure. Mainly the use of biosurfactants can be easily affordable by underdeveloped countries as the bioavailability of the biosurfactants can be increased which can be promising to achieve the desired SDG.

The social and economic reforms of a country are largely influenced by government policies. Setting up infrastructure creates job opportunities that may be considered as a part of economic reforms through which SDG can be achieved. As the sustainability of a project largely depends on financial resources, SDG aims to provide positive cash flow from the projects for a longer time.

So it can be assumed that, ensuring basic services like food, water, and proper sanitation are the basic aims of SDG which might also help in providing social protection as well as reducing climate-related vulnerability. Ensuring good health and well-being is the top priority of the SDG. Poor water quality affects the environment quality, damages the fresh water resources as well as hinders the economic growth of a nation. Hence, developing and implementing SDG policies mainly aims to tackle water pollution by strengthening the system to increase water availability and improving access to safe water for the larger portion of the population in a society.

## 3.8 Recent Findings and Current Challenges

As per the government's projected data, the world population will cross almost 9 billion by 2050 with a projected annual GDP growth rate of 3.5% (Bui and Tseng 2021). To achieve sustainable global growth with a greener world we have to reduce water pollution and future water scarcity problems.

### 3.8.1 Recent Findings

Every year, approximately 300–400 million tons of heavy metal, factory residue, and other effluents enter the world's water system, damaging freshwater resources and harming the water ecosystem. More than 1.7 million people die each year due to hazardous waste materials in drinking water above safety levels. The lack of proper infrastructure and knowledgeable workers causes difficulties in managing factory waste disposal (Bui and Tseng 2021). A good scientific and effective waste treatment and water treatment method requires new sustainable and cost-effective technologies for developing countries. Heavy metals more precisely mercury and microplastics present in drinking water and freshwater, have serious consequences due to prolonged exposure to human health, therefore, reducing the presence of heavy metals and the level of microplastics is another challenging task (Wang et al. 2022). These factory discharges or the disposal along with the pesticides used in agriculture severely damaged the river water bodies and affect marine life which is not limited to fish only but the whole population that leads their life depending on the water resources for use in daily life (Islam et al. 2022).

### 3.8.2 Current Challenges

Due to the lack of a proper municipal water treatment system, ground and surface water levels are reducing dramatically due to wasting and using fresh water. In this situation, global warming has become a global threat to the whole human population worldwide (Neonicotinoid insecticides in the environment 2022). The waste and its hazardous component are damaging the water quality and increasing the risk of a health crisis in the future generation (Neonicotinoid insecticides in the environment 2022). Water contamination is not only limited to urban areas due to improper management systems but also affects rural areas where wastewater treatment facilities are inadequately operated or controlled by the government due to a lack of awareness (Decentralised Wastewater Management Systems in Developing Countries 2022; Kabyl et al. 2022).

Untreated waste generated by the factory and industry every year causes several incidents of accidental toxic chemical spillage, which causes environmental crises leading to damage the public health (Morin-Crini et al. 2022). Rapid urbanization with subsequent use of inorganic fertilizer and waste generated from the agrochemical industry contaminates the surface water and soils (Rudi et al. 2020). Contamination by agrochemicals pollutes is one of the leading causes of damaging the water ecosystem (Holechek et al. 2050; Ray and Iroegbu 2021) (Fig. 3.3).

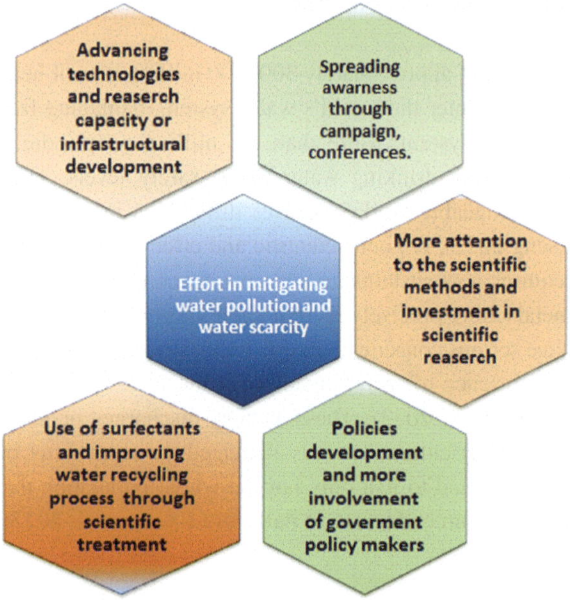


**Fig. 3.3** Representation of different areas of efforts in water pollution and water scarcity mitigation

## 3.9 Future Prospects of Wastewater Management

Government policies need to be reconsidered regarding the condition of water pollution mainly in some countries where infrastructural development requires including new sustainable modern technologies. Therefore, new research areas need to be opened to develop new technologies so that underdeveloped countries can access modern technologies cost-effectively to fight the solid waste-leading water pollution problem. For the treatment of wastewater and reuse, solar energy can be used as it is green and cheap (Datta et al. 2020).

Countries like Japan, Australia, France, and China need to monitor microplastic-containing waste generation and carry out research by a team of experts to mitigate this problem in an eco-friendly way. In this case, many ways are opened for the researcher to work on the plastic degradation and control problem.

The development of new techniques by the experts and research teams to reduce the heavy metal load in the water level and in the environment is another important factor affecting the water quality, which is leading to cause harmful effects on public health. Therefore more information needs to be collected, and the situation should be monitored to select the area for understanding the priority of taking action against the waste accumulation problem.

## 3.10 Conclusive Remarks

Waste management and wastewater treatment are two relevant topics that mainly aim to mitigate water pollution and water scarcity and improve environmental quality by managing the disposed untreated waste materials. The delta-sustainability mainly involves sustainable waste management and water treatment policy by developing infrastructure and managing waste treatment procedures more cost-effectively. In this case, the role of governmental authorities, local stakeholders, and policymakers is monitored for better performance in mitigating water problems. The challenges regarding waste management and wastewater treatment are identified to categorize the issues and draw a possible effective conclusion to mitigate the problems structurally. Microplastic contamination is a big threat to both developed and underdeveloped countries, which directly affects water quality and causes water scarcity. The ecological risk index shows that Asia, Europe, and Africa are going through a difficult situation that establishes a link between the socioeconomic condition and a country's public security infrastructure. The wastes are classified according to the source and types of waste material. Furthermore, it can be concluded that applying different surfactants according to the types of waste materials in the wastewater or water resources gives a promising low-cost water treatment approach. The application of biosurfactants mainly increases bioavailability, has no harmful effects on the environment, and as well as a cost optimization process hence increases the affordability for underdeveloped, poor countries. A comparative idea of the waste management process among developing and underdeveloped countries can be understood based on the amount of micro-plastic, arsenic, and other heavy metals present in the water level in different regions of different countries and continents.

**Acknowledgements** Author (S. Roy) would like to acknowledge 'Scheme for Transformational and Advanced Research in Sciences (STARS)' (MoE-STARS/STARS-2/2023-0175) by Ministry of Education for promoting translational India-centric research in sciences implemented and managed by Indian Institute of Science (IISc), Bangalore, for their support.

**Declarations**

**Conflict of Interest:** The authors declare no competing interests.

## References

Abdel-Shafy HI, Mansour MS (2018) Solid waste issue: Sources, composition, disposal, recycling, and valorization. Egypt J Pet 27(4):1275–1290. https://doi.org/10.1016/j.ejpe.2018.07.003

Akhmouch A, Clavreul D (2016) Stakeholder engagement for inclusive water governance: "Practicing What We Preach" with the OECD Water Governance Initiative. Water 8(5):204. https://doi.org/10.3390/w8050204

Alkhadra MA, Su X, Suss ME, Tian H, Guyes EN, Shocron AN, Conforti KM, de Souza JP, Kim N, Tedesco M, Khoiruddin K, Wenten IG, Santiago JG, Hatton TA, Bazant MZ (2022) Electrochemical methods for water purification, ion separations, and energy conversion. Chem Rev 122(16):13547–13635. https://doi.org/10.1021/acs.chemrev.1c00396

Anuardo RG, Espuny M, Costa ACF, Oliveira OJ (2022) Toward a cleaner and more sustainable world: a framework to develop and improve waste management through organizations, governments and academia. Heliyon 8(4):e09225. https://doi.org/10.1016/j.heliyon.2022.e09225

Bijekar S, Padariya HD, Yadav VK, Gacem A, Hasan MA, Awwad NS, Yadav KK, Islam S, Park S, Jeon BH (2022) The state of the art and emerging trends in the wastewater treatment in developing nations. Water 14(16):2537. https://doi.org/10.3390/w14162537

http://books.google.com/books/about/Public_Private_Partnerships_for_Urban_Wa.html?id=D7YsorFRVLkC

Buenrostro O, Bocco G, Cram S (2001) Classification of sources of municipal solid wastes in developing countries. Resour Conserv Recycl 32(1):29–41. https://doi.org/10.1016/s0921-3449(00)00094-x

Bui TD, Tseng ML (2021) Understanding the barriers to sustainable solid waste management in society 5.0 under uncertainties: a novelty of socials and technical perspectives on performance driving. Environ Sci Pollut Res 29(11):16265–16293. https://doi.org/10.1007/s11356-021-16962-0

Chen Z, Wu Q, Wu G, Hu HY (2017) February). Centralized water reuse system with multiple applications in urban areas: lessons from China's experience. Resour Conserv Recycl 117:125–136. https://doi.org/10.1016/j.resconrec.2016.11.008

Chocat B, Ashley R, Marsalek J, Matos M, Rauch W, Schilling W, Urbonas B (2007) Toward the sustainable management of urban storm-water. Indoor Built Environ 16(3):273–285. https://doi.org/10.1177/1420326x07078854

Correction to Lancet Respir Med (2021) published online April 9. https://doi.org/10.1016/S2213-2600(21)00160-0. (2021, June). Lancet Respir Med 9(6):e55. https://doi.org/10.1016/s2213-2600(21)00181-8

Datta S, Dey A, Singha NR, Roy S (2020) Enhanced performance of dye-sensitized solar cell with thermally stable natural dye-assisted $TiO_2$/$MnO_2$ bilayer-assembled photoanode. Mater Renew Sustain Energy 9(4). https://doi.org/10.1007/s40243-020-00185-3

De S, Debnath B (2016) Prevalence of health hazards associated with solid waste disposal-a case study of Kolkata, India. Procedia Environ Sci 35:201–208. https://doi.org/10.1016/j.proenv.2016.07.081

Dominique K, Money A (2022) Introduction to the water financing landscape: select proven and emerging approaches. Financ Invest Water Secur 249–262. https://doi.org/10.1016/b978-0-12-822847-0.00016-8

Fuller R, Landrigan PJ, Balakrishnan K, Bathan G, Bose-O'Reilly S, Brauer M, Caravanos J, Chiles T, Cohen A, Corra L, Cropper M, Ferraro G, Hanna J, Hanrahan D, Hu H, Hunter D, Janata G, Kupka R, Lanphear B, Yan C (2022) Pollution and health: a progress update. Lancet Planet Health 6(6):e535–e547. https://doi.org/10.1016/s2542-5196(22)00090-0

Geissen V, Mol H, Klumpp E, Umlauf G, Nadal M, van der Ploeg M, van de Zee SE, Ritsema CJ (2015) Emerging pollutants in the environment: a challenge for water resource management. Int Soil Water Conserv Res 3(1):57–65. https://doi.org/10.1016/j.iswcr.2015.03.002

Guerrero LA, Maas G, Hogland W (2013) Solid waste management challenges for cities in developing countries. Waste Manag 33(1):220–232. https://doi.org/10.1016/j.wasman.2012.09.008

Gutberlet J, Uddin SMN (2017) Household waste and health risks affecting waste pickers and the environment in low- and middle-income countries. Int J Occup Environ Health 23(4):299–310. https://doi.org/10.1080/10773525.2018.1484996

Hao R, Huang G, Liu L, Li Y, Li J, Zhai M (2022) Sustainable conjunctive water management model for alleviating water shortage. J Environ Manag 304:114243. https://doi.org/10.1016/j.jenvman.2021.114243

Haseena M, Faheem Malik M, Javed A, Arshad S, Asif N, Zulfiqar S, Hanif J (2017) Water pollution and human health. Environ Risk Assess Remediat 01(03). https://doi.org/10.4066/2529-8046.100020

He C, Liu Z, Wu J, Pan X, Fang Z, Li J, Bryan BA (2021) Future global urban water scarcity and potential solutions. Nat Commun 12(1). https://doi.org/10.1038/s41467-021-25026-3

Holechek JL, et al (2022) A global assessment: can renewable energy replace fossil fuels by 2050?. Sustainability 14(8):4792

Islam MA, Amin SN, Rahman MA, Juraimi AS, Uddin MK, Brown CL, Arshad A (2022) Chronic effects of organic pesticides on the aquatic environment and human health: a review. Environ Nanotechnol Monit Manag 18:100740. https://doi.org/10.1016/j.enmm.2022.100740

Jury WA, Vaux HJ (2007) The emerging global water crisis: managing scarcity and conflict between water users. Adv Agron 1–76. https://doi.org/10.1016/s0065-2113(07)95001-4

Kabyl A, Yang M, Shah D, Ahmad A (2022) Bibliometric analysis of accidental oil spills in ice-infested waters. Int J Environ Res Public Health 19(22):15190. https://doi.org/10.3390/ijerph192215190

Kumar A, Thakur A, Panesar PS (2019) A review on emulsion liquid membrane (ELM) for the treatment of various industrial effluent streams. Rev Environ Sci Bio/Technol 18(1):153–182. https://doi.org/10.1007/s11157-019-09492-2

Li C et al (2023) Interventions of river network structures on urban aquatic microplastic footprint from a connectivity perspective. Water Research, Elsevier BV. p 120418. Crossref. https://doi.org/10.1016/j.watres.2023.120418

Lin L, Yang H, Xu X (2022) Effects of water pollution on human health and disease heterogeneity: a review. Front Environ Sci 10. https://doi.org/10.3389/fenvs.2022.880246

Malik N, Bulasara VK, Basu S (2022) Surfactant induced ultrafiltration of heavy metal ions from aqueous solutions using a hybrid polymer–ceramic composite membrane. Chem Eng Res Des 185:407–417. https://doi.org/10.1016/j.cherd.2022.07.013

Manika N, et al (2023) Towards the Assessment of ecosystem health of ecosystems in asia and promoting nature-based solutions. Anthropocene Sci 2(2). Springer Science and Business Media LLC, pp 172–75. https://doi.org/10.1007/s44177-023-00062-4

Marshall RE, Farahbakhsh K (2013)Systems approaches to integrated solid waste management in developing countries. Waste Manag 33(4):988–1003. https://doi.org/10.1016/j.wasman.2012.12.023

Massoud MA, Tarhini A, Nasr JA (2009) Decentralized approaches to wastewater treatment and management: applicability in developing countries. J Environ Manag 90(1):652–659. https://doi.org/10.1016/j.jenvman.2008.07.001

Mastakar P, Mastakar V, Mukhopadhyay K, Jaju S (2019) Zero waste circular model of MSWM: a success in ward 40, Pune, India. Metamorph: J Manag Res 18(1):36–56. https://doi.org/10.1177/0972622519853420

Morin-Crini N, Lichtfouse E, Liu G, Balaram V, Ribeiro ARL, Lu Z, Stock F, Carmona E, Teixeira MR, Picos-Corrales LA, Moreno-Piraján JC, Giraldo L, Li C, Pandey A, Hocquet D, Torri G, Crini G (2022) Worldwide cases of water pollution by emerging contaminants: a review. Environ Chem Lett 20(4):2311–2338. https://doi.org/10.1007/s10311-022-01447-4

Pathak A, Singh LK (2023) Impact of microplastics and nanoplastics interactions with other contaminants in environment. *Curr Dev Biotechnol Bioeng* 333–359. https://doi.org/10.1016/b978-0-323-99908-3.00015-4

Picó Y, Barceló D (2019) Analysis and prevention of microplastics pollution in water: current perspectives and future directions. ACS Omega 4(4):6709–6719. https://doi.org/10.1021/acsomega.9b00222

Rasheed T, Shafi S, Bilal M, Hussain T, Sher F, Rizwan K (2020) Surfactants-based remediation as an effective approach for removal of environmental pollutants—a review. J Mol Liq 318:113960. https://doi.org/10.1016/j.molliq.2020.113960

Rastogi M, Nandal M, Khosla B (2020) Microbes as vital additives for solid waste composting. Heliyon 6(2):e03343. https://doi.org/10.1016/j.heliyon.2020.e03343

Ray SS, Iroegbu AOC (2021) Nanocellulosics: benign, sustainable, and ubiquitous biomaterials for water remediation. ACS Omega 6(7):4511–4526. https://doi.org/10.1021/acsomega.0c06070

Ritchie H, Roser M (n.d.) Clean water. Our World in Data. Accessed 25 Feb 2023. https://ourworldindata.org/water-access

Roy S, Gupta SK, Prakash J, Habib G, Kumar P (2022) A global perspective of the current state of heavy metal contamination in road dust. Environ Sci Pollut Res 29(22):33230–33251. https://doi.org/10.1007/s11356-022-18583-7

Roy S, Gupta SK, Prakash J, Habib G, Kumar P (2022) A global perspective of the current state of heavy metal contamination in road dust—Environmental Science and Pollution Research. SpringerLink. https://doi.org/10.1007/s11356-022-18583-7

Rudi NN, Muhamad MS, TeChuan L, Alipal J, Omar S, Hamidon N, Abdul Hamid NH, Mohamed Sunar N, Ali R, Harun H (2020) Evolution of adsorption process for manganese removal in water via agricultural waste adsorbents. Heliyon 6(9):e05049. https://doi.org/10.1016/j.heliyon.2020.e05049

Saleh TA, Mustaqeem M, Khaled M (2022) Water treatment technologies in removing heavy metal ions from wastewater: a review. Environ Nanotechnol, Monit Manag 17:100617. https://doi.org/10.1016/j.enmm.2021.100617

Saravanan A, Senthil Kumar P, Jeevanantham S, Karishma S, Tajsabreen B, Yaashikaa P, Reshma B (2021) Effective water/wastewater treatment methodologies for toxic pollutants removal: processes and applications towards sustainable development. Chemosphere 280:130595. https://doi.org/10.1016/j.chemosphere.2021.130595

Sarubbo L, Rocha R, Luna J, Rufino R, Santos V, Banat I (2015) Some aspects of heavy metals contamination remediation and role of biosurfactants. Chem Ecol 31(8):707–723. https://doi.org/10.1080/02757540.2015.1095293

Sasakova N, Gregova G, Takacova D, Mojzisova J, Papajova I, Venglovsky J, Szaboova T, Kovacova S (2018) Pollution of surface and ground water by sources related to agricultural activities. Front Sustain Food Syst 2. https://doi.org/10.3389/fsufs.2018.00042

Shrivastava P (1995) Environmental technologies and competitive advantage. Strateg Manag J 16(S1):183–200. https://doi.org/10.1002/smj.4250160923

Stefania GA, Rotiroti M, Buerge IJ, Zanotti C, Nava V, Leoni B, Fumagalli L, Bonomi T (2019) Identification of groundwater pollution sources in a landfill site using artificial sweeteners, multivariate analysis and transport modeling. Waste Manage 95:116–128. https://doi.org/10.1016/j.wasman.2019.06.010

Surachman EN, Perwitasari SW, Suhendra M (2022) Stakeholder management mapping to improve public-private partnership success in emerging country water projects: Indonesia's experience. Util Policy 78:101411. https://doi.org/10.1016/j.jup.2022.101411

Decentralised wastewater management systems in developing countries: key barriers and potential resource recovery applications (2022) The Italian Association of Chemical Engineering, vol 97. https://doi.org/10.3303/CET2297062

Neonicotinoid insecticides in the environment: a critical review of their distribution, transport, fate, and toxic effects (2022) Neonicotinoid insecticides in the environment: a critical review of their distribution, transport, fate, and toxic effects - ScienceDirect. https://doi.org/10.1016/j.jece.2022.108485

Ugorji OL, et al (2022) The effect of noisome preparation methods in encapsulating 5-fluorouracil and real time cell assay against HCT-116 colon cancer cell line. Heliyon 8(12). Elsevier BV, p e12369. https://doi.org/10.1016/j.heliyon.2022.e12369

Urbina-Suarez NA (2022) Advanced oxidation processes with Uv-h2o2 for nitrification and decolorization of dyehouse wastewater. https://doi.org/10.3303/CET2295040

Wang C, O'Connor D, Wang L, Wu WM, Luo J, Hou D (2022) Microplastics in urban runoff: global occurrence and fate. Water Res 225:119129. https://doi.org/10.1016/j.watres.2022.119129

Wu G, Cao W, Liu L, Wang F (2017) Water pollution management in China: recent incidents and proposed improvements. Water Supply 18(2):603–611. https://doi.org/10.2166/ws.2017.139

Zaman K, et al (2022) Environmental effects of bio-waste recycling on industrial circular economy and eco-sustainability. Recycling 7(4):60

Zanoni A, Fungaroli AA (1973) Potential for ground water pollution from the land disposal of solid wastes. C R C Crital Rev Environ Control 3(1–4):225–260. https://doi.org/10.1080/10643387309381603

# Integrated Waste Management System for Environment Protection in Deltas

4

Leila Soudani, Meriem Chafaa, Moulkheir Selmani, Mohamed Islem Bouacha, Omar Safa, Mhamed Maatoug, and Hermann Heilmeier

**Abstract**

In just a few decades, the increase in municipal solid waste (liquid and solid, hazardous and non-hazardous) has become the most serious environmental problem in the delta regions. This increase was caused by population growth, industrialization, and the rapidity of urbanization. Indeed, the delta's soils and sediments, surface and groundwater, marine environment, atmosphere, and terrestrial ecosystems are all locally adversely and significantly impacted by it. Utilizing integrated waste management allows for the effective reduction, collection, recycling, and disposal of waste products produced by institutional, commercial, industrial, and residential land uses without endangering the environment. It can help improve air and water quality and reduce the spread of disease. It can also help to decrease costly inefficiencies, promote the growth of new markets, and result in the creation of jobs. Integrated waste management also helps society as a whole by reducing unpleasant odors and raising the standard of living for disadvantaged groups like the recycling sector. This chapter focuses on the actions to reduce air and soil pollution from solid waste in deltas, with

L. Soudani (✉) · M. Chafaa · M. Selmani · M. I. Bouacha · O. Safa · M. Maatoug
Laboratory of Agro-Biotechnology and Nutrition in Semi-Arid Areas, Faculty of Nature and Life Sciences, University of Tiaret, Tiaret, Algeria
e-mail: soudani_leila@outlook.fr

H. Heilmeier
AG Biology/Ecology, Institute for Biosciences, TUBergakademie Freiberg, Freiberg, Germany
e-mail: hermann.heilmeier@ioez.tu-freiberg.de

 
M. Nasr and B. Ravindran (eds.), *Solid Waste Management in Delta Region for SDGs Fulfillment*, Deltas of the World, https://doi.org/10.1007/978-3-031-58253-0_4

different strategies to protect water-related ecosystems from releasing hazardous chemicals and materials. Some recommendations to ensure environment protection in deltas using sustainable waste management are provided.

**Abbreviation**

| | |
|---|---|
| EHS | Environmental Health and Safety |
| EIA | Environmental Impact Assessment |
| EM | Environmental Management |
| EMP | Environmental Management Plan |
| EMS | Environmental Management System |
| ISO | International Organization for Standardization |
| GDP | Gross Domestic Product |
| GRI | Global Reporting Initiative |
| NGO | Non Governmental Organization |

## 4.1 Introduction

Different natural environments are affected today by pollution generated by human activity. The anthropization consequences vary according to the landscape configuration of the ecosystem in question: forest, mountain, wetland, sea, delta … etc.). Therefore, the actions to be considered must take into account the characteristics of the physical environment affected (Gomiero et al. 2018).

In this context, the Deltas as a complex geographical unit, as well as the problem of environmental pollution and the actions of treatment and mitigation, are the subject treated. It is a question of identifying innovative solutions in order to reduce the risk of degradation of these environments and to foresee perspectives of sustainable development, integrating at the same time the man, his activities and his environment (Gu et al. 2017).

The deltas are very sensitive to physical pollution. The industrial and agricultural activities as well as the discharges of households in large urban centers are direct sources of pollutants. Although some are transferred to the seas and oceans through currents and water exchanges, solid pollutants tend to settle in these regions through sedimentation processes that are complex to treat and whose longevity is particularly important, and whose impact is on a large spatial scale (Chapman and Wang 2001).

The management of inert and solid wastes has become today the major challenge in Deltas regions. Due to the lack of effective management programs, policies, and regulations; the waste is causing severe health hazard, and environmental impacts. Integrated waste management system, pollution control and environmental protection including

Deltas areas, are the main issues to save our future. Currently, several countries have realized that the way they manage their solid wastes does not satisfy the objectives of sustainable development throughout the world (Seo et al. 2004; Thanh et al. 2011).

However, solid waste management still has gaps due to lack of waste segregation at source level, reuse, recycling, treatment and appropriate disposal. Solid waste management to be effective requires separation of waste at the source level with the implementation of the 3Rs (reduce, reuse and recycling), treatment of organic fractions of wastes at local levels (Tadesse et al. 2008a, b). This chapter seeks to determine the current situation, actions to reduce environmental pollution and also aims to offer a review of the existing strategies that are particularly related to hazardous waste management in Deltas.

## 4.2 Actions to Reduce Soil Pollution from Solid Waste Generation in Deltas

### 4.2.1 Soil Pollution and Generating Activities

Soil pollution is by far the most problematic form of pollution, as soils are non-renewable natural resources and their interest for mankind is no longer to be demonstrated. Moreover, the variability of the pollutants as well as the forms of transfer towards the soil makes the problem one of the most complex to treat (Sheikhavandi 2015). Knowing that soil pollution can affect other natural resources such as groundwater tables, it is obvious that today the urgency is no longer to report but to act. Given the morphological complexity of deltas, the impact of pollution on soils is very serious to treat: deltas ensure at the same time the functions of transfer (water current), accumulation (mouth area and exchange with the land, sea and ocean) and sediment deposition (flow of sediments from the watersheds and surface tributaries) (Meybeck et al. 2004).

The main cause of soil pollution is human activity, in the delta regions the principle activity generating pollution is the hydrocarbon sector: different regions in the world are affected: the Niger Delta in Nigeria being the most sinister example (Raimi et al. 2022). Various reports and studies conducted in the Niger Delta in Nigeria demonstrate the seriousness of the problem with spills of products derived from hydrocarbons that have affected the physical aspect of the delta (landscape denaturation) and chemical water. The reports also alert on the heaviness of the decontamination operation, knowing that the extent of pollution affects even the farmland of the local and riparian populations. Since 2011, the damage caused is estimated at several billions of dollars and the duration of rehabilitation and decontamination is estimated at several decades (Mrozik and Piotrowska-Seget 2010).

Agriculture is also a source of pollution, especially with the use of pesticides and fertilizers on a large scale. The deltas, especially in Europe, have undergone major episodes of degradation, especially through the use of fertilizers and pesticides. The results of a research carried out by a team of the CNRS of Toulouse in France (2022), analyze 600 years of sedimentation of heavy metals in the delta of the Danube and relates the various periods which the delta knew with in particular a report on the seniority of pollution and its continuity with the modern human activities, particularly the development of the delta for the maritime traffic of the countries concerned (Mindrescu et al. 2022).

Urban centers are also an important source of pollution. The extension of cities with high population concentration is another source of pollution, with notably a hazardous management of wastewater discharge systems which include a diversity of pollution: chemical, organic and especially physical. The Indus delta in Asia, which has a surface area of 40,000 $km^2$, is subject to significant pressures that threaten its sustainability and that of the biodiversity in place. The anarchic construction of dams upstream generates physical slash that is deposited upstream in the mangroves and marshy areas at the edge of the Arabian Sea (Kumar et al. 2015), thus preventing the deposition of sediments necessary for the fertilization of the region's soil.

Solid pollutants are the main threat to the degradation of soils and deltas in the South Asian regions. The Ganges River in India, which ends in the Indian Ocean at the mouth of the Bengal Delta, illustrates the situation: a scientific expedition in 2015 estimated that 8 million tons of plastics end up in the Indian Ocean each year after a long process of transfer from the continent through the various rivers that lead to the Bengal Delta (Wilcox et al. 2015). This alarming report perfectly relates the context and the complexity of the pollution generated by urban centers and human activities, as well as their consequences on the deltas and the surrounding soils.

### 4.2.2 Generation, Transfer and Deposition of Solid Pollutants to Deltas

The management of soil pollution and the planning of protection and restoration programs in the affected areas takes into account the knowledge of the processes of production and transfer of pollutants, the deposition and the level of degradation caused by waste, especially solid waste.

The deltas form contacts and exchange zones between the continent, the seas and oceans, the slash and solid waste are therefore transferable via the water currents. Depending on their nature, they can be deposited on the surface (plastic) but also in deep basins, which has the effect of modifying the temporary configuration of the biotope but also of affecting the physical and chemical consistency of the soils through the formation of various heaps of compacted waste.

The main tributaries on the continent are the sewage systems that can contain large quantities of solid and inert waste, not assimilable by the natural environment, and which direct discharged into the rivers feeding the deltas. These waterways are most often used by the riparian populations for fishing but also for agriculture, which has the effect of

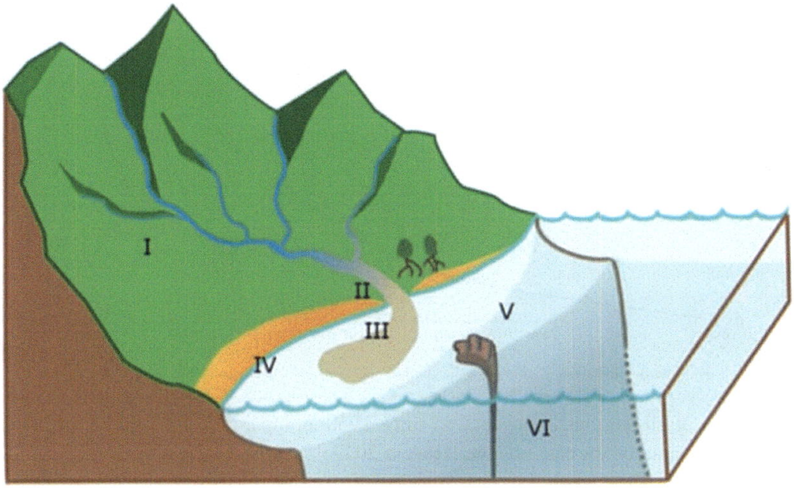


**Fig. 4.1** Main domains/geomorphological units of sediment transport studies. I the watershed; II estuary/delta; III the estuary-littoral transition, including the river plume, the bottom nepheloid layer and the ROFI zone; IV the littoral zone; V the continental shelf; VI canyon and deep sea (Ouillon 2018)

degrading the agricultural soils and other fertile land on the banks of the waterways but also on the swamps and potential mangroves in the delta (Okonkwo et al. 2015).

Soil pollution by solid waste discharges in deltas is made problematic by the complexity of deposition and sedimentation processes of transferable loads in water (Fig. 4.1): the various solid wastes, particularly plastic, will form sedimentary layers that are difficult to treat and whose repercussions on the sustainability of the natural environment, the perpetuation of biodiversity and ecosystem services are greatly affected. The actions to be planned must take into account the impact of the pollution and its consequences in order to set up a prioritized intervention plan (Naess 2017).

### 4.2.3 Degradation and Consequences

The solid discharges in nature and, particularly in the deltas accentuate more and more the irreversible degradation of the natural environments, in a demanding economic framework and although innovative, the human activity generates more and more solid waste difficult to treat and whose repercussions are both short and long term.

Plastic waste is a typical example of the problem that affects both the physical consistency of the deltas and the soils, the increasing economic development promoted by the use of plastic in different industries, and in the daily household consumption induced large quantities of solid waste. The versatility of the material and its low cost make its

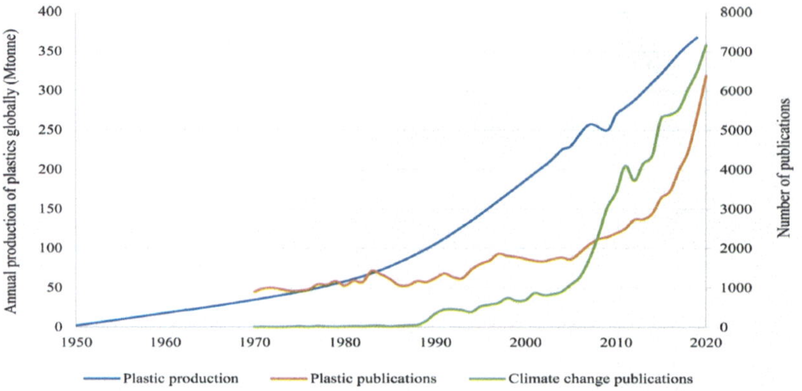


**Fig. 4.2** Volume of global plastic production since 1950, compared to the number of publications on plastics and, separately, climate change (Horton 2022)

use important and unavoidable, however, considering its durability in time, the ecosystems in the world are unable to assimilate the plastic quickly. Its degradation in nature is estimated at several hundred years (Fig. 4.2).

In a report of the FAO on the plastic pollution of the soils, the agriculture sector is the 1st consumer sector of plastic in the world, on average it is about 12.5 million tons which are used in the agricultural production, Asia is the continent which consumes the most plastic for its agriculture, and in front of the absence of viable and sustainable alternative of management of these wastes, it is the same quantities of plastics which are rejected in the nature and which affect directly the agricultural soils and the fertile grounds (FAO and UNEP 2021).

The six main types of plastics are now present everywhere around us, especially in soils and surface waters, their effects and consequences are devastating both on natural resources and on human health and activity.

Once rejected in nature, this waste undergoes a cycle of transfer by the effect of water or wind and is subjected to different processes of degradation, mechanical by the effects of solar radiation and temperature, chemical and biological, which has the effect of releasing in the soil different chemical compounds. These solid wastes are not assimilated, this fragmentation leads to tiny fragments inferior to 1 mm which infiltrate into the soil, are absorbed by the microorganisms of the soil, transferable towards the water table and can be present in the food we consume. This long chain has many consequences and it is still difficult today to judge with exactitude the impact that this pollution can have on the various compartments of the environment, although appreciable through biological indicators, the scientific community is slow to estimate and quantify the damage caused by the pollution of soils by plastic (Oliviera and Almeida 2019). However, the consequences of this pollution can have a direct impact on the living environment through;

- Soil degradation: loss of fertility, decrease of microbial activity in the soil, secondary pollution of groundwater.
- Loss of faunal biodiversity (microorganisms in the soil, invertebrates, fungi, plants) and consequent ecological imbalance of the trophic chain.
- Landscape deformation, fragmentation of habitats and reduction of arable land.

The pollution of deltas by solid waste has as a consequence to damage the aquatic biodiversity but also contributes to the degradation of these solid wastes in tiny fragments that disperse and integrate into the process of deposition and sedimentation. The direct impact is of ecological and sanitary aspect. The chain of degradation of macro debris into micro and nano debris leads to growing concerns about their toxicological impacts on the health of wildlife and humans. The animals that are trapped end up dying from exhaustion, the ingestion of micro debris causes internal injuries that lead to their death (Roman et al. 2021).

Agricultural soils are directly exposed by pollution, micro debris infiltrates into the soil, lead to the degradation of the biology in place including nematodes and invertebrates necessary for the degradation and mineralization of organic matter. Their presence in the soil contributes to a chemical alteration of the soil solution, the pH becomes acidic and the soil consequently becomes unfit for cultivation. These often-irreversible processes contribute to the loss of useful agricultural surface in the world, which consequently accentuates the economic crises related to the food of the populations. Soil pollution in the world affects 3.4 billion people or 40% of the world population. It is therefore urgent to put in place intervention strategies to inhibit the thresholds of soil degradation by solid and inert waste (FAO and UNEP 2021).

### 4.2.4 Actions to Reduce and Manage Solids Waste and Soil Ecosystem Restoration

Today, the international community is taking into consideration the magnitude of the problem and its environmental, social and economic consequences. The technical processes of decontamination and elimination of waste being expensive and complicated to set up, everywhere in the world; policies of sustainable development of the deltas and the fight against the pollution and its consequence on the grounds, are adopted. In a broader vision, political commitments are agreed upon, in particular to work towards so-called green economies, whose main niche is the reduction of waste in nature. The use of the deltas and the bordering biotopes knows a regulatory organization in a multidimensional framework that takes into account their vulnerabilities.

In this sense, the actions undertaken in the world to reduce soil pollution by solid waste in deltas can be enumerated in different essential points, moving from universal actions to localized actions, adapted to the problems at local scale.

#### 4.2.4.1 Regulatory and Legal Actions

The awareness of the need to protect soils and deltas against pollution and solid waste has led to the implementation of strategies based essentially on the establishment of regulatory and legal texts, notably for:

- **Classify the deltas as protected sites:** considering the richness and biological diversity of the deltas, several sites have been classified as World Biosphere Reserves, in Europe in particular with the first resolutions for the protection of the Danube basin and delta (UNESCO World Heritage Committee 1991). The interest of classification opens the way to a new vision of exploitation and use of the deltas in a sustainable economic perception, it is essentially a question of promoting the sustainability and durability of these living environments, the continuity of ecosystem services but beyond that, to limit the factors of degradation in particular the discharges of solid and chemical waste.
- **Regional treaties for the exploitation and sustainable use of deltas:** in particular for a better use of deltas for navigation and maritime traffic, such as the Danube in Europe.

#### 4.2.4.2 Limiting Industrial and Household Solid Waste

Actions are undertaken at the level of emission sources, especially industrial activities that generate untreated solid waste that is usually discharged into the environment. For this purpose, a standardization is implemented in a context of environmental protection, which consists in obliging the big industries to implement approaches and production processes involving an inhibition and reduction of solid discharges into nature, as well as the installation of specific treatment lines that allow the recovery of waste and bring back solids at the level of the production units in this sense, the main rules dedicated to the problem are:

- Standard ISO 13.030.01: Waste in general;
- Standard ISO 13.030.10 Solid waste;
- Standard ISO 13.030.30 Special waste;
- Standard ISO 13.030.40 Installation and equipment for waste removal;
- Standard ISO 13.030.50 Recycling including the corresponding equipment (ISO 2015).

#### 4.2.4.3 Reducing Emissions and Changing Consumption Patterns

Among the actions undertaken at the level of the main sources of emissions and pollution, the recovery, which consists in collecting recyclable solid waste such as plastic for their reuse (Eriksson et al. 2005).Organized and structured processes in various phases aim at reducing solid waste in nature. Three essential steps are implemented:

1. **Pre-treatment:** at the level of the source of emission (industry, household) the waste is treated in such a way as to filter out non-recoverable and non-recyclable waste with a proven level of dangerousness.
2. **Collection and transport:** the waste likely to be recovered and recycled in transformation units is treated and recovered by the local communities in adapted centers for their re-use.
3. **Treatment:** transformation and use for the production of raw materials and derived products.

Recovery and recycling are the main actions adopted, especially in developing countries with archaic waste management infrastructures. The concept of waste management is therefore shifting towards waste recovery. This allows the development of a parallel economic activity which is expanding at a lower rate compared to the amount of waste produced in the world. The percentage of recycled waste in the world is estimated at 09% while the quantity produced has practically doubled in 20 years. In a context of continuous demographic growth, it seems necessary to adopt new modes of consumption: to consume while protecting the environment by favoring products, waste, said biodegradable and recyclable (Al Maaded et al. 2012).

#### 4.2.4.4 Wastewater Treatment and In-Situ Waste Recovery

Wastewater systems are the major contributors of solid waste that accumulates at the delta level. Urban and industrial discharges contain large quantities of biosolids, sludge and other inert wastes that are most often discharged directly into the environment via waterways. Wastewater is the main source of solid waste pollution. Wastewater treatment and management is one of the key actions adopted for the reduction of solid waste in nature, through the installation of wastewater treatment plants that treat wastewater before it is discharged to the outlets. This has the advantage of limiting the discharge of solid waste as much as possible, although the problem is complex due to the continuous problem of sewage sludge that contains high levels of heavy metals, which ensures the continuity of the pollution (Cheremisinoff and Cheremisinoff 1993).

However, the problem cannot be solved in such a simplified manner, knowing that the fragmentation of solid debris is the main characteristic of solid waste transferred and transferable via water vectors (Hui et al. 2006). In disadvantaged countries, in the absence of wastewater treatment plants, some rudimentary but effective actions are advocated, such as the installation of recovery nets at the outlet of the drains, which has the effect of recovering macro debris.

#### 4.2.4.5 Integrated Soil Management Options for Polluted Soils

Environnemental and scientific experts analyze the baseline concentration of soil contaminants in the field and then evaluate the remediation of soil contamination. Remediation of oil-contaminated soil and sediments in Deltas has been practiced using a variety of technologies:

- **Controlling pollutant entry into the soil** by creating composts from source-separated biodegradable municipal solid wastes to reduce heavy metal entry and increase agricultural land productivity;
- **Decontaminating contaminated soil using physical and chemical methods**, including soil excavation, thermal extraction for volatile metals, solidification/stabilization, electro kinetics, chemical oxidation, vitrification and soil flushing with chemical extracts;
- **Phytoremediation** has grown in importance as a viable ecoremediation technique. The soil type and the climatic and agro-ecological characteristics of the remediation site restrict the use of phytoremediation procedures. The development of plants may be constrained by extremes of environmental toxicity, therefore phytoremediation takes longer than other treatments (Shankar and Shikha 2017).

## 4.3 Strategies to Protect Water-Related Ecosystems in Deltas from Releasing Hazardous Chemicals and Materials

Urbanization, agriculture, and industrial discharges are the main causes of surface water and groundwater contamination. Physical aspects of the environment can also contribute significantly to contamination. Pollutants released by agricultural activities include metals, pesticides, pathogens, nutrients and salts also influence the quality of water resources. Moreover, untreated and partially treated sewage, construction waste and solid/liquid waste factors contain hazardous substances emitted into the river water by urban activities (Akhtar et al. 2021).

In Deltas areas, elimination of waste generation and treatment of hazardous wastes are important environmental challenges of chemical industry in order to fulfill legal emission limits and comply with sustainability targets (Papadokonstantakis et al. 2013). For the time being, many studies describe the management of hazardous waste in Deltas, in terms of treatment, prevention, and the ideal procedures of management.

Couto et al. (2013) has determined that all the involvement of all the stakeholders, including waste generators, regulators, decision-makers, waste processors, and the formal and informal sectors, must be included for effective management of hazardous waste. According to Fan et al. (2013), in China, hazardous waste has been classified into three types, household hazardous waste, industrial hazardous waste, and medical waste. However, other developing countries including Deltas areas have implemented several regulations and guidelines based on the international safety standards in combination with the Deltas situation. This suggests that integrated strategies for hazardous wastes may reduce the disposal rate of these wastes in Deltas regions. These are adopted in order for waste producers to minimize waste in fuel production and fuel cycles, materials classification and purification and also to guarantee a high volume reduction (Fan et al. 2013).

This suggests recovery are needed that may reduce the disposal rate of these wastes along the Deltas coastline. For this reason, an integrated approach should be adopted to design programs, and agencies to lead and coordinate a wide range of prevention initiatives to reduce the potential environmental and public health impacts (Olukunle and Okonkwo 2015). Official environmental protection policies in the world especially UK and the European Community have changed in the last few years, coinciding with a growing level of public anxiety about the prospects for a sustainable economy and environment. New techniques has a critical role in the final treatment of hazardous wastes, integrated solid waste management strategy is suggested through technology and policy interventions. One of these innovative processes is:

- Bioremediation and Phytoremediation techniques that involve the use of microorganism and plant to prevent and repair environment from hazardous wastes, including restoration of degraded soil, and conservation (Shankar and Shikha 2017).These techniques are actually used in different Deltas regions like China, India, Pakistan, etc. (Mmereki et al. 2016).However, according to Sheng et al. (2008), scientific research has found that phytoremediation has a limitation of long lasting and of low efficiency. Is just one example of strategies contributing to a reduction of hazardous waste generated level in the Deltas coastline Other formal strategies include the following:
- Adopting alternative strategies (e.g., physico-chemical treatment) (Menad et al. 1999).
- Treatment with or without heat recovery. For instance, in Chile, used oil and spend solvents are used as raw materials for alternative fuels production (Olukunle and Okonkwo 2015).
- Making new regulations where necessary and appropriate (Menad et al. 1999). In Nigeria Deltas (Ogoniland) records show the existence of well over 25 significant environmental laws since 1956 when the Oil Pipelines Act (OPA) was declared (amended in 1965) (Yakubu 2017).In Nigeria the declaration of the Environmental Guidelines and Standard for the Petroleum Industry (EGASPIN) in 2002 (Ite et al. 2013).
- Ensuring that other government departments and public bodies fulfill their roles and responsibilities identified in hazardous waste management (Menad et al. 1999). It is important to note that policy makers, regulators, product producers, generators, and holders of hazardous waste need to play a vital role in ensuring that such materials are prevented, minimized, collected, and treated properly in accordance with the waste hierarchy (Duan et al. 2008).
- Namely the three Rs, the first and most important option is source reduction. Source reduction is any activity that reduces or eliminates either the generation of waste at the source or the release of a contaminant from a process. The next preferred option is recycling. Recycling is the reclamation of the useful constituents of a waste for

reuse, or the use or reuse of a waste as a substitute for a commercial feedstock or as a feedstock in an industrial process (Onwukwe and Nwakaudu 2012).

- Facilitating a two way communication with sectorial and stakeholder interests.
- Municipalities fulfill their important role in providing small scale collection services and generally raising awareness in hazardous waste management.
- Prevention: engaging with priority sectors in hazardous waste prevention waste activities through cleaner technologies and better compliance with regulation.
- Devising future new policies on the management of hazardous waste by also taking into consideration the generation rates of these types of wastes and potential savings from the recovery of these wastes (Menad et al. 1999). In the Pearl River Delta region, China Administration and higher level Environmental Protection Bureaus (Li 2006). About 350 industries are discharging their sewage-water either directly into the Nile, the river Nile and the co-ordinated canals receive enormous amounts of biological and chemical pollutants. Based on this concept, in Egypt, they receive policy mandates from the State Environmental Protection appropriate law was issued (Abdel-Shafy et al. 2002). Government of India has formulated Solid Waste Management Rule, 2016 (SWM 2016).
- Integrate more investment in environmental protection the case of China where over 1300 billion (US $156.6 billion) invest in environmental protection between 2006 and 2010, more than 1.5% of the country's Gross Domestic Product (GDP). But, compared to other developed coastal countries such as the USA, the UK and Germany, who have invested more than 2.0% of their respective GDPs. China's investment in environmental protection still falls short of the required amount (Zhouet al. 2010).
- Pollution Prevention Techniques, ISO on Environmental Management Systems, defines pollution prevention as the adoption of processes, practices, techniques, materials, products, services, substances or energy that avoid or minimize or control (separately or in combination) the creation, emission or discharge of any type of pollutants or wastes thus reducing adverse environmental impacts (ISO 2015). Both the public and private sectors are required to give prior consideration to the environmental effects of any activity before it is embarked upon. Pollution prevention can be achieved by process or procedure modification, redesigning equipment and technology. Improvements in maintenance, inventory control, training programs and housekeeping are all pollution prevention efforts (Yakubu 2017).
- Environmental management (EM) according to Thompson (2002) who suggests that EM is the system that visualizes and in advance prepares for, and circumvents, or provides solutions to environmental and resource sustentation problems (Fig. 4.3).The objectives of integrated EM are as follows:
  - Prevent, reduce and control deleterious change in the environment, likely to impact on quality of life, thereby maintaining and improving its life support and productivity capacities;

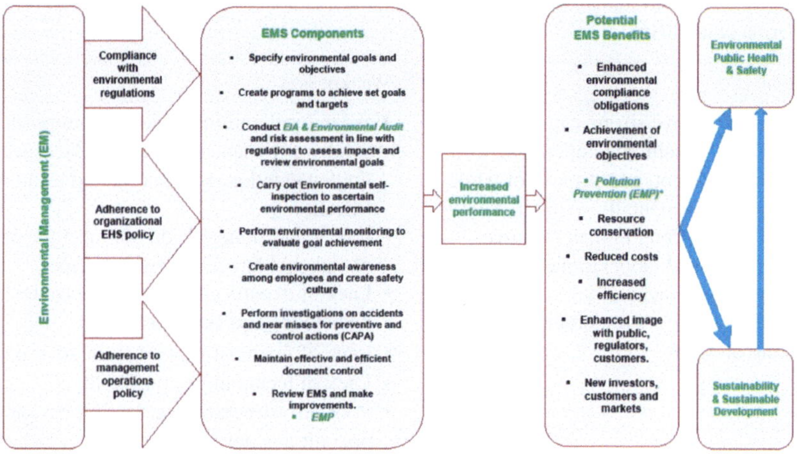


**Fig. 4.3** EM/EMS Model. EM-environmental management; EMS—environmental management system; EIA—environmental impact assessment; * EMP—environmental management plan; EHS—environmental health and safety (Yakubu 2017)

- Develop and increase the potential of living resources to meet human nutritional needs, as well as social, economic and development goals;
- Promote the integrated management and sustainable development of terrestrial, freshwater and coastal marine environments (Moore et al. 2004).

Further support the municipal solid waste MSW plan, we analyzed gaps and SWOT (strengths, weakness, opportunities, and threats) to be incorporated in the action plans for efficient management of hazardous waste in deltas. SWOT is a tool to identify the positive and negative points of the performance of any process, organization, project, and company (Verma et al. 2016). We applied to SWOT analysis for the MSW management as shown in Table 4.1.

- Green development: actually many countries have made great efforts to create a sustainable future through developing green development approaches, policies and actions. For examples, United States stimulated green development by investing in clean energy, and European Union (EU) developed green economy by building "green industry". South Korea proposed "green growth" economic revitalization strategy, while Japan promoted the growth of the green economy by building "low-carbon society" (Skea and Nishioka 2008).Other developing countries like Cambodia and South Africa have also developed a green economic strategy plan (UNEP 2012) proposed the concept of "green index" (Fig. 4.4).

**Table 4.1** SWOT (Strengths, Weaknesses, Opportunities, and Threats) analysis of MSW management (Verma et al. 2016)

| Strengths | Weaknesses |
|---|---|
| • Private sectors have strong capacity to manage MSW, including collection, transport, treatment, and dispose of while applying environmentally sound technologies<br>• Availability of strong human resources (including manual labor) to meet the demand of MSW management<br>• Informal sectors play important role in recycling of MSW | • Weak legal and institutional framework for solid waste management (SWM), especially implementation, monitoring and evaluation (M&E)<br>• Policy and standards do not meet the ability of stakeholders (e.g. private sectors)<br>• Lack of experts and staff who have technical knowledge and skills in SWM<br>• Lack of financial support from government<br>• Lack of technologies for SWM<br>• Poor infrastructure (e.g. landfill design and operation is not efficient<br>• Weak environmental protection in landfills<br>• Does not take advantages of the private sector role in SWM<br>• Lack of waste separation at source due to lack environmental awareness of local people |
| Opportunities | Threats |
| • City location is good for attracting investment from outside foreign and internal organizations<br>• Better engagement of private sectors in SWM by changing economic structures and policies<br>• Being a biggest city, the national government has focused on city development and MSW plan will be supported by the government<br>• With high fraction of organic waste (65–90%) there are opportunities and potential of converting waste to energy or waste to compost | • Rapid economic growth and changing life style increased consumption of resources which in turn increasing waste generation<br>• Increasing population due to migration from other part of the country resulting increased waste generation<br>• Harsh climate as a logistical constraint |

## 4.4 Actions to Reduce Air Pollution from Solid Waste Generation in Deltas

Accelerated urbanization and industrialization have led to an increase in the number of waste treatment facilities. Landfills are one of the most common methods of municipal and industrial waste disposal worldwide. In addition to the potential hazards of landfills to soil and groundwater, gases and odors are also emitted by anaerobic decomposition of solids (Adesinaet al. 2018). Gases and odors Dincer et al. (2006) state that most pollutants released from landfills are malodorous and have a negative impact on the environment and health.

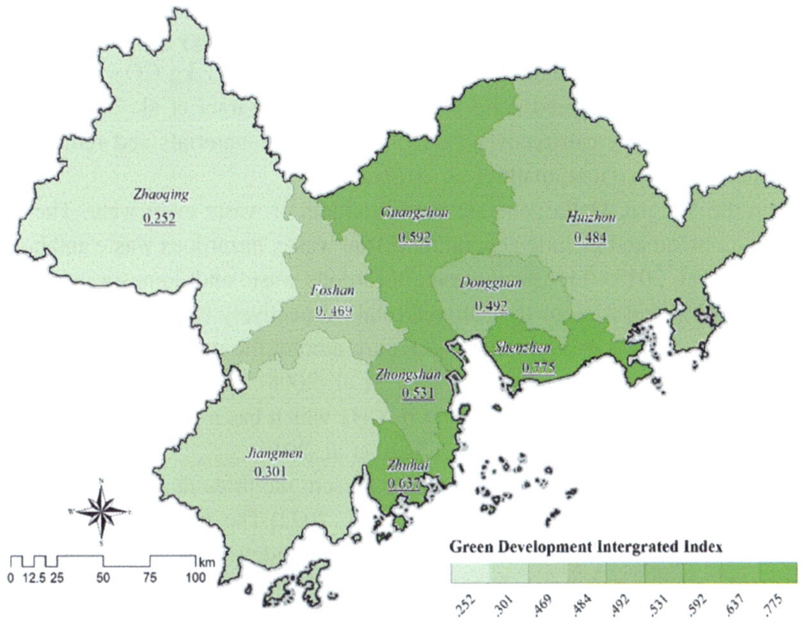


**Fig. 4.4** Distribution of green development integrated index in cities within the Pearl River Delta, China in 2015 (Wang et al. 2018)

Air pollutants emitted from waste treatment plants include carbon monoxide, CO, nitrogen oxides ($NO_X$); sulfur dioxide ($SO_2$); particulate matter (PM); hydrocarbons (HC) and Volatile Organic Compounds (VOCs).

An association between air pollution and high morbidity and mortality from hypertension and cardiovascular problems has been reported. In addition to the stench, the landfill produced foul-smelling gas. Previous studies have shown that they pose respiratory, neurotoxic, carcinogenic, and teratogenic risks to humans, especially those living close to host communities (Sonibare et al. 2019).

Ejuveyerome (2020) states that pollution prevention is the reduction of pollution at the source resulting in less waste to be managed treated or eliminated. Pollution control improves health, reduces economic losses, protects the environment and promotes economic growth through efficient production in industry.

### 4.4.1 Air Pollution Sources

The entire waste management process Air pollution and its environmental impacts such as respiratory diseases, smog, acid rain, and global warming are major threats to people and the environment (Testa et al. 2014).

Produces greenhouse gas emissions, including collection, transportation, composting, digestion, incineration, and landfill, which affect the urban energy use and carbon cycles (Zhou et al. 2015). It is estimated that approximately 4,984.71 kg $CO_2$ is produced per 8,500 tonnes of fully processed solid waste (Nabavi-Pelesaraei et al. 2017). Improving waste disposal efficiency can recover valuable recyclable materials and reduce negative environmental impacts (Gundupalli et al. 2017).

Cities in the Yangtze Delta generate huge amounts of waste every year. The types of solid waste are broad and include general industrial waste, hazardous waste and household waste (Gupta et al. 2015). Industrial waste, hazardous waste and domestic waste account for about 9%, 17% and 21% of the national total, respectively (Chen et al. 2014).

Methane ($CH_4$) and carbon dioxide ($CO_2$) are among the most important greenhouse gases (GHG) emitted from landfills (Bogner et al. 2008). Anaerobic decomposition of wastes in landfills promotes the production of $CH_4$ which has a global warming potential of about 21 multiples of that of $CO_2$ (Sonibare et al. 2019).

Transportation, energy sources and industry present the three main sectors responsible for $CO_2$ and air pollutant emissions (Chen et al, 2022).The transport emits very high concentrations of carbon monoxide (CO), nitrogen dioxide ($NO_2$), sulfur dioxide ($SO_2$), and hydrocarbons (Echendu et al. 2022).

The Combustion of biomass and waste in the Niger Delta region cause air pollution and releases carbonaceous aerosols with strong radiative effects. Aerosols generally affect climate through their effects on radiation and clouds. These aerosols, nitrogen oxides, nitrogen, volatile organic compounds and also carbon monoxide, affect climate by perturbing ozone and methane concentrations and producing secondary aerosol particles (Knippertz et al. 2015).

Giwa et al. (2019) declare that gas flares are a major source of pollution in the Niger Delta. The negative impact on ecosystem health is due to the production of crude oil-related natural gas, which releases gaseous, particulate, and caloric species into the atmosphere. Flare gases are a major contributor to global warming and climate change as they are sources of greenhouse gases, volatile organic compounds, precursor gases, toxins (including benzene, hydrogen sulfide, toluene, etc.) and black carbon. All of these are harmful because that cause ecosystem degradation and destruction (Giwaet al. 2019).

Roughly, the amount of environmental toxins and carbon dioxide released will increase, accounting for one-third of total emissions, and nitrogen oxides and methane will flow into the delta ecozone. Others may contain fluorine and black carbon, as shown in Fig. 4.4 (Okpara and Oruche 2019).

One of the main contributors to the greenhouse effect is solid waste from emissions or incineration, which reduces the amount of trees available to hold carbon dioxide ($CO_2$), and the release of stored carbon dioxide has contributed to global warming in recent years, the temperature is rising (Akinbami et al. 2003) (Fig. 4.5).

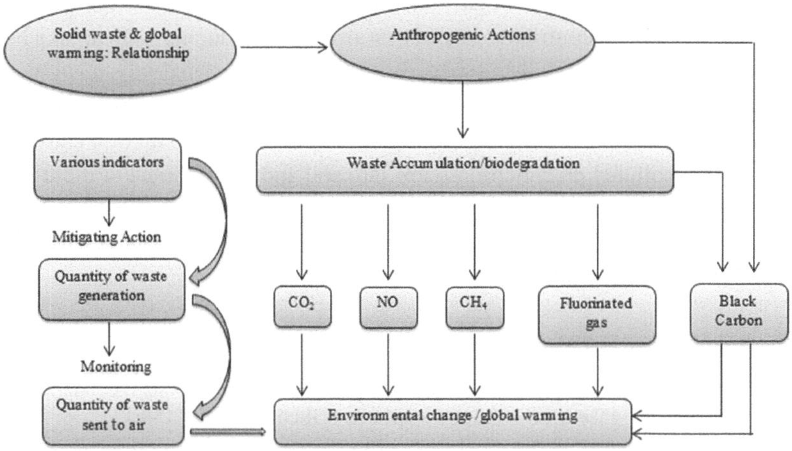


**Fig. 4.5** Relationship between climate change and solid waste (Okpara and Oruche 2019)

## 4.4.2 Impact of Air Pollution

### 4.4.2.1 Adverse Effects of Air Pollutants on Health

Emissions of air pollutants as a by-product of gas flaring and solid waste in the Niger Delta have shown that exposure to high concentrations of air pollutants can have detrimental effects on human health through exposure–response studies in different regions of the world. It is of great concern because it has been shown to be highly sensitive (Pope et al. 2002).

Studies have shown that high levels of particulate matter (PM2.5) cause respiratory problems, cerebrovascular disease, ischemic heart disease, and lung cancer (Burnett et al. 2014). Chronic obstructive pulmonary disease is also associated with elevated ozone concentrations (Chuwah and Santillo 2017).

### 4.4.2.2 Impact on Plants

It is known from field experiments that high air pollutant concentrations can adversely affect plants (De Bock et al. 2011).

For example, the study of Dung et al. (2008) found that gas flares and associated air pollutants in the Niger delta slowed the growth of plants growing closer than farther away from the point of the flare. Their results also show that air pollutants from gas flares can degrade the nutritional value of harvested crops. These air pollution impacts on crops mean that increased food demand will be met by consolidation of existing land or expansion of cultivated land, leading to land-use change and other related environmental problems (Chuwah et al. 2015).

Furthermore, air pollutants from gas flares adversely affect natural vegetation, thereby reducing biodiversity and contribute to global warming by reducing $CO_2$ uptake by plants (Chuwah and Santillo 2017).

Air pollutants produced during flaring cause ecosystem damage by adversely affecting the quality of soil and water resources through precipitation. Air pollution causes a chain reaction in which polluted precipitation seeps into soil and water. Some of these pollutants are also released into the atmosphere. Acid rain, a common result of polluted air, alters soil chemistry, resulting in various effects on plant growth and water quality, he was experienced in the Niger Delta (Echendu et al. 2022).

#### 4.4.2.3 Impact on Local Climate

Gas flares contribute directly to global warming by acting as sources of $CO_2$ and $CH_4$, two major greenhouse gases, into the atmosphere (Okeagu et al. 2006).

Studies in various parts of the world have shown that air pollution can have a significant impact on local climates. Due to the short residence time of air pollutants in the atmosphere, their spatial distribution is uneven, usually with higher concentrations near the emission source. This suggests that changes in air pollutant emissions have significant impacts on local climate (Chuwah and Santillo 2017), and in some cases through remote atmospheric connections.

High concentrations of pollutants such as methane and carbon dioxide released into the atmosphere by human activities cause an energy imbalance between the atmosphere and the surface, resulting in temperature changes that disrupt the normal chemical composition of the atmosphere (Echendu et al. 2022).

Air pollutants interact with the atmosphere, causing warming and environmental degradation. Burning fossil fuels contributes to greenhouse gas emissions that contribute to climate change. This change contributes to wildfires becoming more frequent and severe by creating higher temperatures that allow more dry fuel to burn. Fires can intensify and spread (Pausas and Keeley 2021). So there is a cycle in which climate influences fires, and fire activity influences climate change by releasing more pollutants into the atmosphere.

#### 4.4.2.4 Impact on the Environment

Dry or wet deposition of nitrates and sulfates is recognized as a major problem for the inhabitants of the Niger Delta. Studies have shown that wet and dry deposition of acid and acid particles in the Niger Delta affects ecosystems, especially aquatic and soil microbes (Ekpoh and Obia 2010). The same authors found a direct link between gas flaring, associated acid rain, and zinc corrosion in parts of the Niger Delta.

Airborne polycyclic aromatic hydrocarbons (PAHs) are produced by the combustion, when deposited in water, it interferes with the dissolution of atmospheric oxygen, affects circulation and limits oxygenation of aquatic organisms. It also impedes the penetration of light into the water, affects the photosynthetic activity of primary producers such as

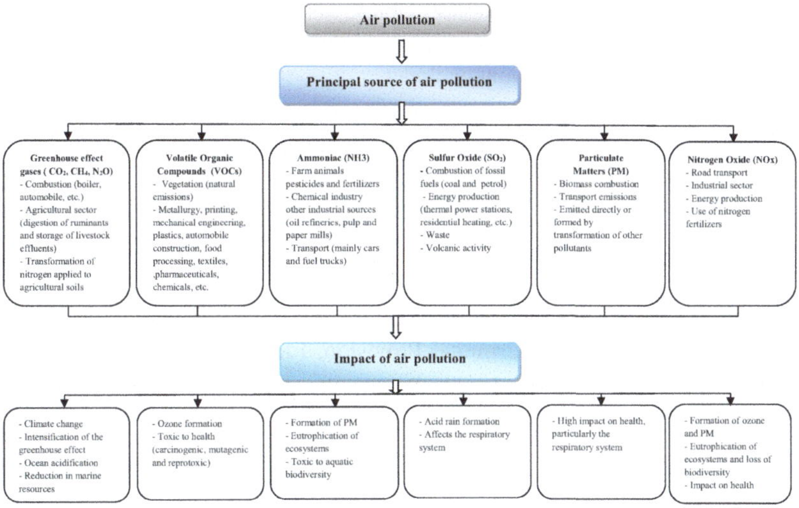


**Fig. 4.6** Sources of air pollution and their impacts on the environment

phytoplankton, and causes imbalances in food webs and ecosystems (Echendu et al. 2022) (Fig. 4.6).

## 4.5 Different Actions Used in Deltas to Reduce Air Pollution from Solid Waste

The experience of joint control of air and river basin pollution in the Yangtze River Delta provides a cooperative foundation for the collaborative governance of solid waste. The region has made some valuable explorations about this pollution under the mode of trans-regional and multi-entity's participation (Wang et al. 2020). The process also introduced scientific instruments of interregional cooperation, such as joint leadership meetings, regional environmental cooperation bodies, regional environmental cooperation agreements, and a system of horizontal ecological compensation mechanisms (Zhang and Zhu 2022).

The same authors stated that Germany's Rhein-Ruhr metropolitan area has established a waste exchange system that aggregates trade intention information from various regions to meet the demand and supply of solid waste in various markets. In order to reduce the discharge of air pollutants, the surrounding areas of Tokyo have formed a clear industrial chain of solid waste disposal, which jointly eliminates the solid waste generated in the region through the division of labor. There are also quite a few attempts domestically.

According to Abonyi and Eleje (2020), some other domestic statutes enacted by the Nigerian legislature have been documented to guide waste management in the Delta and the nation. Some of these legislations include:

- The Federal Environmental Protection Agency (FEPA) Act of 1988. This was established by Decree 55 of December 30th 1988. The Act vested FEPA with the statutory responsibility for overall protection of the environment. It requires FEPA to issue environmental guidelines and standards for the abatement and control of all forms of pollution. In 1991, FEPA established an inspectorate and enforcement department with divisions for standard regulation, chemical tracking and compliance monitoring.
- The Harmful Waste (Special Criminal Provision, Etc.) Act 1988, essentially a penal legislation, this Act was enacted with the specific object of prohibiting the carrying, depositing and dumping of hazardous wastes on any land, territorial waters and matters relating thereto.

Establishment of private companies under the control of the local People's Committee (PC) or Department of Natural Resources and Environment (DONRE) in Mekong Delta region which are directly responsible for the collection, transportation and processing of municipal solid waste (MSW) in the provinces/cities to monitor and protect the environment from pollution. The national regulations and standards, governmental decrees, joint circular and local regulations and decisions are more and more strict in order to turn environmental protection into positive effect (Hoang 2012). Qualified scientific personnel were trained abroad to develop a solid waste management system in the MD region.

Realizing that there is increase in waste generation and change in waste composition, they have been promoting the 3Rs (reduce, reuse and recycling) since 2007 in order to increase material recovery and to reduce waste disposed in landfill. The new recycling policy was issued by the Ministry of Environment in 2008 (Meidiana and Gamse 2010).

Landfill is currently the only treatment option for municipal solid waste (MSW). Low-plain deltas face the most serious environmental challenges in the near future due to the effects of climate change and sea level rise. A landfill should not be considered for permanent use in this area. Appropriate MSW treatment options should be applied along with reuse, recycling and other solutions for MSW reduction and pollution prevention. An integrated MSW management approach to climate change adaptation is considered necessary.

## 4.6 Recommendations to Ensure Environment Protection in Deltas Using Sustainable Waste Management Practices

Conservation of the delta is a prerequisite for biodiversity conservation. Balancing economic development with measures to avoid or mitigate ecological risks remains a challenge in the Delta. Sustainable development should no longer be seen as a declarative goal. Because its implementation may be the only effective way to ensure continued access to natural resources for present and future generations.

A variety of endemic flora and fauna, migratory bird species (some of which are rare or endangered) and some traditional populations depend on the delta ecosystems and are vulnerable to irreversible habitat change. The preservation of the delta is of paramount importance because it is fragile.

Deltas are environments of exceptionally rich biodiversity, home to thousands of plant and animal species, often rare or threatened with extinction on a global scale, and therefore enjoy international protection.

Protection of the biological, ecological, and scientific values of the delta is a concern because delta ecosystem habitats are particularly sensitive to environmental change. Some deltas are under pressure from human activities and are under severe threat. Environmentally sound management of hazardous waste is becoming a major concern in developing countries due to the diversity of waste streams and the toxic substances they contain and the negative environmental and public health impacts they cause. Therefore, some practical recommendations are suggested, such as (Mmereki et al. 2016):

- Raise public awareness of the recycling potential of hazardous waste.
- Capacity building and staff development for hazardous waste recycling.
- Monitoring and evaluation of hazardous waste management systems and reporting programs.
- Development of appropriate infrastructure, technical knowledge and expertise.
- Strengthening and reforming existing regulatory frameworks.
- Developing decision aids to identify appropriate technologies for treating hazardous waste.
- We provide funding to identify the best new and cutting edge technologies.
- Development of a regional treatment system for hazardous waste.

For the successful implementation of hazardous waste management best practices, there is an urgent need to consider the experiences of developed countries and combine this with the socioeconomic situation of developing countries (Lu et al. 2013).Waste minimization through source reduction, reuse and recycling must be effectively implemented to reduce the amount of hazardous waste generated and disposed of. Achieving this will require a fundamental reform of the current regulations in developing countries (Mmereki et al. 2014).

We therefore need sound environmental policies backed by robust mechanical science. Environmental elements for sustainable industrial development must include (Moore et al. 2004):

- Key environmental, health and sustainability indicators;
- Broad support for the objective of minimizing human impact on the environment;
- An effective environmental policy framework that minimizes environmental impacts based on an ecosystem approach;
- Cleaner industrial production and pollution control;
- Environmental emission and discharge standards (with the aim of minimizing these);
- Enforced environmental protection and waste management;
- Ecotoxicology to assess the environmental impacts of pollution and resource overexploitation;
- Environmental modeling for policy making;
- Risk assessment, mitigation and management;
- Integration process with socio-economic conditions and governance issues;
- Broad public support for environmental action.

According to Thonart et al. (2005), the importance of waste sorting at the base and at the time of the burial makes it possible to consider zones with different and less expensive arrangements according to the type of waste, that implies:

- Political decisions and socio-economic aspects through the establishment and management of landfills (legislation, regulation, control, etc.);
- Installation of pumps and treatment plants for leachate and biogas;
- Environmental monitoring (groundwater and surface water, gas infiltration, diffusion control of odor, the achate and biogas treatment plants and intervention in case of problems).

It will be preferable to adopt the anaerobic waste digestion method, this method can handle several types of waste, ranging from wet to dry and from clean organics to gray waste (Eunomia 2015).

The Anaerobic digestion process provides a renewable source of energy when food is wasted It can be broken down into biogas (a mixture of methane and carbon dioxide). Used for energy production. Biogas can be used to generate electricity, he three ways of power generation. On-site devices and surplus electricity can be fed into the national grid. Another by-product of this process is a biofertilizer rich in nutrients such as nitrogen. Phosphorus and other elements essential for healthy plant growth and fertile soil (Halkos and Kleoniki 2016).

It will be necessary to use recycling as a waste treatment process in developing countries; this process leads to many benefits (EPA 2016):

- Reduces the amount of waste sent to landfills and incinerators;
- Protect natural resources such as wood, water and minerals;
- Prevents pollution by reducing the need to gather new resources;
- Save energy;
- Reduce greenhouse gas emissions that contribute to global climate change;
- Help protect the environment for future generations;
- Help create new high-paying jobs in recycling and manufacturing.

As the issue of municipal solid waste management is prevalent in Vietnam, especially in the Mekong Delta, many studies will focus on sustainable integrated municipal solid waste management due to the impact of climate change in the coming years. Opportunities and potential challenges were identified to guide an integrated MSW management system as a new approach. Open landfills, along with integrated solutions for MSW management activities such as source separation, size reduction, waste management, recycling, composting, pyrolysis and incineration, state promotion or subsidies, and public participation, should be transferred to the bioreactor landfill (Nguyen 2012).

Quality bulletins and research reports confirm increased land-based pollution, eutrophication, mangrove deforestation, depletion of marine fisheries, loss of marine biodiversity, and increased occurrence of red tides. Understanding the drivers of environmental degradation in the Pearl River Delta region (PRD) will help raise public awareness. The general public recognizes their role in protecting the environment. Inadequate government institutions, inefficient frameworks of maritime institutions, and low environmental awareness among the general public contribute to environmental degradation. In this context, it is necessary to (Zhou and Cai 2010):

- Chinese government must balance economic development and environmental goals protection. The evaluation and reward system for government officials needs to be reformed, and social and environmental indicators need to be linked so that they can better track sustainable development. It is also highly recommended to promote financial ties between her EPB and the government at the same level. The EPB can function more efficiently with support from her EPB above her or from the new integrated governing body.
- Deficiencies in the environmental legal and institutional framework must be rectified. Regarding the regulatory framework, regional differences in the policymaking at the central level should be taken into account. This allows policies to be flexible enough to take into account important regional differences (social, environmental and economic). The policy should keep both national and local environmental regulations up to date with the latest environmental changes (Table 4.2). A comprehensive institution should be established in the Pearl River Delta.

**Table 4.2** Waste management authority and its responsibility for pollution reduction in Deltas

| Waste management authority | Responsibility |
|---|---|
| The government, regulatory organizations and machinists in the petroleum sector in Nigeria deltas | The creation of appropriate environmental laws and regulations should reassure everyone that the environment is being conserved (Emoyan 2008). The government of Nigeria in 1989, the National Environmental Policy (NEP) was broadened. In 1999 the policy was updated to reflect contemporary and developing environmental concerns and the revised policy aimed to achieve sustainable development goals in Nigeria Deltas. This might be achieved by preserving biodiversity, restoring and maintaining the ecosystem and ecological progressions and promoting sustainable natural resources (National Environmental 2010) |
| Industrial sector in Nigeria deltas | Many rural areas of the deltas are subject of industrial ecological techniques that promote energy and material conservation (such as recycling activities) (Dieu 2003) |
| The new ministry of natural resources and environment (created in 2002) its predecessor was the ministry of science, technology and the environment (set up in 1993) in Vietnam | Vietnam's institutional framework for environmental protection has lately been reorganized. The Ministry's main environmental responsibilities are assigned to the Vietnam Environmental Protection Agency; It is also tasked with carrying out the general environmental protection law. Responsibility of the environmental management is in the hands of various departments of natural resources and environment functioning on the red river delta of northern Vietnam, which are a significant entities for the execution of environmental laws in general, and for pollution control, including industrial discharges, in particular (Konstadakopulos 2008) |

(continued)

**Table 4.2** (continued)

| Waste management authority | Responsibility |
|---|---|
| Vietnam's voluntary pollution reduction program | The red river delta of northern Vietnam's environmental policy has been impacted by the collaboration of its public institutions with transnational governmental and non-governmental organizations (Weidner 2002) |
| Several ministries in Egypt are involved in water waste management:<br>– Ministry of water resources and irrigation<br>– State ministry for environmental affairs<br>– Ministry of industry<br>– Ministry of interior | The functions of the three primary control mechanisms for waste management as it is performed by the various Egyptian ministries and institutions: legislation, monitoring and pollution control. The important national policies and action plans that have an impact on Waste management are (Abdel-Dayem 2011):<br>– National water resources plan 2017: water for the future<br>– National environmental action plan (NEAP) 2002<br>– Agricultural policy<br>– National policy for inhabitation of deserts |

a. To manage the natural resources and environmental issues of this unique habitat and important estuary;
b. Improve the efficiency of government and thereby defuse the situation. This overarching body should have final authority and overall responsibility for the integrated management of the PRD's natural resources and coastal environment.

– Further efforts are needed to raise public awareness and increase investment in environmental protection. More protected demonstration plots and nature a park should be set up and open to the public. Environmental awareness needs to spread across schools, businesses and communities. Speaking of in addition to investing in environmental protection, the Pearl River Delta needs to increase investment from non-governmental sources such as the private sector and foreign investors, in addition to most of the funds coming from the national treasury (Cai et al. 2001). Already polluted areas require investment from government funds to fund scientific research and ecosystem restoration projects. In late 2005, the "Shenzhen Bay Fengtang Estuary Mangrove Restoration Demonstration Area" was opened. Aimed at estuary water purification and mangrove restoration projects, the project has invested a total of RMB 240 million (Cai et al. 2007).

The Spanish legislation aims to progressively protect the Ebro Delta by limiting the development of cities and towns and by applying the GRI-environmental measures to reduce fertilizers and pesticides used in rice cultivation. The Spanish regional government, especially at the request of the European Union (Natura 2000 network), plans to expand its natural parks (currently 11,500 hectares, including 7,800 hectares of natural parks) to better protect them (Platvoet 2005).

To create an institution for the Ebro Delta that houses both the administrative and economic sectors involved, with the necessary competencies and financial resources to implement comprehensive development and management plans for the Delta (Platvoet 2005).

To develop an effective action plan to successfully manage MSW, the following recommendations can be included in the action plan (Verma et al. 2016).

*Awareness campaign*
Information sharing campaigns can be organized to raise local awareness of waste separation at source and encourage 3Rs (reuse, reduce, recycle) practices. Organizing a public participation program is important for good results. Household interviews can be conducted to explore community attitudes towards waste separation at source. You can set goals. For example, the amount of waste generated can be reduced by 5–10% per year. Young minds learn quickly, so schools are a great place to start an awareness campaign. At school level, 3R-related activities can be implemented as students will play an important role in the future. This activity raises awareness not only for the students but also for the families of the students.

*Resource recovery*
It is important that the proposed anaerobic digestion facility meets the criteria of the chosen environment market area. Providing 'work guidelines' to landfill workers and workers is very important to reduce as much as possible Potential environmental risk.

*Build a transport station*
The first activity proposed to be accomplished is the preparation of the building location, design, standards, regulations and especially the finance for transportation stations. The need of conducting this activity is due to the lack of transportation stations within the city, which leads to difficulties during the waste collection and transportation stages.

*Staff development*
It is also necessary to improve the skills of staff in order to properly manage municipal solid waste. Capacity development for Environmental officers has an important role to play and should be concerned. Employees can be encouraged to do this Attend waste management training courses. It is possible to implement training and evaluate after training. To give environmental workers the opportunity to develop their technical skills through

**Table 4.3** Cost estimation of waste management for pollution reduction in deltas

| Programs | Cost estimation |
|---|---|
| The Niger Delta Development Commission (NDDC) | In 2000, the Niger Delta Development Commission (NDDC) was officially established. Its main objective was to explore about, plan, and implement projects and programs for the development sustainability of the region. In accordance with the NDDC act, there was to take their finance from four different sources which are 15% of the monthly Federal Statutory Allocations (FSA), due to petroleum producing states, 3% of the operational budgets of petroleum firms, 50% of the Ecological Fund (EF) due to the states in the NDRN and any additional contributions from the private sector or foreign aid (Okonkwo and Etemire 2017) |
| Some states in the Niger Deltas Regions | Use a portion of the resource revenues provided to them to create development plans or join groups that that allocate a percentage of their derivation income to help areas impacted by solid waste and oil (Kimenyi et al. 2014) |

capacity building as part of urban exchanges, study abroad is a valuable resource as it allows workers to showcase themselves and learn new skills and techniques always beneficial. Personal and financial preparation for school trips is important, study abroad performance indicators are based on staff performance.

*Public-private partnerships in solid waste management*

Public–Private Partnerships (PPPs) play an important role in protecting the environment. Forming a stakeholder network should be the first activity that helps identify all private sectors relevant to the solid waste sector. A stakeholder agreement must then be developed to ensure the responsibilities, obligations, and benefits of each stakeholder in this collaboration.

*MSW management fee*

Governments can not only promote his 3Rs, but also persuade locals to be willing to pay for waste disposal services and encourage them to generate less household waste. Imposing a waste management fee can lead to a reduction in waste generation at the household level (Table 4.3).

## 4.7 Conclusion

The deltas, through their ecological and economic interests, are at the center of the attention of decision-makers and local development actors. Economic development undeniably depends on better management of cuttings and the resulting pollution. Moreover, the complexity of the forms of pollution and the growing economic stakes make it all the more important to protect the natural environments of the delta regions.

The consequences of solid waste disposal lead to serious forms of degradation which affect different compartments of the environment, in particular the non-renewable natural resources which will be used by man at the advent. Although attempts are made to ensure the sustainability of the rate of economic growth, relying on technological innovation as the main asset to solve the problem of waste, the fact remains that the high cost of use as well that their new features are not a guarantee. Utilizing integrated waste management in any perspective aimed at reducing the quantity of solid waste, generating income by recovering recyclable and reusable waste is one of the actions adopted by several delta countries. The objective being to encourage consumers to sort their waste at the household level, to collect what can be reused and recycled. It also means encouraging people to collect waste from disaster areas. The problem is not only under the responsibility of the governmental authorities, the associative movement as well as the NGOs is mobilized and involved in this process which seems to be an effective and saving solution for a better management of solid waste.

Integrating humans into the protection of their own environment appears to be the ultimate long-term solution, even if it is difficult for current generations to change their consumption patterns by using recyclable and biodegradable products; Awareness of long-term behavior change in particular.

In terms of economic development and exploitation of deltas, current societies are obliged to take into consideration the diversity of ecosystem services provided by these biotopes. Totalitarian visions integrating deltas as a landscape unit called to assimilate all kinds of dejections are no longer in the majority. The notions of sustainable planning and ecologically sustainable economic development impose new management, exploitation practices of these resources in order to ensure the sustainability and to prepare future generations in better conditions for the challenges to come.

**Acknowledgements** We would like to acknowledge the help of all the people involved in this work and, more specifically, to the authors who contributed their time and expertise to this chapter book.

We are deeply grateful to Dr. Mahmoud Nasr for graciously accepting us to contribute to this book. We also thank him for his availability and for his clarifications on our chapter.

**Declaration of Competing Interest** The authors declare that they have no known competing financial interests or personal relationships that could have appeared to influence the work reported in this paper.

## References

Abdel-Dayem S (2011) Water quality management in Egypt. Int J Water Resour Dev 27:181–202

Abonyi NN, Eleje JN (2020) Industrial waste management and environmental pollution in the Niger delta region of Nigeria: issues and realities. J Soc Sci 2:73–83

Adesina OA, Sonibare JA, Diagboya PN, Adeniran JA, Yusuf RO (2018) Spatiotemporal distributions of polycyclic aromatic hydrocarbons close to a typical medical waste incinerator. Environ Sci Pollut Res 25:274–282

Akhtar N, Ishak MIS, Bhawani SA, Umar K (2021) Various natural and anthropogenic factors responsible for water quality degradation: a review. Water 13:2660

Akinbami J, Salami A, Siyanbola W (2003) An integrated strategy for sustainable forest-energy-environment interactions in Nigeria. J Environ Manage 69:115–122

Al-Maaded M, MadiKahramanRamazan NK, Hodzic A, Ozrkan NG (2012) An overview of solid waste management and plastic recycling in Qatar. J Polym Environ 20:186-194

Burnett RT, Pope CA, Ezzati M (2014) An integrated risk function for estimating the global burden of disease attributable to ambient fine particulate matter exposure Environ Health Perspect. 122:397–403

Bogner J, Pipatti R, Hashimoto S, Diaz C, Mareckova K, Diaz L (2008) Mitigation of global greenhouse gas emissions from waste: conclusions and strategies from the intergovernmental panel on climate change (IPCC) fourth assessment report, Working group III (mitigation). Waste Manag Res 26:11–32

Cai LZ, Zheng TL, Lin JD (2001) Spatial and temporal distributions of Denronereispinnaticirris and Neanthesglandicincta (Polychaeta: Nereididae) in the organically-enriched mudflat of Deep Bay, China. Asian Mar Biol 18:25–33

Cai LZ, Lin P, Li HM (2007) Restoration process of macrofaunal community at Luomazhou Section of the Shenzhen River after pollution interception. Chin J App Environ Biol 13:497–500

Chapman PM, Wang F (2001) Assessing sediment contamination in estuaries. Environ Toxicol Chem 20:3–22

Chen WM, Kim H, Yamaguchi H (2014) Renewable energy in eastern Asia: Renewable energy policy review and comparative SWOT analysis for promoting renewable energy in Japan, South Korea, and Taiwan. Energy Policy 74:319–329. https://doi.org/10.1016/j.enpol.2014.08.019

Chen W, Li H, Zhu1 Y, Jang J, Lin CJ, Chiang PC, Wang S, Xing J, Fang T, Li1 J, Yang Q, Zheng K (2022) Impact Assessment of Energy Transition Policy on Air Quality over a Typical District of the Pearl River Delta Region, China. Aerosol and Air Quality Research 22: 1-14

Cheremisinoff NP, Cheremisinoff PN (1993) Water treatment and waste recovery: advanced technology and applications. Englewood Cliffs, USA

Chuwah C, Van Noije T, Van Vuuren DP, Stehfest E, Hazeleger W (2015) Global impacts of surface ozone changes on crop yields and land use. Atmos Environ 106:11–23

Chuwah C, Santillo D (2017) Air pollution due to gas flaring in the Niger Delta: a review of current state of knowledge. Greenpeace Res Lab Tech Rep (rev) 04:1–9

Couto N, Silva V, Monteiro E, Rouboa A( 2013) Hazardous waste management in Portugal: an overview. Terra Green13 international conference. Energy Procedia 36:607–611

De Bock M, Op de Beeck M, De Temmerman L, Guisez Y, Ceulemans R, Vandermeiren K (2011) Ozone dose-response relationships for spring oilseed rape and broccoli. Atmos Environ 45:1759–1765

Dieu TTM (2003) Greening food processing industry in Vietnam: putting industrial ecology to work. Wageningen university and research.

Dincer F, Odabasi M, Muezzinoglu A (2006) Chemical characterization of odorous gases at a landfill site by gas chromatography–mass spectrometry. J Chromatogr A 1122:2229

Duan H, Huang Q, Wang Q, Zhou B, Li J (2008) Hazardous waste generation and management in China: a review. J Hazard Mater 158:221–227

Echendu AJ, Okafor HF, Olayinka L (2022) Air Pollution, climate change and ecosystem health in the Niger delta. Soc Sci 11:1–17

Ejuveyerome OW (2020) Novel metric for managing the protection of humanity and the environment against pollution and its adverse effects. Heliyon 6:1–8

Ekpoh IJ, Obia AE (2010) The role of gas flaring in the rapid corrosion of zinc roofs in the Niger Delta Region of Nigeria. Environmentalist 30:347–352

Emoyan OO (2008) The oil and gas industry and the Niger Delta: implications for the environment. J Appl Sci Environ Manag 12:29–37

EPA (2016) Recycling basics. US Environmental Protection Agency. http://www.epa.gov/recycle/recycling-basics

Eriksson O, Reich Carlsson M, Frostell B et al (2005) Municipal solid waste management from a systems perspective. J Clean Prod 13:241–252

Eunomia (2015) Economic analysis of options for managing biodegradable municipal waste, final report to the european commission. http://ec.europa.eu/environment/waste/compost/pdf/econanalysis_appendices.pdf

Fan Z, Liu Y, Wang J, Ren G, Le WE (2013) China: experience of radioactive waste (RAW) management. Processes, technologies and international experience A volume. Woodhead Publ Ser Energy 697–722:723e–725e

FAO and UNEP (2021) Global assessment of soil pollution-summary for policy makers. Rome, FAO. https://doi.org/10.4060/cb4827en

Giwa S, Collins N, Sidikat K, Kayode O (2019) Gas flaring attendant impacts of criteria and particulate pollutants: a case of Niger Delta region of Nigeria. J King Saud Univ-Eng Sci 31:209–217

Gomiero A, Strafella P, Fabi G (2018) Frommacroplastic to microplastic litter: occurrence, composition, source identification and interaction with aquatic organisms. Experiences from the Adriatic Sea. In: Plastics in the environment. IntechOpen, London

Gu B, Jiang S, Wang H, Wang Z, Jia R, Yang J, ... Cheng R (2017) Characterization, quantification and management of China's municipal solid waste in spatiotemporal distributions: a review. Waste Manag. 61:67–77

Gundupalli SP, Hait S, Thakur A (2017) A review on automated sorting of source-separated municipal solid waste for recycling. Waste Manag 60:56–74. https://doi.org/10.1016/j.wasman.2016.09.015

Gupta N, Yadav KK, Kumar V (2015) A review on current status of municipal solid waste management in India. J Environ Sci 37:206–217. https://doi.org/10.1016/j.jes.2015.01.034

Halkos G,Kleoniki N (2016) Efficient waste management practices: a review. MPRA Paper No 71518, p 36. https://mpra.ub.uni-muenchen.de/71518/

Hoang NX (2012) Integrated municipal solid waste management approach in adaptation to climate change in Mekong Delta. J Viet Env 3:19–24. https://doi.org/10.13141/jve.vol3.no1.pp19-24

Horton AA (2022) Plastic pollution: when do we know enough?. J Hazard Mater 422:126885. https://doi.org/10.1016/j.jhazmat.2021.126885

Hui Y, Li'ao W, Fenwei S et al (2006) Urban solid waste management in Chongqing: challenges and opportunities. Waste Manag 6:1052–1062

International Organization for Standardization (ISO) (2015) Environmental management systems—requirements with guidance for use, 3rd edn. ISO, Geneva, Switzerland

Ite AE, Ibok UJ, Ite MU, Petters SW (2013) Petroleum exploration and production: past and present environmental issues in the Nigeria's Niger Delta. Am J Environ Prot 1:78–90

Kimenyi MS, Deressa TT, Pugliese JE, Onwuemele A, Mendie M (2014) Analysis of community-driven development in Nigeria's Niger delta region: use of the institutional analysis and development (IAD) framework

Konstadakopulos D (2008) Environmental and resource degradation associated with small-scale enterprise clusters in the Red River Delta of Northern Vietnam. Geogr Res 46:51–61

Knippertz P, Mat JE, Paul R, Field PR, Andreas H, Liousse C, Marsham H (2015) The possible role of local air pollution in climate change in West Africa. Nat Clim Chang 5:815–822

Kumar P, Kaushal RK, Nigam AK (2015) Assessment and management of Ganga river water quality using multivariate statistical techniques in India. Asian J Water Environ Pollut 12:61–69

Li WX (2006) Looking at the Songhua River incident from an environmental regulatory governance perspective: a long standing issue in China. Perspectives 7:21–35

Lu JW, Chang NB, Liao L (2013) Environmental informatics for solid and hazardous waste management: advances, challenges, and perspectives. Crit Rev Environ Sci Technol 43:1557–1656

Meidiana C, Gamse T (2010) Development of waste management practices in Indonesia. Eur J Sci Res 40(2):199–210

Menad N, Kanari N, Allain E, Gaballah I (1999) Thermal treatment of industrial waste in control atmospheres for the elimination of as, hg, cd, se and concentration of pb, cu and zn. Resour Conserv Recycl 25:233–254

Meybeck M, Vogler J, Turner K et al (2004) River/coast integration in European regional seas. EUROCAT project (www.iiacnr.unical.it/EUROCAT)

Mîndrescu M, Haliuc A, Zhang W et al (2022) A 600 years sediment record of heavy metal pollution history in the Danube Delta. Sci Total Environ 823:153702

Mmereki D, Baldwin A, Hong L, Li B (2016) The management of hazardous waste in developing countries. Manag Hazard Wastes 6:39–50

Mmereki D, Li B, Meng L (2014) Hazardous and toxic waste management in Botswana: practices and challenges. Waste Manage Res 32:1158–1168

Moore MN, Depledge MH, Readman JW, Leonard DP (2004) An integrated biomarker-based strategy for ecotoxicological evaluation of risk in environmental management. Mutat Res/fundam Mol Mech Mutagen 552:247–268

Mrozik A, Piotrowska-Seget Z (2010) Bioaugmentation as a strategy for cleaning up of soils contaminated with aromatic compounds. Microbiol Res 165:363–375

Nabavi-Pelesaraei A, Bayat R, Hosseinzadeh-BandbafhaH AH, Berrada A (2017) Prognostication of energy use and environmental impacts for recycle system of municipal solid waste management. J Clean Prod 154:602–613. https://doi.org/10.1016/j.jclepro.2017.04.033

Naess A (2017) The shallow and the deep, long-range ecology movement. A summary. In: The ethics of the environment. Routledge, Lonon, pp 115–120

National Environmental (2010) Economic and development study for climate change in Nigeria (NEEDS), pp 1–43. https://www.semanticscholar.org/paper/FEDERAL-MINISTRY-OF-ENVIRONMENT-ABUJA%2C-NIGERIA-AND/47b935e8e23b825ca878d38fda4e4acc5a4ba8ae

Nguyen XH (2012) Integrated municipal solid waste management approach in adaptation to climate change in Mekong Delta. J Viet Env 3:19–24

Okeagu JE, Okeagu JC, Adegoke AO, Onuoha CN (2006) The environmental and social impact of petroleum and natural gas exploitation in Nigeria. J Third World Stud 23(1):199–218

Okonkwo CNP, Kumar L, Taylor S (2015) The Niger delta wetland ecosystem: what threatens it and why should we protect it? Afr J Environ Sci Technol 9:451–463

Okonkwo T, Etemire U (2017) Oil injustice in Nigeria's Niger delta region: a call for responsive governance. J Environ Prot 8:42–60

Okpara DA, Oruche HC, Bassey OM (2019) Emerging solid waste leachate pollutants and brewing effluent on delta eco-zones; Impact on environmental resources sustainability, a case of Niger Delta Nigeria. Int Conf Resour Environ Sci. https://doi.org/10.1088/1755-1315/398/1/012006

Oliveira M, Almeida M (2019) The why and how of micro (nano) plastic research. TrAC Trends Anal Chem 114:196–201

Olukunle OI, Okonkwo OJ (2015) Concentration of novel brominated flame retardants and HBCD in leachates and sediments from selected municipal solid waste landfill sites in Gauteng Province, South Africa. Waste Manage 43:300–306

Onwukwe SI, Nwakaudu MS (2012) Drilling wastes generation and management approach. Int J Environ Sci Dev 3:252–257

Ouillon S (2018) Why and how do we study sediment transport? Focus on coastal zones and ongoing methods. Water 10(4):390. https://doi.org/10.3390/w10040390

Papadokonstantakis S, Rerat C, Hungerbühler K (2013) Integrated waste management in batch chemical industry based on multi-objective optimization. J Air Waste Manag Assoc. 63(3):349–366

Pausas JG, Keeley JE (2021) Wildfires and global change. Front Ecol Environ 19:387–395

Platvoet L (2005) La protection des deltas européens. Rapport de la Commission de l'environnement, de l'agriculture et des questions territoriales. http://assembly.coe.int/nw/xml/XRef/X2H-Xref-ViewHTML.asp?FileID=10868&lang=fr

Pope CA, Burnett RT, Thun MJ, Calle EE, Krewski D, Ito K, Thurston GD (2002) Lung cancer, cardiopulmonary mortality, and long-term exposure to fine particulate air pollution. J Am Med Assoc 287:1132–1141

Raimi OM, Sawyerr OH, Ezekwe CI, Gabriel S (2022) Many oil wells, one evil: comprehensive assessment of toxic metals concentration, seasonal variation and human health risk in drinking water quality in areas surrounding crude oil exploration facilities in rivers state, Nigeria. Int J Hydrol 6:23–42

Roman L, Schuyler Q, Wilcox C, Hardesty BD (2021) Plastic pollution is killing marine megafauna, but how do we prioritize policies to reduce mortality? Conserv Lett 14:e12781

Seo S, Aramaki T, Hwang Y, Hanaki K (2004a) Environmental impact of solid waste treatment methods in Korea. J Environ Eng 130:81–89

Shankar S, Shikha (2017) Management and remediation of problem soils, solid waste and soil pollution. In: Singh RL (ed),Principles and applications of environmental biotechnology for a sustainable future, applied environmental science and engineering for a sustainable future. Springer, Singapore, pp 143–171

Sheikhavandi T (2015) Soil contamination. In: Handbook of research on advancements in environmental engineering. IGI Global, pp 173–199

Sheng XF, He LY, Wang QY, Ye HS, Jiang CY (2008) Effects of inoculation of biosurfactant-producing Bacillus sp. J119 on plant growth and cadmium uptake in a cadmium amended soil. J Hazard Mater 155:17–22

Skea J, Nishioka S (2008) Policies and practices for a low-carbon society. Clim Policy 8:5–16

Sonibare OO, Adeniran JA, Bello IS (2019) Landfill air and odour emissions from an integrated waste management facility. J Environ Health Sci Eng 17:13–28

SWM (2016) Solid waste management rule, Government of India. http://www.moef.gov.in/sites/default/files/SWM%202016.pdf

Tadesse T, Ruijs A, Hagos F (2008) Household waste disposal in Mekelle city, Northern Ethiopia. Waste Manag 28:2003–2012

Tadesse T, Ruijs A, Hagos F (2008b) Household waste disposal in Mekelle city, Northern Ethiopia. Waste Manage 28:2003–2012

Testa F, Daddi T, De Giacomo M, Iraldo F, Sant'Anna S (2014) The effect of integrated pollution prevention and control regulation on facility performance. J Clean Prod 64:91–97
Thanh NP, Matsui Y, Fujiwara T (2011) Assessment of plastic waste generation and its potential recycling of household solid waste in Can Tho City, Vietnam. Environ Monit Assess 175:23–35
Thompson D (2002) Tools for environmental management: a practical introduction and guide. New Society Publishers, Gabriola, BC
Thonart P, Diabaté SB, Hiligsmann S, LardinoisM (2005) Guide pratique sur la gestion des déchets ménagers et des sites d'enfouissement technique dans les pays du Sud. p 146
UNEP (2012) Green Economy Indicators-brief Paper. p 24
UNESCO World Heritage Committee (1991) Fifteenth session (Carthage, 9-13 December 1991). Doc. SC-91/CONF.002/15, 12 December 1991
Verma RL, Borongan G, Memon M (2016) Municipal solid waste management in Ho Chi Minh City, Viet Nam, current practices and future recommendation. Procedia Environ Sci 35:127–139
Wang MX, Zhao HH, Cui JX, Fan D, Lv B, Wang G, Zhou GJ (2018) Evaluating green development level of nine cities within the Pearl River Delta, China. J Clean Prod 174:315–323
Wang YJ, Liu Z Y, Huang L, Lu GB, Gong YG, Yaluk E et al (2020) Development and evaluation of a scheme system of joint prevention and control of PM2.5 pollution in the Yangtze River Delta region, China. J Clean Prod 275:122756. https://doi.org/10.1016/j.jclepro.2020.122756
Weidner H (2002) Capacity building in national environmental policy: a comparative study of 17 countries. Springer Science and Business Media. p443
Wilcox C, Van Sebille E, Hardesty BD (2015) Threat of plastic pollution to seabirds is global, pervasive, and increasing. Proc Natl Acad Sci 112:11899–11904
Yakubu OH (2017) Addressing environmental health problems in Ogoniland through implementation of United Nations environment program recommendations: environmental management strategies. Environments 4:28
Zhang J, Zhu M (2022) Collaborative governance of municipal solid waste in urban agglomerations: the case of Yangtze River Delta. Front Environ Sci 10:999120. https://doi.org/10.3389/fenvs.2022.999120
Zhou C, Huang H, Cao A, Xu W (2015) Modeling the carbon cycle of the municipal solid waste management system for urban metabolism. Ecol Model 318:150–156. https://doi.org/10.1016/j.ecolmodel.2014.11.027
Zhou X, Cai L (2010) Coastal and marine environmental issues in the Pearl River Delta region, China. Int J Environ Stud 67:137–145. https://doi.org/10.1080/00207231003683549

# Eco-utilization and Sustainable Management of Solid Wastes of Steel Industries

# 5

Trinath Biswal and Pratap Kumar Swain

**Abstract**

Today, huge quantities of solid waste of various kinds are produced globally; some of these are non-hazardous, but a major fraction of them are non-biodegradable and hazardous. The landfill dumping of these huge quantities of industrial solid waste is the cause of environmental degradation, specifically groundwater contamination through leaching, surface water contamination due to rain, and soil contamination, which is continuously increasing throughout the world. Hence, the eco-utilization and sustainable management of industrial wastes are highly essential to save our environment and biological community from drastic environmental contamination. Various kinds of solid wastes, such as fly ash, bottom ash, slags, concrete, bottom ash, toxic solvents, chemicals, scrap metal, masonry, etc., are released from integrated steel industries. Now the 7Rs cradle-to-cradle rule, such as regulation, recycling, reduction, rethinking, reuse, recovery, and renovation, is the best way to achieve sustainability in the management of solid wastes in the steel industry. Fly ash is one of the major solid wastes generated in the steel industry that can be reused eco-friendly in a variety of ways, including bricks, zeolite, road sub-base, cost-effective adsorbents for the removal of organic materials, mine backfill, controlling fire, and most recently in gardening by mixing with cow dung. Fly ash is the major raw material used to produce cement. Since the properties of the slag are comparable to those of natural stones, it can be used as civil

T. Biswal (✉)
Department of Chemistry, Veer Surendra Sai University of Technology, Burla 768018, Odisha, India
e-mail: biswal.trinath@gmail.com

P. K. Swain
Department of Chemistry, Government Autonomous College, Rourkela 769004, Odisha, India



M. Nasr and B. Ravindran (eds.), *Solid Waste Management in Delta Region for SDGs Fulfillment*, Deltas of the World, https://doi.org/10.1007/978-3-031-58253-0_5

engineering materials such as roadbed, asphalt track bed, coarse aggregate for concrete, and raw material fused for the manufacturing of cement. The huge quantities of mill scale obtained from steel industries can be eco-utilized for the manufacturing of mortars and some construction materials. SMS slag has been used successfully as pavement and road material, filling materials, soil conditioners, cement, brick manufacturing, and mine filling. The major aim of this chapter is to explore clear explanations regarding the eco-utilization and sustainable management of huge solid wastes generated from steel industries throughout the globe and the production of clean and green steel with zero waste generation. The future eco-utilization of huge solid wastes generated by steel industries is the development of an eco-industrial park.

## 5.1 Introduction

Sustainable waste management is defined as the best approach for protecting public health when it involves prevention, reduction, reuse, recovery of energy, recycling, and ultimately the treatment and disposal of solid waste. According to the 1976 Resource Conservation and Recovery Act (RCRA), any discarded material or sludge from industries, mines, the agricultural sector, wastewater treatment plants, water treatment plants, and community activities is called solid waste (Kumar 2014). After the industrial revolution, each country in the world put emphasis on industrial development, and now the industry is the major determinant of the country's growth. The steel industry is one of the most important heavy industries and makes a greater contribution to the development of any country. The steel industry produces a wide range of products that can be utilized in various sectors such as agriculture, electrical, domestic, automobile, defence, aerospace, and some other important industries (Shipa 2020). In an integrated steel industry, various complex operations are employed to produce steel from iron ore, where coke, limestone, and dolomite function as fluxes by using the basic oxygen process in a blast furnace. Besides these, the use of scrap materials, such as sponge iron, direct reduced iron, and hot briquette iron, in electric arc furnaces and induction furnaces, is the cause of the generation of appreciable quantities of solid waste. During the process of producing waste, a variety of pollutants are generated in all three states (solid, liquid, and gas). These wastes have a very detrimental effect on the environment, and the generation of these wastes is directly related to the demand for industrial products and the expansion of industrial sectors (Swain et al. 2020; Tiwari et al. 2016). As per the data of the World Steel Association in 2019, the global demand for steel was 1,735 million metric tonnes, and in Asian countries, the steel demand increased by 6.5% and 6.4% in 2019 and 2020, respectively, excluding China. According to the WSA Report for 2019, the demand for steel all over the world will touch 1,735 Mt in 2019. In 2018, global steel demand increased by 2.1%, which is less than that of 2019. Steel demand in China is steadily declining as the country's economic situation deteriorates. In developed countries, steel demand decreased from 3.1% in 2017 to 1.8% in 2018. It is further expected that steel demand will further decrease to 0.3% in 2019 and 0.7% in 2020. It is forecasted by the WSA that the steel demand in Asian countries except China will increase by 6.5 and 6.4% in 2019 and 2020, respectively. Due to increased demand for steel and steel-finished products, huge quantities of solid, liquid, and gaseous waste are produced in various steel industries around the globe. Gaseous wastes are released into the atmosphere, and liquid wastes are discharged into water bodies. The solid wastes are continuously dumped in the open fields demarcated by the steel industries. These solid waste dump yards now look like a high hill, which is the cause of soil, air, and water contamination either directly or indirectly (Nikiforova 2018; Ravazzolo and Vespignani 2020).

## 5.2 Steel Industry Scenario of the World

In the whole world, generally, two major types of steel industries are found: integrated steel plants and secondary producers. Integrated steel plants use iron ore, coke, and fluxing materials like dolomite and lime stones in blast furnaces, whereas secondary producers use scrap materials, direct reduced iron (DRI), and briquetted iron in electric arc furnaces and induction furnaces to produce a variety of steel products. In the whole world, around 2.01 billion metric tonnes of solid waste are generated, and the contribution of the steel industry to total global solid waste generation is about 6.8%. During the process of producing steel, various kinds of discarded products, such as SMS slag, BF slag, dolo dust, flue dust, fly ash, some refractory materials, sludge, mill scale, and some debris products, are generated as solid waste. The huge quantities of solid waste generated from steel plants are not being scientifically disposed of, managed, or utilized in an eco-friendly manner. Thus, these wastes have a catastrophic effect on the environment and degrade the quality of air, water, and soil. Environmental regulatory authorities like the state pollution control board (SPCB) and the central state pollution control board (CPCB) specify norms for the utilization of these wastes (Lopez et al. 2022; Sarkar and Mazumder 2015a). They put an emphasis on zero waste to minimise the pressure on our ecosystem and reduce environmental pollution. Thus, the eco-utilization of solid waste is now gaining importance to achieve zero waste in the industrial sector. The term waste is replaced by co-product because efforts were continuously made to reutilize the solid wastes generated from several industrial complexes in an economical and sustainable manner (Lopez et al. 2022). Thus, technocrats, scientists, and researchers are continuously focusing on utilizing the solid waste in an efficient, economical, and eco-friendly manner. The quantitative description of steel produced from steel plants around the world in different years is given below. Figure 5.1 shows the production of steel in various production pathways in India and basic oxygen furnace route accounts for the largest amount of steel production.

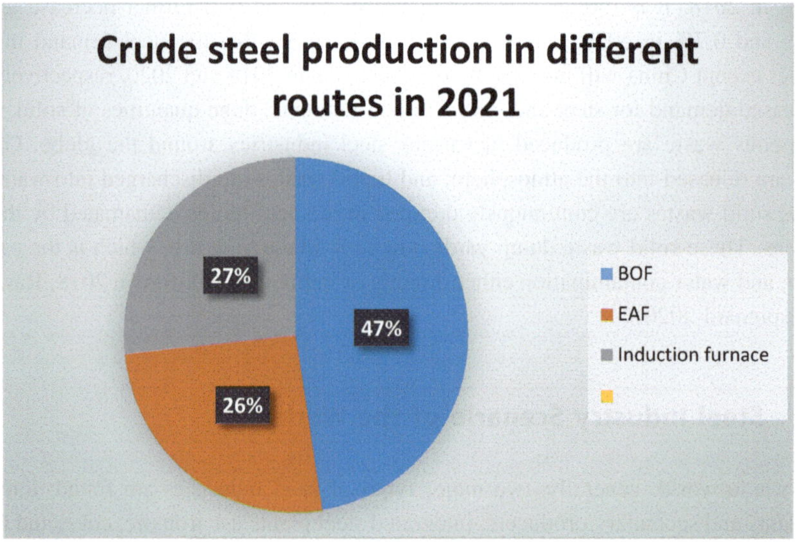


**Fig. 5.1** Steel production through different production routes in India

Figure 5.2 indicates the world steel production in MT; and Fig. 5.3 shows the steel production (MT) in India. From Figs. 5.2 and 5.3 it is clear that the quantity of steel production goes on increasing in India and the world which in turn increases the generation of solid wastes.

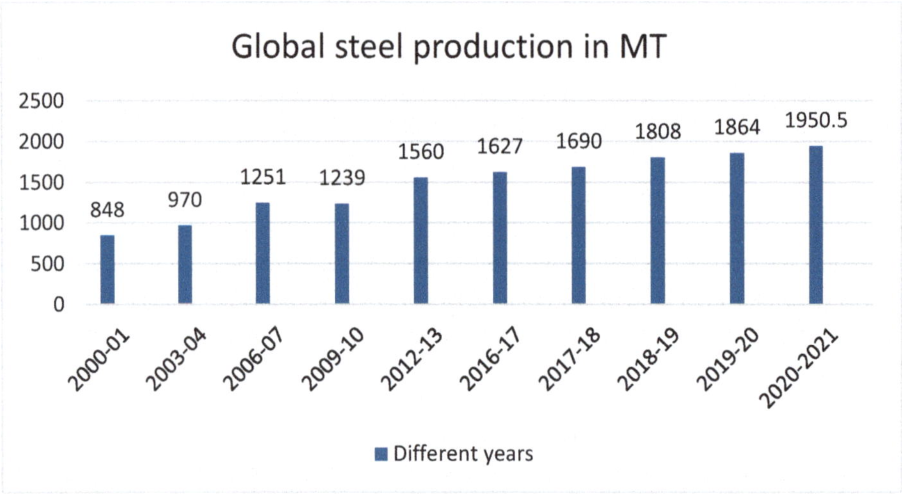


**Fig. 5.2** World steel production in MT. *Source* World Steel Association Report-2021

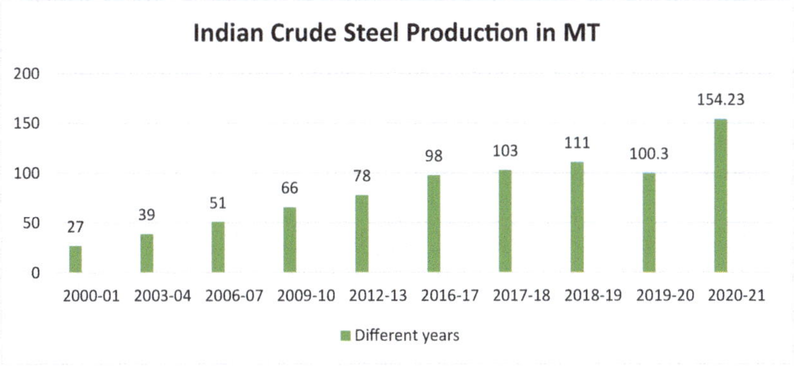


**Fig. 5.3** Indian steel production in MT. *Source* World Steel Association Report-2021

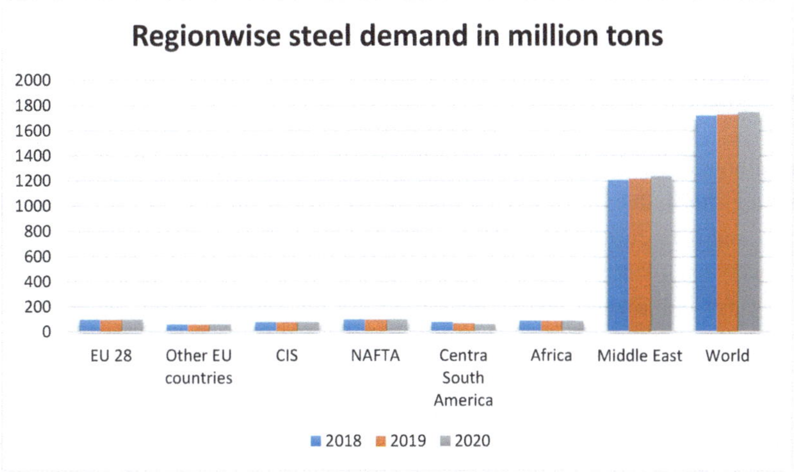


**Fig. 5.4** Region wise steel demand in million tons as per WSA Report 2020

Figure 5.4 indicates the region-wise production of steel in the whole world (Sarkar and Mazumder 2015b; Biswal et al. 2019). From Fig. 5.4 it is clear that the region-wise steel demand increases which helps in increasing the production of steel globally.

## 5.3 Categories of Solid Waste Produced from a Steel Plant

Solid wastes produced from a steel plant are broadly classified into two categories viz.

- Wastes from production units
- Wastes from pollution control devices.

In a steel plant, solid waste is produced from the blast furnace, SMS, hot rolling mill, coke ovens, cold rolling mill, captive power plant, and lime and dolomite brick plant. The major solid waste produced from a steel plant is some refractory materials, such as fine powders of dolomite. The various kinds of solid waste generated from integrated steel plants are blast furnace slag, SMS slag, BF flue dust, blast furnace sludge, SMS sludge, mill scale, etc. (Sarkar and Mazumder 2015b). Generally, the solid wastes produced are classified into the following categories:

i. Fly ash
ii. Ferruginous solid waste
iii. Non-ferruginous solid waste

The ferruginous wastes are the iron-containing waste materials, which include mill scale, steel melting shops, flue dust, sludges obtained from blast furnaces at gas cleaning plants, SMS slag, and blast furnace slag. The wastes belong to the non-ferruginous kind, which includes acetylene plant sludge, lime fines, broken fire clay bricks, and broken refractory bricks. Fly ash is the solid waste generated from the captive power plants of the steel industry. In developed countries, about 550 kg of solid waste are produced per tonne of crude steel production, whereas in India, the generation of solid waste for the same condition is about 1100 kg because of the use of low-grade raw materials. Figure 5.5 illustrates the various steps of steel production and the generation of BF and SMS slag, and Fig. 5.6 illustrates the various kinds of solid waste generated from a steel plant (Krishna et al. 2020; Lobato et al. 2015; Khunte 2018).

**Problems Associated with the Solid Wastes Generated from Steel Industries**

In steel industries, normally huge resource materials like iron ores, limestone, coal, air, water, electric power, etc. are used as per requirement. For the production of 1 tonne of crude steel, approximately five tonnes of raw materials, 15 tonnes of coal, and 60 tonnes of water are used, (Rourkela Steel Plant panorama page no-29) which may vary based on the composition, quality of the raw materials, technology adopted, the capacity of the production units, etc. (Agrawal et al. 2004). During the process of production, a large volume of liquid waste is discharged from byproduct plants, coke ovens, and pickling operations. These wastes are directly discharged into the rivers, and about 60 gaseous wastes are emitted into the atmosphere in the form of gases, dust, suspended particulate

**Fig. 5.5** Process of steel production and generation of BF slag and SMS slag

**Fig. 5.6** Different types of solid wastes generated from a steel plant

matter, fumes, etc., which are very harmful not only to the biological community but also to the whole world. Besides this, the solid waste generated from different units is continuously piled in the open, causing serious soil contamination. These wastes are carried away by the wind and rainwater because of the degradation of the water quality index of both groundwater and surface water. Surface water contamination causes fatal diseases in aquatic animals and, indirectly or directly, in other animals of various trophic levels and humans. The soil contamination leads to infertility in the soil. It also inhibits the growth of useful microorganisms and the earthworms present in the soil. The solid waste debris piled on open land also helps in increasing several insects like flies, mosquitoes, and other disease-causing microbes that seriously affect human health through diseases like malaria, elephantasis, dengue, diarrhoea, encephalitis, etc. The fine particles of the solid waste carried by the wind degrade the air quality and cause several respiratory diseases like asthma, bronchitis, pneumonia, tuberculosis, etc. (Biswal et al. 2019; Krishna et al. 2020). Moreover, a number of vehicles engaged in the transportation of raw materials, finished products, and solid waste contribute to noise and air pollution. Disposal of solid waste requires huge revenue and more land area. As the land area decreases due to the dumping of solid waste, it has a detrimental effect on agricultural production. Now the United Nations Organisation blows the alarm to all the nations of the world regarding the serious deterioration of air and water quality and issues stringent guidelines for achieving the sustainable development goals by conserving all the natural resources for posterity (Kumar and Agrawal 2020; Das et al. 2007).

**Characterization of Steel Plant Solid Waste**

Solid waste generated from steel plants causes serious environmental degradation and health hazards. To ascertain the detrimental effects of these wastes, physico-chemical and mineralogical characterization is absolutely necessary (Singh et al. 2017). The physical properties of solid waste generated from a steel plant are represented in Table 5.1.

**Table 5.1** Physical properties of solid wastes generated from a steel plant (Pappu et al. 2007)

| Physical property | BF slag | SMS slag | Mill scale | Fly ash | BF flue dust | Wet ash | SMS sludge |
|---|---|---|---|---|---|---|---|
| Colour | White colour | Blackish | Bluish black | Greyish | Yellowish | Blackish brown | Blackish |
| $p^H$ | 8.7 | 11.3 | 11.1 | – | 5.4 | 4.5 | 5.1 |
| Elastic property | No elastic | No plastic property | No elastic | No elastic | No elastic | No elastic | No elastic property |
| Bulk density in gm. $cm^{-3}$ | 2.3 | 3.6 | 2.9 | 1.3 | 2.1 | 1.06 | 1.87 |
| Specific gravity | 2.7 | >2.9 | 4.96 | 1.8 | 1.92 | 2.1 | 1.8 |
| % of moisture | <0.14 | <0.12 | 1.3 | 5.3 | <0.12 | 3.4 | 3.6 |
| % of absorption | 5.2 | 6.0 | 21.2 | 21.0 | 17.3 | 16.2 | 16 |
| Relative density | 2.6 | 2.7 | 1.9 | 1.3 | 1.7 | 2.1 | 2.3 |
| Dielectric Strength | 1.5 | 1.7 | 2.0 | 1.9 | 1.75 | 1.4 | 2.2 |

**Mineralogical Properties of Solid Waste**

The mineralogical study includes SEM and XRD analysis. The shape, size, and morphology of the particles in the solid waste were ascertained from SEM analysis. An XRD study determines the mineralogical phase of the crystalline materials present in the different solid wastes. From chemical analysis, the chemical composition of the solid wastes can be determined by XRF analysis. Figures 5.7, 5.8, 5.9, 5.10, 5.11, and 5.12 represent the SEM images of fly ash, BF slag, SMS slag, dolodust, flue dust, and mill scale. Figures 5.13, 5.14, 5.15, and 5.16 represent the XRD patterns of BF slag, fly ash, mill scale, and dolodust, respectively. Table 5.2 illustrates the chemical analysis of fly ash, and Table 5.3 illustrates the micro- and macronutrients present in fly ash. Table 5.4 represents the different concentrations of heavy metals present in fly ash. Table 5.5 illustrates the chemical composition of BF slag, while Table 5.6 illustrates the chemical composition of SMS slag. Table 5.7 illustrates the chemical composition of BF flue dust, while Table 5.8 shows the chemical composition of mill scale (Wei et al. 2011; Kirby and Rimstidt 1993). From Table 5.2, it is evident that fly ash contains oxides of aluminum 22–28%, silicon 52–62%, and calcium 5–8% etc. So it has good pozzolanic characteristics. Table 5.3 clearly indicates that Al, B, Ca, K, P, S, Fe, Na etc. present in fly ash can assist in plant growth as nutrients. Table 5.4 shows the presence of heavy metals like Pb, Co, Ni, Cd, As etc. in trace amount which has no appreciable toxicity towards the plant growth. Table 5.5 shows the chemical composition of blast furnace slag. It is evident that the blast furnace slag has the oxides of silicon, aluminium, magnesium and calcium as the chief constituents. So the slag has good pozolanicity due to presence of these oxides. Besides this as the slag is alkaline, so it can be utilized to amend the acidic soil. Table 5.6 shows that SMS slag contains oxides of silicon, aluminium, calcium and iron as the chief constituents. So it can be utilized as construction materials. Its high phosphorous content makes it suitable for the manufacturing of fertilizer. Its oxide content shows the alkaline characteristics which helps in modifying the acidity of the soil. Table 5.7 shows the chemical composition of the blast furnace flue dust contains the oxides of magnesium, aluminium, silicon. These oxides have good pozolanic properties. So it can be utilized in cement industries. Its iron content in the form of oxide also indicates that it can be reutilized in the production process of steel as sinter. Table 5.8 shows the chemical composition of mill scale and from the table it is evident that mill scale contains higher percentage of iron. So the recovery of iron can be done by adopting the direct reduction process.

**Ecoutilization of Solid Wastes Generated from Steel Plants**

Eco-utilization of solid waste is based on the 4 R concepts, i.e., recycling, reduction, reuse, and recovery. It helps with the conservation of resources and energy. It also reduces the cost of disposing of the waste materials. It makes the environment clean and green. It strengthens the laws enacted by the government (Li et al. 2022).

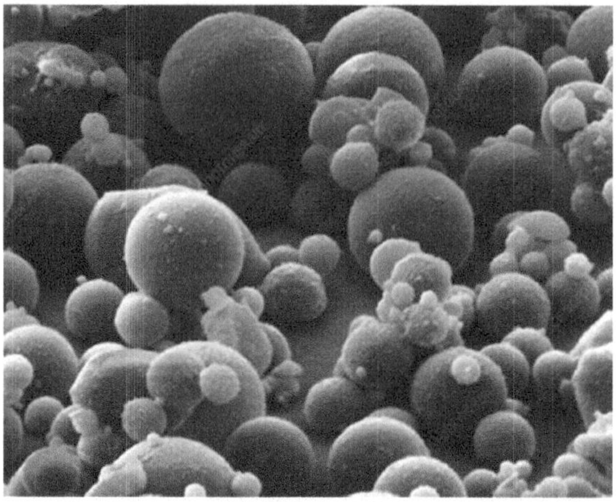

**Fig. 5.7** SEM image of fly ash

**Fig. 5.8** SEM image of BF slag

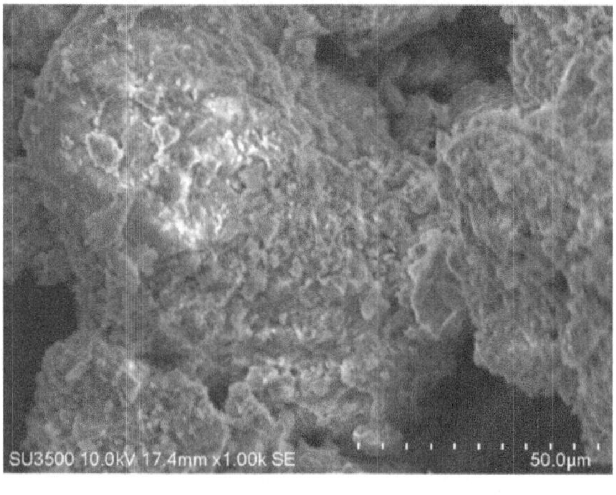


**Fig. 5.9** SEM image of SMS slag

**Fig. 5.10** SEM image of dolo dust

**Fig. 5.11** SEM image of flue dust

**Fig. 5.12** SEM image of mill scale

**Fig. 5.13** XRD graph of BF slag

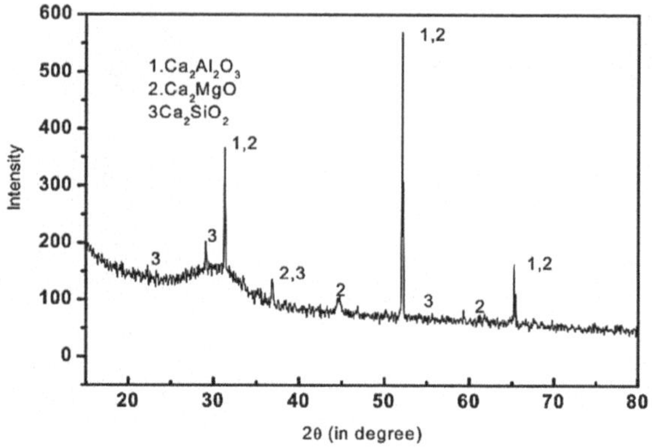


**Fig. 5.14** XRD graph of fly ash

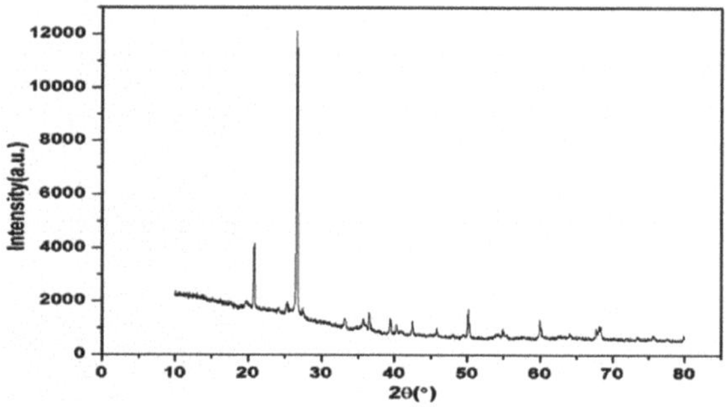


**Fig. 5.15** XRD graph of mill scale

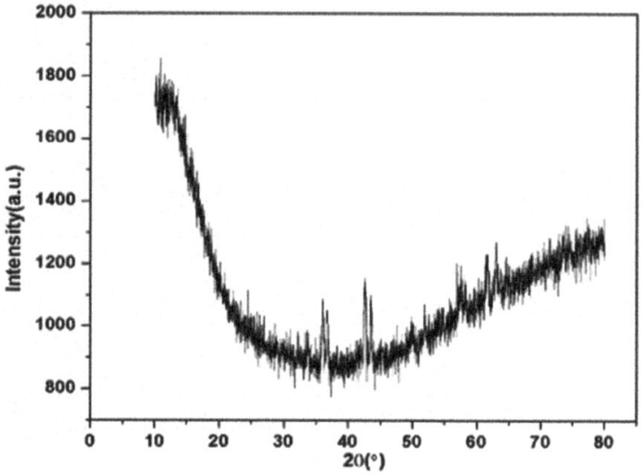


**Ecoutilization of BF Slag in Brick Manufacturing**

From the mineralogical analysis, it is evident that the particles of BF slag are subround to sub-angular in shape. The major mineralogical phase is gehelinite, which contains trace amounts of Mg. It also has other mineralogical phases like quartz, rankinite, wustite, and

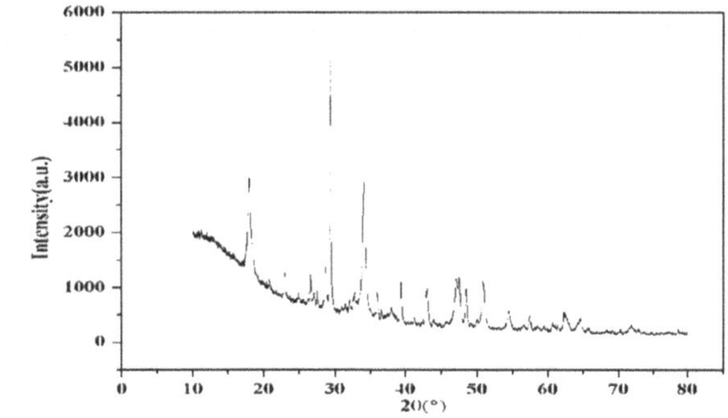


**Fig. 5.16** XRD graph of dolodust

**Table 5.2** Chemical analysis of fly ash (Swain et al. 2019; Zhou et al. 2013)

| Composition | Percentage of weight (w/w%) |
|---|---|
| $SiO_2$ | 52.6±9.7 |
| $Al_2O_3$ | 22.9±5.3 |
| $Fe_2O_3$ | 7.6±4.2 |
| CaO | 5.0±2.8 |
| MgO | 1.2±0.6 |
| $Na_2O$ | 1.0±1.0 |
| $K_2O$ | 1.4±0.7 |

**Table 5.3** Micro and macronutrients present in fly ash (Swain et al. 2019; Yu et al. 2019)

| Elemental composition | In ppm |
|---|---|
| Al | 11,520–144,200 |
| B | 10–3005 |
| Ca | 5450–177,000 |
| Cl | 14–25,000 |
| Co | 31–3025 |
| Cu | 5–1498 |
| Mn | 31–4405 |
| P | 600–2490 |
| K | 1538–14,005 |
| S | 0.10–0.27 |
| Fe | 7805–289,003 |
| Na | 1185–20,301 |
| Ni | 11.9–8001 |

**Table 5.4** Heavy metals present in fly ash (Swain et al. 2019; Singh et al. 2010)

| Parameters | Total | Available (in ppm) |
|---|---|---|
| Se | 0.7–2.7 | 0.2–0.5 |
| Cr | 50.5–151 | 0.4–0.6 |
| Pb | 10.2–70.5 | BDL |
| Co | 10.1–50.3 | 0.06–0.16 |
| Ni | 51–152 | 0.16–0.26 |
| Cd | 5.2–10.1 | 0.04–0.07 |
| As | 1.01–4.03 | BDL |
| Hg | BDL | BDL |

**Table 5.5** Chemical composition of BF slag (Swain et al. 2019; Escalante-García et al. 2003)

| Composition of BF slag | % composition (mean value) | Range |
|---|---|---|
| CaO | 41.0 | 34–48 |
| $SiO_2$ | 36.0 | 31–45 |
| $Al_2O_3$ | 13.0 | 10–17 |
| MgO | 7.0 | 1–15 |
| Oxides of iron | 0.5 | 0.1–1.0 |
| MnO | 0.8 | 0.1–1.4 |
| Sulphur | 1.5 | 0.9–2.4 |

**Table 5.6** General chemical composition of SMS slag (Swain et al. 2019; Kim et al. 2012)

| Chemical constituent | % composition |
|---|---|
| CaO | 41–51 |
| $Al_2O_3$ | 1–3 |
| $SiO_2$ | 10–19 |
| $P_2O_5$ | 0.5–1.0 |
| FeO | 10–40 |
| S | <0.1 |
| MnO | 5–8 |
| MgO | 5–10 |
| Fe | 0.5–1.0 |

**Table 5.7** General chemical composition of BF flue dust (Swain et al. 2019; Das et al. 2002)

| BF flue dust | % composition | Range % |
|---|---|---|
| C | 29.9 | 20–31 |
| $Fe_2O_3$ | 51.1 | 50–63 |
| $SiO_2$ | 6.31 | 5–8 |
| $Al_2O_3$ | 5.12 | 3–6 |
| CaO | 4.90 | 4–6 |
| MgO | 0.88 | 1–4 |
| Pb | 0.024 | 0–0.5 |
| Zn | 0.042 | 0–0.5 |
| MnO | 0.58 | 0–1 |
| $K_2O$ | 1.22 | 0–1.5 |
| $Na_2O$ | 0.47 | 0–0.5 |
| Fe (T) | 35.73 | 0–37 |

**Table 5.8** Chemical composition of mill scale (Swain et al. 2019; Ghods et al. 2011)

| Elements | % by weight |
|---|---|
| $Fe_{Total}$ | 67.4 |
| $SiO_2$ | 4.1 |
| $Al_2O_3$ | 1.84 |
| CaO | 0.7 |
| MgO | 0.25 |
| MnO | 0.65 |
| $Na_2O$ | 0.41 |
| $K_2O$ | 0.25 |
| LOI | 6.8 |
| *Trace elements* | *% wt.* |
| P | 0.07 |
| Ti | 0.08 |
| Cu | 0.14 |
| Zn | 0.01 |
| Co | 0.026 |
| Ni | 0.038 |
| Cr | 0.033 |
| Pb | 0.033 |

hematite. The chief constituents of BF slag are calcium oxide, silica, aluminium oxide, magnesium oxide, manganese oxide, and oxides of iron in traces. Due to its low iron content and the good cementitious properties of calcium oxide and magnesium oxide, BF slag can be used effectively as a construction or structural material. By taking the mixture of slag and lime in different proportions, slag lime brick can be manufactured, which possesses good compressive strength and water swelling capability. It also showed better results in the water absorption test and the efflorescence test. The use of slag lime brick can replace the burnt clay brick and save the top fertile soil cover of the land, which is massively used in fire clay brick manufacturing. Slag brick is an eco-friendly, efficient, and profitable building material (Balaji et al. 2022).

**Ecoutilization of BF Slag for Production of Cement**

Since BF slag contains calcium and magnesium oxide, the slag can be utilized to manufacture cement, which replaces Portland cement. Slag cement actually reduces $CO_2$ emissions by around 60% around the globe. Slag cement used in the manufacture of concrete saves around 40% of the energy used. Slag cement has low heat of hydration, good durability, promotes controlling the alkali-silica reaction, prevents acid attack, and reduces the cost of cement production (Li et al. 2016).

**Ecoutilization of BF Slag in Sub-base Construction of the Road**

Slag is whitish and possesses a specific gravity of 2.7 kg/m$^3$ with a bulk density of around 2.3 g/cm$^3$. So it can be utilized successfully in road construction as sub-base material (Nicolae et al. 2007).

**For the Manufacture of Ceramic Materials and Glass**

BF slag contains different crystalline phases like gehelinite, wustite, hematite, and barium aluminium silicate, but at high temperatures the slag can be converted into a glassy state by taking the mixture of slag and calcia in a 0.1–0.3 ratio, where the maximum mechanical strength of the ceramic tile was noticed (Khater 2002).

**As Adsorbent**

Kostura et al. reported that from the contaminated water, phosphate can be adsorbed by the BF slag, which is why it is helpful for water treatment and purification. The adsorption agrees with the Langmuir adsorption isotherm and follows second-order kinetics (Kostura et al. 2005).

**Preparation of Silica Gel**

Seggiani et al. reported that blast furnace (BF) slag, when leached with concentrated $H_2SO_4$, silica gel, and gypsum, can be precipitated out at pH 3.2 (Seggiani and Vitolo 2003).

**Utilization of BF Slag as Soil Amender and Fertilizer**

BF slag contains oxides of calcium, manganese, magnesium, and aluminium, which are similar to the texture of the soil due to their almost similar composition. The concentrations of heavy metals like Pb, Hg, As, Cd, etc. are below the detection limit. It can be used as a good nutrient source for the plants and can fulfil the iron chlorosis in iron-deficient calcarious soil. It can also be used as a complex NPK fertilizer by modification with phosphoric acid, ammonium hydroxide, and ammonium nitrate (Lim et al. 2022).

**Ecoutilization of BF Flue Dust**

**In Cement Manufacturing**

BF flue dust contains oxides of magnesium, iron, aluminium, and silica as the main constituents. So blast furnace flue dust exhibits cementitious properties. It can be utilized in the cement industry for the manufacture of cement. As flue dust contains oxides of Mg and Al, it can be effectively used as a co-processing material. Thus, it reduces the carbon footprint of the steel and cement plants, saves the natural resources from exhaustion, and maintains the sustainability of the environment (Baidya et al. 2019).

**As Sinter**

BF flue dust is also being utilized as sinter material in the blast furnace due to the high concentration of iron oxides in it. But it also causes some problems when taken back into the blast furnace as sinter. This is because it contains a high percentage of alkalies, which reduce the melting and softening temperatures. The alkali present in the flue dust causes permeability of the load and creates cracks in the refractory materials. Zinc present in the flue dust increases coke consumption. Alkali cyanides are also found in the flue dust, which causes environmental problems and hazards to the biological community. Using different benefication and effective techniques, alkali content can also be minimized as per requirements (El-Hussiny and Shalabi 2010).

**Recovery of Zn from Flue Dust**

BF flue dust contains Zn along with the oxides of aluminium, magnesium, and silica. Besides these elements, it also contains alkalies, chromium, cadmium, and arsenic. It can be regarded as a hazardous waste material due to the presence of heavy metals like cadmium, chromium, and arsenic. So, its use as sinter is not suitable. By the process of leaching at a low concentration of sulphuric acid, it can be used as sinter (Zeydabadi et al. 1997).

**Eco-utilization of SMS Slag**

From the chemical composition, it is evident that SMS slag contains a higher percentage of iron oxide. So, it can be utilized in the following ways:

***As Rail Ballast***

Due to its high iron oxide content, it cannot be used in cement production. It is very hard with a high density and acts as a good abrasion and attrition-resistant material. Owing to these unique properties, it can be used as rail ballast (Sarkar and Mazumder 2015b).

***As Subbase in Road Construction***

SMS slag can be used as a sub-base in road construction because it is a rigid material with high compactness. It can replace the stone chips and save huge amounts of revenue for the government. In this way, it acts as an eco-friendly construction material without causing any damage to our environment (Barala et al. 2022).

**As a Component in Fertilizer**

The fine particles can be used as a major component in fertilizer as they are alkaline in nature and have a high phosphorous content (Sarkar and Mazumder 2015b).

**For Brick Manufacturing**

SMS slag in granulated form is mixed with an appropriate proportion of fly ash, and the bricks are manufactured, which have comparable compressive strengths to burnt clay bricks. It is an economic, eco-friendly, and efficient building material, which helps in maintaining a clean and green environment. It helps to prevent the loss of fertile top land cover, which is traditionally utilized to manufacture burnt clay brick, causing serious damage to the environment (Gencel et al. 2021).

**Soil Conditioner**

SMS slag contains oxides of aluminium, calcium, and iron. Thus, it can be used as a good soil conditioner for tea and coffee plantations. It also neutralizes the pH of the water and wastewater (Branca et al. 2014).

**Flooring Tiles Manufacturing**

As the SMS slag contains the oxides of calcium, aluminium, silicon, manganese, and iron, it exhibits good abrasion characteristics. Due to these abrasion characteristics, SMS slag can be suitably used to manufacture tiles for the floor (Khater et al. 2022).

**Eco-utilization of Fly Ash**

**For the Manufacture of Cement**

From the data based on chemical composition, fly ash contains lime, silica, and alumina as the major components. So, it shows good pozzolanic properties and can be used to replace Portland cement. Generally, 20–30% replacement of Portland cement with fly ash can act as a good binder. Thus, concrete made from fly ash exhibits much more strength and durability than normal concrete. Fly ash concrete is resistant to corrosion and can be used in coastal areas to prevent saline water (Ahmaruzzaman 2010).

**Used for Filling Low-Lying Areas**

The low-lying marshy lands can be filled with fly ash with the addition of a tap layer of soil, and the areas can be utilized for constructional activities in the future. It can also be used for the reclamation of land in coastal regions (Gorai and Ash 2018).

**Construction of Road and Embankment**

Fly ash is a lightweight material used for the construction of embankments. As it has low compressibility, it can therefore be used for the construction of sub-base materials (Martin et al. 1990).

**For Manufacturing Bricks and Tiles for Pavement**

Fly ash contains oxides of calcium, magnesium, aluminium, and silica. It has cementitious properties, and CaO has good binding abilities. Thus, it can be used in manufacturing good-quality bricks when mixed with gypsum or lime in a definite proportion. The fly ash can be used along with sand and lime to manufacture good quality bricks. The fly ash bricks are lightweight and have comparable compressive strengths with the fire clay bricks. Fly ash can also be used to manufacture tile, which can be used for constructing pavements, courtyards, footpaths, and vehicle parking areas. The fly ash used for making bricks and the flooring tiles will conserve the land area burrowed for burnt clay brick production and maintain the fertility of the soil as the top cover of the soil is preserved. It also prevents soil erosion and reduces the depletion of firewood required for burning conventional clay bricks (Yousuf et al. 2020).

**In Agriculture**

Fly ash contains oxides of calcium, magnesium, aluminium, and silica. Thus, it is usually alkaline in nature. In addition to this, it contains macro- and micronutrients like Ca, K, P, Al, Cu, Na, Fe, etc. but is devoid of organic carbon and nitrogen. The presence of these micronutrients helps improve water retention capacity and soil aeration. So, it can be effectively used on farmland to increase crop production. By burying the fly ash around the crops in the farmland, it partially exhibits the properties of hydrogel, which helps in extending the gap between two water cycles. Fly ash can be used as a soil amendment for saline alkali soil in an economical manner. Fly ash can also be applied to the acidic soil to rectify the acidity and neutralize the soil. It acts as a good soil conditioner. It has been experimentally verified that the mixture of 70% soil, 20% fly ash, and 10% cow dung can improve the production of crops and flowers and therefore can be used for domestic gardening. It can also be applied as an insecticide, and if used along with bio-wastes, it can boost the production of various crops (Ansari et al. 2011). Table 5.9 illustrates the yield of various crops with variations in the percentage of soil, compost, and fly ash.

**Table 5.9** Production of different crops in different plots with varied composition of fly ash, soil and cow dung

| Plot No | Condition | Types of seed cultivated | Amount of yield per ha in tons |
|---|---|---|---|
| 1 | Control (100% soil) | Beet, maize, potato, cucumber | 3.25 |
| 2 | Soil 75% + fly ash 25% | Beet, maize, potato, cucumber | 3.74 |
| 3 | Soil 70% + fly ash 30% | Beet, maize, potato, cucumber | 3.85 |
| 4 | Soil 50% + fly ash 50% | Beet, maize, potato, cucumber | 3.86 |
| 5 | Soil 70% + fly ash 20% + compost 10% | Beet, maize, potato, cucumber | 4.17 |

*Source* Swain et al. (2019)

**Application of Fly Ash in Mine Filling**

Fly ash can be utilized to fill the excavated mines. In this process, 25% of the fly ash generated in the world can be effectively used in an eco-friendly manner. By introducing a large quantity of fly ash in the excavated mines, it will save the river sand from depletion. Other metals like Fe, Mn, Se, etc. can be successfully extracted from it (Dutta and Kumar 2016).

**Metallurgy**

Fly ash contains about 20–27% of $Al_2O_3$ and thus aluminium can be extracted from it. It was estimated that 1.5 tonnes of aluminium and 1.25 tonnes of pozzolanic cement can be produced from 1 tonne of fly ash and 4 tonnes of additives like gypsum and lime. In this process, metals like manganese, zinc, and selenium can be successfully recovered by using a suitable methodology or procedure (Han et al. 2018).

**Ecoutilization of Mill Scale for the Production of Sponge Iron**

A SEM study (Fig. 5.12) reveals that the millimetre scale contains both elongated and globular structures. From XRD studies (Fig. 5.15), it was found that mill scale mainly contains mineralogical phases such as hematite and magnetite. From physical and chemical properties, mill scale is found to contain the highest percentage of iron. Thus, it can be classified as non-dangerous waste. Thus, mill scale can be used to recover iron through the process of reduction. The materials required for this purpose are mill scale, coal, and lime fines, and these are crushed into the 50–400 ASTM mesh size specification. As coal constitutes around 27% of fixed carbon, it can therefore be used to reduce the mill scale. For this purpose, the ratio of mill scale and coal, along with lime fines, is 1:1. Then the mixture is heated at different temperatures, i.e., 700–920 °C in a muffle furnace. It was

**Table 5.10** Chemical analysis of mill scale after reduction at 700 °C temperature and different time interval (Sista et al. 2019)

| Sl. No | Particle of different ASTM sieve size | Residence time in hour | Wt. % of Fe |
|---|---|---|---|
| 1 | Up to 50 | 0.5 | 61.24 |
| 2 | | 1.0 | 62.53 |
| 3 | | 1.5 | 65.87 |
| 4 | 50–100 | 0.5 | 71.35 |
| 5 | | 1.0 | 74.43 |
| 6 | | 1.5 | 78.76 |
| 7 | 150–200 | 0.5 | 65.68 |
| 8 | | 1.0 | 69.56 |
| 9 | | 1.5 | 74.23 |
| 10 | 250–350 | 0.5 | 68.73 |
| 11 | | 1.0 | 71.22 |
| 12 | | 1.5 | 74.19 |
| 13 | Up to 400 | 0.5 | 70.63 |
| 14 | | 1.0 | 72.82 |
| 15 | | 1.5 | 74.73 |

experimentally verified that about 86.43% of metallic iron was obtained from 50 to 100 ASTM-size milliscales if heated at 920 °C for 1.5 h. Table 5.10 illustrates the chemical analysis of mill scale after reduction at 700 °C at various time intervals. Table 5.11 illustrates the chemical analysis of mill scale after reduction at 800 °C with various time interval. Table 5.12 illustrates the chemical analysis of mill scale after reduction at 920 °C temperature at various time intervals. Figure 5.17 shows the plot of different size of mill scale and their % of metallization in different residence time at 920 °C (Gautham and Sasmal 2022). Table 5.11 shows the experimental result of reduction of mill scale of different ASTM size heated at different temperatures for different time intervals. It shows that the metallization increases with the increase in temperature for each ASTM size. Figure 5.17 shows percentage of metallization in graphical representation. It is evident that as the temperature and time interval increases, the metallization percentage also increases.

Thus, the mill scale sample with 50–100 ASTM sieve sizes undergoing reduction at a furnace temperature of about 920 °C yields about 86.43%.

**Sustainable Development Goals**

Eco-utilization of solid waste will achieve the sustainable development goals viz. good health and well-being, clean water and sanitation, decent work and economic growth, industry, innovation and infrastructure, responsible consumption and production and climate action.

**Table 5.11** Chemical analysis of mill scale after reduction at 800 °C temperature with various time interval (Joshi and Dhokey 2015)

| Sl. No | Particle of different ASTM sieve size | Residence time in hour | Temperature (°C) | Wt. % of Fe |
|---|---|---|---|---|
| 1 | Up to 50 | 0.5 | 800 | 64.56 |
| 2 | | 1.0 | | 65.84 |
| 3 | | 1.5 | | 68.93 |
| 4 | 50–100 | 0.5 | 800 | 74.23 |
| 5 | | 1.0 | | 77.57 |
| 6 | | 1.5 | | 80.54 |
| 7 | 150–200 | 0.5 | 800 | 68.46 |
| 8 | | 1.0 | | 71.64 |
| 9 | | 1.5 | | 77.37 |
| 10 | 250–350 | 0.5 | 800 | 71.45 |
| 11 | | 1.0 | | 73.87 |
| 12 | | 1.5 | | 76.53 |
| 13 | Up to 400 | 0.5 | 800 | 73.38 |
| 14 | | 1.0 | | 75.56 |
| 15 | | 1.5 | | 77.86 |

**Table 5.12** Chemical analysis of mill scale after reduction at 920 °C temperature at various time interval (Sen and Pandel 2020)

| Sl. No | Particle of different ASTM sieve size | Residence time in hour | Temperature (°C) | Wt. % of Fe |
|---|---|---|---|---|
| 1 | Up to 50 | 0.5 | 920 | 68.13 |
| 2 | | 1.0 | | 69.68 |
| 3 | | 1.5 | | 73.54 |
| 4 | 50–100 | 0.5 | 920 | 77.83 |
| 5 | | 1.0 | | 78.73 |
| 6 | | 1.5 | | 86.43 |
| 7 | 150–200 | 0.5 | 920 | 71.98 |
| 8 | | 1.0 | | 74.56 |
| 9 | | 1.5 | | 80.97 |
| 10 | 250–350 | 0.5 | 920 | 74.92 |
| 11 | | 1.0 | | 76.72 |
| 12 | | 1.5 | | 79.13 |
| 13 | Up to 400 | 0.5 | 920 | 76.51 |
| 14 | | 1.0 | | 78.72 |
| 15 | | 1.5 | | 80.64 |

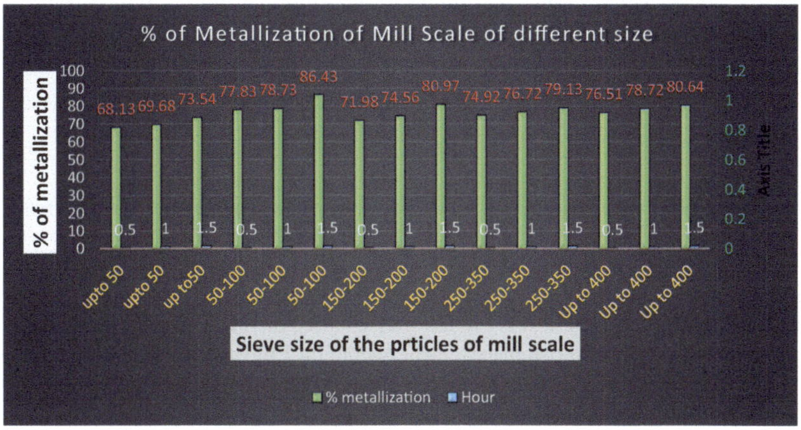


**Fig. 5.17** Plot of different size of mill scale and their % of metallization in different residence time at 920 °C

## 5.4 Conclusion

Today, the steel industry places emphasis on adopting clean, green, and sustainable technology to reduce the contaminants generated from it. To achieve the development of sustainable technology in the steel industry, the waste management system plays a significant role. By the way, we must implement suitable methods and technology for 100% recycling, reuse, and eco-utilization of huge quantities of solid waste (mill scale, BF slag, SMS slag, fly ash, dolo dust, flue dust, etc.) generated from steel plants throughout the world. This concept of sustainable waste management is the reason for the achievements in the steel industry in manufacturing clean and green steel in the global steel market. In the sinter plant unit, waste and other by-product materials combine and form an improved quality of input materials, which is the proper substitute for virgin iron ore and thus can save useful natural resources. The steel industries, builders, and ready-mix concrete units emphasized using the huge quantities of waste generated in an eco-friendly manner, such as in cement manufacturing, brick and slag manufacturing, agriculture, mine filling, road construction, etc. The eco-utilization of solid wastes not only produces useful resource materials with the reduction of wastes but also saves energy, saves the environment from degradation, and reduces the emission of $CO_2$ gas, thereby reducing the global warming effect.

**Acknowledgements** We acknowledge the support of our institution 'Veer Surendra Sai University of Science and Technology', Burla for providing facility of design, development of this work.

**Conflict of Interest** The corresponding author on behalf of all authors declared that there is no financial and non-financial conflict of interest.

## References

Agrawal A, Sahu KK, Pandey BD (2004) Solid waste management in non-ferrous industries in India. Resour Conserv Recycl 42(2):99–120

Ahmaruzzaman M (2010) A review on the utilization of fly ash. Prog Energy Combust Sci 36(3):327–363

Ansari FA, Gupta AK, Yunus M (2011) Fly-ash from coal-fed thermal power plants: bulk utilization in horticulture—a long-term risk management option. Int J Environ Res 5(1):101–108

Baidya R, Kumar Ghosh S, Parlikar UV (2019) Blast furnace flue dust co-processing in cement kiln—a pilot study. Waste Manage Res 37(3):261–267

Balaji G, Pavan PS, Ashok P, Suganth V, Vetturayasudharsanan R (2022) Utilization of lignite coal ash and steel slag in fly ash brick manufacturing: a review. IOP Conf Seri: Earth Environ Sci 1125(1):12015–12027

Barala D, Mohanta R, Sahoo B, Dalai R, Mallik PK (2022) Dielectric properties of steel melting shop (SMS) slag. Mater Today: Proc 49:1905–1909

Biswal T, Swain PK, Panda RB (2019) Solid waste generation in the Rourkela Steel Plant, its recycling and eco-utilization. Pollut Res 38(2):438–446

Branca TA, Pistocchi C, Colla V, Ragaglini G, Amato A, Tozzini C, Romaniello L et al (2014) Investigation of (BOF) converter slag use for agriculture in Europe. Metall Res Technol 111(3):155–167

Das B, Prakash S, Reddy PSR, Biswal SK, Mohapatra BK, Misra VN (2002) Effective utilization of blast furnace flue dust of integrated steel plants. EJMP & EP (Eur J Miner Process Environ Prot) 2(2):61–68

Das B, Prakash S, Reddy PSR, Misra VN (2007) An overview of utilization of slag and sludge from steel industries. Resour Conserv Recycl 50(1):40–57

Dutta RK, Kumar V (2016) Suitability of flyash-lime-phosphogypsum composite in road pavements. Period Polytech Civ Eng 60(3):455–469

El-Hussiny NA, Shalabi MEH (2010) Effect of recycling blast furnace flue dust as pellets on the sintering performance. Sci Sinter 42(3):269–281

Escalante-García JI, Fuentes AF, Gorokhovsky A, Fraire-Luna PE, Mendoza-Suarez G (2003) Hydration products and reactivity of blast-furnace slag activated by various alkalis. J Am Ceram Soc 86(12):2148–2153

Gautham S, Sasmal S (2022) Nano-scale fracture toughness of fly ash incorporated hydrating cementitious composites using experimental nanoindentation technique. Theoret Appl Fract Mech 117:103180

Gencel O, Munir MJ, Kazmi SMS, Sutcu M, Erdogmus E, Velasco PM, Quesada DE (2021) Recycling industrial slags in production of fired clay bricks for sustainable manufacturing. Ceram Int 47(21):30425–30438

Ghods P, Isgor OB, McRae GA, Li J, Gu GP (2011) Microscopic investigation of mill scale and its proposed effect on the variability of chloride-induced depassivation of carbon steel rebar. Corros Sci 53(3):946–954

Gorai S, Ash F (2018) Utilization of fly ash for sustainable environment management. J Mater Environ Sci 9(2):385–393

Han G, Yang S, Peng W, Huang Y, Wu H, Chai W, Liu J (2018) Enhanced recycling and utilization of mullite from coal fly ash with a flotation and metallurgy process. J Clean Prod 178:804–813
Joshi C, Dhokey NB (2015) Study of kinetics of mill scale reduction: for PM applications. Trans Indian Inst Met 68:31–35
Khater GA (2002) The use of Saudi slag for the production of glass-ceramic materials. Ceram Int 28(1):59–67
Khater GA, Nabawy BS, El-Kheshen AA, Abdel Latif MAB, Farag MM (2022) Use of arc furnace slag and ceramic sludge for the production of lightweight and highly porous ceramic materials. Materials 15(3):1112–1133
Khunte M (2018) Process waste generation and utilization in steel industry. Int J Ind Manuf Syst Eng 3(1):1–5
Kim K, Asaoka S, Yamamoto T, Hayakawa S, Takeda K, Katayama M, Onoue T (2012) Mechanisms of hydrogen sulfide removal with steel making slag. Environ Sci Technol 46(18):10169–10174
Kirby CS, Rimstidt JD (1993) Mineralogy and surface properties of municipal solid waste ash. Environ Sci Technol 27(4):652–660
Kostura B, Kulveitova H, Leško J (2005) Blast furnace slags as sorbents of phosphate from water solutions. Water Res 39(9):1795–1802
Krishna RS, Mishra J, Meher S, Das SK, Mustakim SM, Singh SK (2020) Industrial solid waste management through sustainable green technology: case study insights from steel and mining industry in Keonjhar, India. Mater Today: Proc 33:5243–5249
Kumar A (2014) Utilisation of wastes from integrated steel plant with special reference to India. Int J Eng Res Technol (IJERT) 3(1):47–52
Kumar A, Agrawal A (2020) Recent trends in solid waste management status, challenges, and potential for the future Indian cities—a review. Curr Res Environ Sustain 2:1–19
Li Y, Liu Y, Gong X, Nie Z, Cui S, Wang Z, Chen W (2016) Environmental impact analysis of blast furnace slag applied to ordinary Portland cement production. J Clean Prod 120:221–230
Li Y, Li S, Pan X, Zhao X, Guo P (2022) Eco-utilization of steel slag: preparation of Fe-based calcium silicate hydrate and its application in As (V) removal. Appl Surf Sci 597:153763
Lim JY, Kang YG, Sohn KM, Kim PJ, Galgo SJC (2022) Creating new value of blast furnace slag as soil amendment to mitigate methane emission and improve rice cropping environments. Sci Total Environ 806:150961
Lobato NCC, Villegas EA, Mansur MB (2015) Management of solid wastes from steelmaking and galvanizing processes: a brief review. Resour Conserv Recycl 102:49–57
Lopez G, Farfan J, Breyera C (2022) Trends in the global steel industry: evolutionary projections and defossilisation pathways through power-to-steel. J Clean Prod 375:1–25
Martin JP, Collins RA, Browning JS, Biehl FJ (1990) Properties and use of fly ashes for embankments. J Energy Eng 116(2):71–86
Nicolae M, Vîlciu I, Zaman F (2007) X-ray diffraction analysis of steel slag and blast furnace slag viewing their use for road construction. UPB Sci Bull 69(2):99–108
Nikiforova VA (2018) World steel industry: current challenges and development trends (analytical overview). Econ Ind 1(81):86–114
Pappu A, Saxena M, Asolekar SR (2007) Solid wastes generation in India and their recycling potential in building materials. Build Environ 42(6):2311–2320
Ravazzolo F, Vespignani J (2020) World steel production: a new monthly indicator of global real economic activity. Can J Econ/rev Can D'économique 53(2):743–766
Sarkar S, Mazumder D (2015a) Solid waste management in steel industry—challenges and opportunities. Int J Soc, Behav, Educ, Econ, Bus Ind Eng 9(3):884–887
Sarkar S, Mazumder D (2015b) Solid waste management in steel industry—challenges and opportunities. Eng Technol Int J Soc Behav Educ Econ Bus Ind Eng 9(3):978–981

Seggiani M, Vitolo S (2003) Recovery of silica gel from blast furnace slag. Resour Conserv Recycl 40(1):71–80

Sen R, Pandel U (2020) Closed crucible reduction of lump powdered mill scale or iron ore by coal: the sequential methodology and mechanism for optimization of process parameters. Adv Powder Technol 31(9):3760–3773

Shipa SK (2020) Utilization of metallurgical wastes in Indian steel industries. Int J Eng Res Technol 9(8):82–184

Singh R, Singh DP, Kumar N, Bhargava SK, Barman SC (2010) Accumulation and translocation of heavy metals in soil and plants from fly ash contaminated area. J Environ Biol 31(4):421–430

Singh PK, Lava KA, Katiyar PK, Maurya R (2017) Agglomeration behaviour of steel plants solid waste and its effect on sintering performance. J Mater Res Technol 6(3):289–296

Sista KS, Dwarapudi S, Nerune VP (2019) Direct reduction recycling of mill scale through iron powder synthesis. ISIJ Int 59(5):787–794

Swain PK, Biswal T, Panda RB (2019) Physico chemical and minerological properties of fly ash of Rourkela Steel Plant, Odisha and its eco-utilization. Pollut Res 38(2):144–152

Swain PK, BiswaL T, Panda RB (2020) A short review on solid waste generations, recycling and management in the present scenario of India. J Ind Pollut Control 34(1):2008–2014

Tiwari MK, Bajpai S, Dewangan UK (2016) Steel slag utilization—overview in Indian perspective. Int J Adv Res 4(8):2232–2246

Wei Y, Shimaoka T, Saffarzadeh A, Takahashi F (2011) Mineralogical characterization of municipal solid waste incineration bottom ash with an emphasis on heavy metal-bearing phases. J Hazard Mater 187(1–3):534–543

Yousuf A, Manzoor SO, Youssouf M, Malik ZA, Khawaja KS (2020) Fly ash: production and utilization in India—an overview. J Mater Environ Sci 11(6):911–921

Yu CL, Deng Q, Jian S, Li J, Dzantor EK, Hui D (2019) Effects of fly ash application on plant biomass and element accumulations: a meta-analysis. Environ Pollut 250:137–142

Zeydabadi BA, Mowla D, Shariat MH, Kalajahi JF (1997) Zinc recovery from blast furnace flue dust. Hydrometallurgy 47(1):113–125

Zhou Y, Ning XA, Liao X, Lin M, Liu J, Wang J (2013) Characterization and environmental risk assessment of heavy metals found in fly ashes from waste filter bags obtained from a Chinese steel plant. Ecotoxicol Environ Saf 95:130–136

# Valorization of Deep Eutectic Solvent Pretreated Lignocellulosic Biomass for Improved Biogas Production

6

Akinola David Olugbemide, Ikhazuagbe Hilary Ifijen, Chandan Mahata, Filipa A. Vicente, and Blaž Likozar

**Abstract**

Chemical pretreatment is an important method in the valorization of biomass for bioenergy production. The recalcitrant nature of lignocellulosic materials makes them resistant to microbial attacks, which reduces or impairs their usefulness as feedstock for bioenergy production. Although chemical pretreatment has been proven useful and efficient in fractionating lignocellulosic materials/biomass for the production of biofuels and bio-based chemicals, the effects of wastewater generated from the process on the environment have been an issue of great concern. The need for greener solvents to arrest the situation has encouraged significant research efforts in this direction. Chemicals such as ionic liquids (ILs) have been extensively studied, and research is still ongoing. However, some of the undesirable properties of the ILs, which limited their

A. D. Olugbemide (✉)
TETFUND Centre of Excellence in Creative-Innovation and Entrepreneurship Dynamics, Auchi Polytechnic, Auchi, Edo State, Nigeria
e-mail: akinolaore@auchipoly.edu.ng

Department of Basic Sciences, Auchi Polytechnic, Auchi, Edo State, Nigeria

I. H. Ifijen
Department of Research Outreach, Rubber Research Institute of Nigeria, Benin City, Edo State, Nigeria

C. Mahata
Biological Systems Engineering, Virginia Polytechnic Institute and State University, Blacksburg, VA 24061, USA

F. A. Vicente · B. Likozar
Department of Catalysis and Chemical Reaction Engineering, National Institute of Chemistry, Hajdrihova 19, 1001 Ljubljana, Slovenia



M. Nasr and B. Ravindran (eds.), *Solid Waste Management in Delta Region for SDGs Fulfillment*, Deltas of the World, https://doi.org/10.1007/978-3-031-58253-0_6

applications, have led to the quest for better solvents, which eventually led to the use of deep eutectic solvents (DESs). Numerous studies have been conducted on the applications of different classes of DESs, including catalysis, extraction, gas adsorption/absorption, purifying and manufacture of biodiesels and metal electrodeposition; however, the use of DESs to pretreat biomass for biogas production via anaerobic digestion (AD) has been sparsely explored. This review seeks to reveal the extent of work done so far in this respect, address the likely challenges, identify untapped possibilities and future research opportunities. Overall, this review provides valuable insights into the use of DESs for lignocellulosic biomass pretreatment and their potential to revolutionize the production of renewable energy from biomass.

## 6.1 Introduction

Anaerobic digestion (AD) is an environmentally friendly process that converts biodegradable organic materials to biogas, mainly methane (the only energy-rich component) and carbon dioxide in the absence of oxygen. In addition to biofuel generation, AD technology is a sustainable waste management process that can mitigate some of the human and environmental hazards caused by the indiscriminate handling of organic wastes (Uddin et al. 2021; El Asri 2023).

Biogas provides clean energy that can be used for electricity, heat and automobile fuels while the residue generated from the process known as digestate can be used as a soil conditioner or biofertilizer. Additionally, methane and the digestates can serve as platforms for the production of value-added chemicals that are required in industries. The production of these chemicals from an environmentally friendly process, such as AD, would further reduce the negative impacts of producing them from fossil fuels (World Biogas Association 2018; Petravić-Tominac et al. 2020).

Four stages are usually involved in AD: hydrolysis, acidogenesis, acetogenesis and methanogenesis. Each stage has its unique nutrient and environmental requirements, which, to a large extent, determine the success of the AD process. It is worth mentioning that AD can adequately address twelve of the seventeen UN Sustainable Development Goals (SDG) which are: SDG 1, SDG 2, SDG 3, SDG 5, SDG 6, SDG 7, SDG 9, SDG 11, SDG 13, SDG 14, SDG 15 and SDG 16 (Ngan et al. 2020; Obaideen et al. 2022).

Anaerobic digestion, as an "omnivorous" system, can be fed with streams of organic materials such as agricultural wastes, municipal sewage sludge, food wastes, energy crops, micro and macroalgae and effluents or wastewater from industries. Among the aforementioned streams of organic materials, lignocellulosic materials stand out due to their abundance and availability in different parts of the world. It has been estimated that 181.5 billion tonnes of lignocellulosic biomass are annually produced (Khan and Ahring 2019; Dahmen et al. 2019). Lignocellulosic biomass has been reported as a sustainable feedstock that will contribute immensely to the implementation of a circular bioeconomy in the near future. However, its recalcitrant nature creates a bottleneck in harnessing their full potential as a reliable feedstock, necessitating pretreatment to make them more amenable to enzymatic attack for optimal results (Morone and Pandey 2014; Ubando et al. 2020).

Different pretreatment methods exist for fractionation or delignification of biomass for biogas production via AD process, namely physical, biological and chemical pretreatments as well as a combination of these methods. Chemical pretreatment methods include the use of acids, alkaline, organic solvents (organosolvs), ionic liquids (ILs) and deep eutectic solvents (DES). Among these, DESs have attracted considerable attention due to their fascinating properties, for instance their designer solvent character, lower cost, higher biocompatibility, possibility of using natural compounds and solubility of both hydrophobic and hydrophilic compounds. These unique properties can be used to enhance the use of lignocellulosic materials as viable substrates for biogas production (Abraham et al. 2020; Sumiati and Suryadi 2021). Other chemical pretreatment methods have been explored to improve the chemical composition and, morphology of lignocellulosic materials for biogas production (Mirmohamadsadeghi et al. 2021) but little has been done using DESs. A Web of Science search using the keywords "anaerobic digestion" and "biogas production" and "lignocellulosic biomass pretreatment" and "deep eutectic solvents" only yielded 8 hits. This further confirms that this emerging research area needs to be fully explored to push the frontiers of knowledge on biogas enhancement using green solvents like DES. Therefore, this study aims to provide:

- a comprehensive review of the application of DESs in pretreating feedstocks for biogas production;
- an evaluation of the impacts of DES pretreatment on lignocellulosic biomass, including alterations in lignin and hemicellulose structures, heightened accessibility of cellulose, and diminished recalcitrance;
- a correlation between enhanced biogas production due to DES and achievement of the UN SDGs;
- a highlight on the challenges linked with DES pretreatment, such as reuse of DES, sustainability of DES production, and the variability of biomass properties, and future prospects.

### 6.1.1 A Brief History on Deep Eutectic Solvents

DESs are a relatively new class of green solvents that were first discovered in 2003 by Abbott and his colleagues (Abbott et al. 2003). DESs are mixtures of pure compounds, a hydrogen bond acceptor (HBA) and a hydrogen bond donor (HBD), for which the eutectic point temperature is far below an ideal liquid mixture (cf. Fig. 6.1). In more detail, a DES is a eutectic mixture of Lewis or Brønsted acids and bases, in which it is not only required the presence of a hydrogen bond between a donor and an acceptor but there also must be a

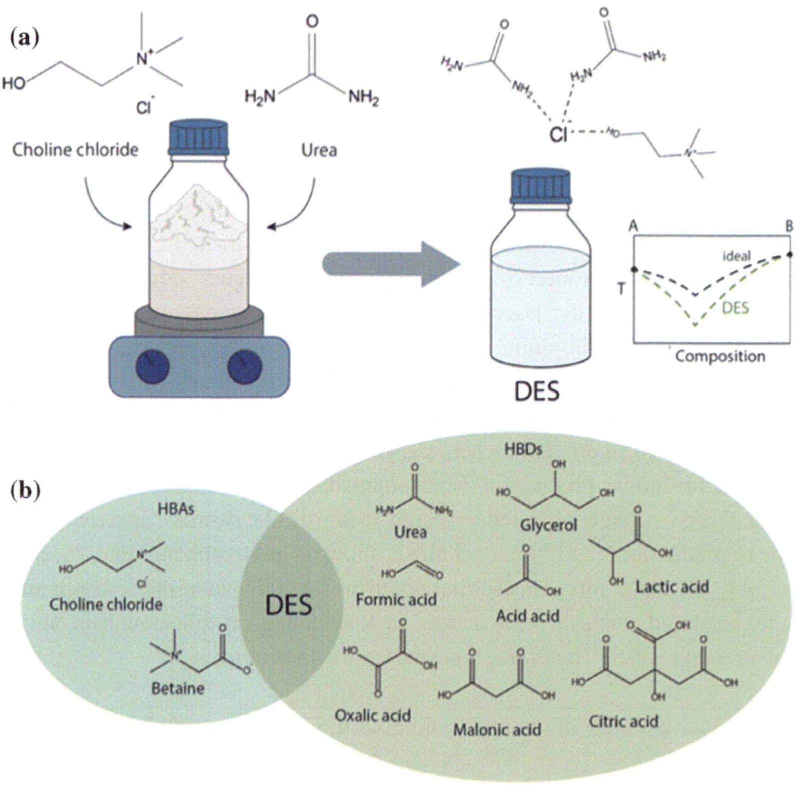


**Fig. 6.1** **a** Schematic representation of DES formation with the phase diagram representing the thermodynamic behaviour of the system, namely the melting points of the ideal mixture and the experimental mixture. **b** Examples of hydrogen bond donors (HBDs) and hydrogen bond acceptors (HBAs) usually used to prepare natural DES

small difference in acidity between the HBD and the HBA (Abbott et al. 2004; Abranches and Coutinho 2023). The first DES discovered was a mixture of choline chloride and urea, which was found to be a highly effective solvent for a wide range of applications (Smith et al. 2014). Since then, researchers have identified and developed many other types of DESs with various properties and potential applications (Wang et al. 2020). One of the main advantages of DESs is that they are usually composed of inexpensive, abundant, and environmentally friendly compounds. In general, DESs are biodegradable, non-toxic, and have low volatility, making them safer and more sustainable than many traditional organic solvents (Dheyab et al. 2021).

Owing to their unique properties, DESs have been explored in a wide range of applications, including green chemistry, electrochemistry, materials science, and pharmaceuticals. They have been used as solvents, electrolytes, catalysts and even as reaction media for chemical reactions (Svigelj et al. 2021). Despite the many potential benefits of DESs, there are still challenges that need to be addressed, such as high viscosity, low thermal stability, difficult recovery of DESs and high cost of this process after biomass pretreatment (Yang et al. 2016). Nevertheless, the field of DES is rapidly expanding and holds great promise for sustainable and innovative solutions in various industries (Satlewal et al. 2018; Radošević et al. 2018).

### 6.1.2 Effects of Different DES Mixtures on Lignocellulosic Materials

The effects of different DES mixtures on lignocellulosic biomass pretreatment have been investigated in several studies. These studies have shown that the composition of the DES can have a significant impact on the pretreatment efficiency, biogas production, lignocellulosic composition, and enzymatic hydrolysis efficiency (Miftah et al. 2022; Mamilla et al. 2019; Suopajärvi et al. 2020).

Table 6.1 shows the effect of different DES mixtures on the removal of lignin and hemicellulose from different feedstocks. Lignin-carbohydrate complexes (LCCs) have been identified as responsible for the recalcitrant nature of lignocellulosic biomass, which is a result of cross-linking between lignin and carbohydrates, especially hemicellulose. The cleavage of the LCCs leads to hemicellulose removal which in turn enhances microorganisms' access to the substrates for improved biogas yields (Satlewal et al. 2018). Some of the DESs (Table 6.1) were effective in removing hemicellulose to a large extent, whereas other were incapable of achieving this. The level of lignin removal was also significant whereas cellulose remained intact in most cases, which is beneficial for AD process.

### 6.1.3 Chemical Pretreatment Methods

Pretreatment of lignocellulosic materials is an important step in their conversion to biofuels and other valued-added chemicals. Generally, the goals of any pretreatment are to: (a) avoid degradation or loss of carbohydrates, (b) prevent or minimize formation of inhibitors, (c) minimize energy consumption, (d) produce less residues and (e) be cost effective (Muthu et al. 2023; Yogalakshmi et al. 2023). As aforementioned, chemical pretreatment methods are commonly employed, however several researchers have reported on both the antagonistic and synergistic effects of these on biogas production except for DES pretreatment, which has not been adequately studied in comparison to the other chemical pretreatment methods. Therefore, in this section, we shall take a look at some of the research endeavours on these chemical pretreatment methods for the enhancement of biogas yields and, ultimately, focus on what has been done so far on biogas production from DES-pretreated biomass.

**Table 6.1** Lignocellulosic biomass pretreated with DES for biogas production (**A**) and for other purposes (**B**), DES pretreatment conditions, and percentage removal of lignin, hemicellulose and cellulose

| Feedstock | DES | Molar ratio | Temperature and time conditions | Lignin (%) | Hemicellulose (%) | Cellulose (%) | Biomass for biogas production | References |
|---|---|---|---|---|---|---|---|---|
| **A** | | | | | | | | |
| Halophyte (*Atriplexcrassifolia*) | ChCl:LA | 1:2 | 120 °C, 4 h | 89.5 | – | – | Yes | Nawaz et al. (2023) |
| Sugarcane leaves | ChCl:monoethanolamine | 1:6 | 120 °C, 3 h | 86.4 | 74.0 | – | Yes | Miftah et al. (2022) |
| Willow | ChCl:LA | 1:10 | 160 °C, 15 min | 80 | – | – | Yes | Kang et al. (2023) |
| Rice straw | Betaine:LA | 1:2.5 | 130 °C, 3 h | 81.3 | – | – | Yes | Basak et al. (2022) |
| Corn stover | Ammonium thiocyanate:urea | | 100 °C, 1 h | 23.5 | – | – | Yes | Olugbemide et al. (2021) |
| *Roystonearegia* leaves | ChCl-various natural compounds* | 2:1 (w/w) | – | 53.6 | – | | Yes | Yu et al. (2019) |
| **B** | | | | | | | | |
| Switchgrass | ChCl:EG | 1:2 | 130 °C, 30 min | 74.41 | – | – | No | Chen et al. (2018) |
| Oil palm fronds | ChCl:urea | 1:2 | 120 °C, 4 h | 16.31 | – | – | No | New et al. (2019) |
| Corn stover | ChCl:formic acid | 1:2 | 130 °C, 2 h | 23.8 | 66.2 | – | No | Xu et al. (2016) |

(continued)

**Table 6.1** (continued)

| Feedstock | DES | Molar ratio | Temperature and time conditions | Lignin (%) | Hemicellulose (%) | Cellulose (%) | Biomass for biogas production | References |
|---|---|---|---|---|---|---|---|---|
| Switchgrass | Guanidine hydrochloride:EG:p-toluenesulfonic acid | 1:1.94:0.06 | 120 °C, 6 min | 82 | 79 | – | No | Chen et al. (2019a, b) |
| Wood | ChCl:OA | 1:1 | 130 °C, 30 min | 85 | – | – | No | Kohli et al. (2020) |
| Coconut coir | ChCl:LA | 1:4 | 150 °C, 20 min (with microwave) | 82 | – | – | No | Mankar et al. (2022) |
| Wheat straw | ChCl:BA:PEG-200 | 1:1:1.5 | 120 °C, 4 h | 88.4 | 84.4 | – | No | Ci et al. (2020) |
| Switch grass | ChCl:LA | 1:2 | 130 °C, 30 min | 83.4 | 80.4 | – | No | Chen et al. (2019a, b) |
| Sugar cane bagasse | ChCl:Gly | 1:10 | 121 °C, 15 min (with ultrasonic waves) | – | – | – | No | Sharma et al. (2021) |
| Sorghum bagasse | ChCl:Gly | 1:2 | 180 °C, 40 min | – | – | – | No | Ai et al. (2020) |
| Maritime pine sawdust | ChCl:LA:TA | 1:4:1 | 175 °C, 1 h | 95 | – | – | No | Fernandes et al. (2021) |

(continued)

**Table 6.1** (continued)

| Feedstock | DES | Molar ratio | Temperature and time conditions | Lignin (%) | Hemicellulose (%) | Cellulose (%) | Biomass for biogas production | References |
|---|---|---|---|---|---|---|---|---|
| Rice straw | ChCl:AA | 1:3.6 | 126 °C, 2.5 h | 83.1 | 82.1 | – | No | Maibam and Goyal (2022) |
| *Pinusbungeana* wood | ChCl:LA | 1:10 | 120 °C for 8 min (with microwave) | 42.8 | 74.7 | 10.1 | No | Li et al. (2019) |
| Bamboo residues | ChCl:OA | 1:1 | 130 °C, 6 h | – | 91.1 | – | No | Li et al. (2022) |
| Raw poplar sawdust | $ZnCl_2$:LA | 1:10 | 140 °C, 3 h | 97.5 | – | – | No | Bai et al. (2022) |

*Notes* Choline chloride—ChCl, lactic acid—LA, ethylene glycol—EG, oxalic acid—OA, boric acid—BA, polyethylene glycol-200—PEG-200, glycerol—Gly, tartaric acid—TA, acetic acid—AA. * Herein DES is formed of added ChCl and various compounds present in the biomass

### 6.1.4 Biogas Production from Acid-Pretreated Lignocellulosic Biomass

Acid pretreatment involves the use of concentrated and diluted mineral and organic acids for fractionation or modification of lignocellulosic materials (LMs) for biogas production. Some of the commonly used acids include $H_2SO_4$, $HNO_3$, HCl, $CH_3COOH$, HCOOH, oxalic acid, etc.

The use of dilute acids is more prevalent compared to concentrated acids owing to the cost effectiveness, mildness, low pretreatment severity, safety concerns, and reduction in the formation of inhibitors (Taherzadeh and Karimi 2008; Gomes et al. 2022). Compared with organic acid pretreatment, more research efforts have been carried out on mineral acid pretreatments with varying degrees of success (Mirmohamadsadeghi et al. 2016; Sarto et al. 2019). However, mineral acid pretreatments have some inherent disadvantages such as high cost, formation of inhibitory substances and loss of sugar which can be mitigated by using organic acids. The use of organic acids for biomass pretreatment has increased in the last few decades because of some obvious advantages they have over mineral acids such as, reduction in the formation of inhibitory compounds, lower sugar degradation, and less corrosive nature in comparison to mineral acids (Gönen et al. 2021). Some promising results have been reported in the literature on pretreatment of biomass with organic acids. Tantayotai et al. (2019) compared the effect of organic acid pretreatment on water hyacinth for biogas production. Two organic acids, citric and oxalic acids were used with percentages increase of 75.62 and 61.87% compared to the untreated sample. In this same work, the authors further stated that the water hyacinth pretreated with hydrochloric acid performed poorly with yields lower than the control. In another study, Nugraha et al. (2018) reported a significant increase (about 35.67% in comparison to the untreated sample) in biogas production when rice husk soaked in 3% acetic acid for 24 h was used in the solid-state AD process.

### 6.1.5 Biogas Production from Alkaline-Pretreated Lignocellulosic Biomass

Alkali pretreatment method is one of the leading pretreatments due to its several advantages over other methods which include: (a) solubilization of lignin, (b) neutralization of various acidic productions and (c) prevention of drop in pH during subsequent acidogenesis process (Kumar et al. 2020). Alkaline pretreatment makes use of solutions such as calcium, potassium, sodium and ammonium hydroxide. The solutions are mixed with the substrates at a suitable temperature for a certain period of time using either wet or solid-state pretreatment method (Badiei et al. 2014; Olugbemide et al. 2020).

Jankovičová et al. (2022) reported an increase of 159% biogas yield from rapeseed straw pretreated with 0.5% NaOH at a temperature range of 90–100 °C for two hours.

In another study by Gu et al. (2015), extruded rice straw was pretreated with calcium hydroxide with 8 and 10% concentrations, being the most promising of the chemical concentrations under consideration producing 34.3 and 36.7% higher biogas than the untreated sample.

### 6.1.6 Biogas Production from Organosolv-Pretreated Lignocellulosic Biomass

Organosolv pretreatment method is based on the pretreatment of lignocellulosic materials with an organic or aqueous organic solvent at temperatures between 120 and 200 °C. The process relies on the chemical breakdown of the lignin macromolecules by cleavage of ether linkages and their subsequent dissolution. The solvents most commonly employed within this process are methanol, ethanol, acetone and ethylene glycol. Methanol and ethanol are often preferred because of their low boiling points and ease of recovery by simple distillation, which implies low energy requirements for their recycling (Oliva et al. 2021). Organosolv pretreatment is usually catalyzed using a salt, an acid or a base. The commonly used catalysts due to their effectiveness in enhancing digestibility are sulphuric acid and sodium hydroxide (Orlando and Borja 2020). The goal of organosolv pretreatment is to fractionate lignocellulosic materials into their individual main fractions, by depolymerizing hemicelluloses and delignifing the lignocellulosic matrix (Oliva et al. 2021; Chu et al. 2021).

In the work by Olugbemide et al. (2016), biogas was produced from rice husk using acid-catalyzed ethanol-organosolv pretreatment method. Three concentrations of ethanol were used namely 1, 3 and 5% at two pretreatment temperatures of 100 and 120 °C. The 3% pretreated substrate produced the highest biogas of 5.55 L, which was about 122% higher than the control. Production started on day one in all the organosolv pretreated digesters, which was an indication of the rapid acclimatization (as a result of low or absence of inhibitors) of the microorganisms (Abu-Reesh 2014).

### 6.1.7 Biogas Production from Ionic Liquid-Pretreated Lignocellulosic Biomass

ILs are preferred to conventional solvents owing to their low melting point, high thermal stability, high reaction rates and low volatility. ILs have been used in various fields of research such as catalysis, nanotechnology, separation process and in pretreatment of biomass prior to anaerobic digestion (Arumugam et al. 2018; Pérez-Pimienta et al. 2020). Some of the commonly used ILs in AD are N-methylmorphine-N-oxide (NMMO), 1-ethyl-3-methylimidazolium acetate ([EMIM]-[OAc]) and 1-buthyl-3-methylimidazolium acetate ([BMIM]-[OAc]). Shafiei et al. (2014) pretreated wood chips and powder with 85% NMMO at 120 °C for 1, 3 and 15 h prior to anaerobic digestion. The anaerobic digestion conducted at mesophilic temperature of 37 °C for a retention time of 45 days increased methane yields by 3.5, 5.4 and 6.8 folds compared to raw wood chips while the treated wood powder increased by 2.6, 3.0 and 3.4 folds in comparison to the untreated wood powder. IL pretreatment has been reported to reduce cellulose crystallinity, partially remove lignin and hemicellulose thus increasing the digestibility of the pretreated biomass for biogas production (Allison et al. 2016). In another work, Padrino et al. (2018) reported 28 and 80% increase in methane yield when barley pretreated at 105 °C for 7.5 h with [EMIM]-[OAc] IL was anaerobically digested at mesophilic and thermophilic temperatures (37 and 55 °C) for 35 days.

### 6.1.8 Biogas Production from DES-Pretreated Lignocellulosic Biomass

As earlier stated, production of biogas or biomethane from DES-pretreated biomass is an emerging research area awaiting exploration by researchers in the field of renewable energy. In this section, we shall discuss the few pioneering research works reported so far that the authors believe will stimulate interest among more researchers in taking this relatively new research area to the next level. Figure 6.2 is a simplified representation of biogas and biofertilizer production from DES-treatment lignocellulosic biomass.

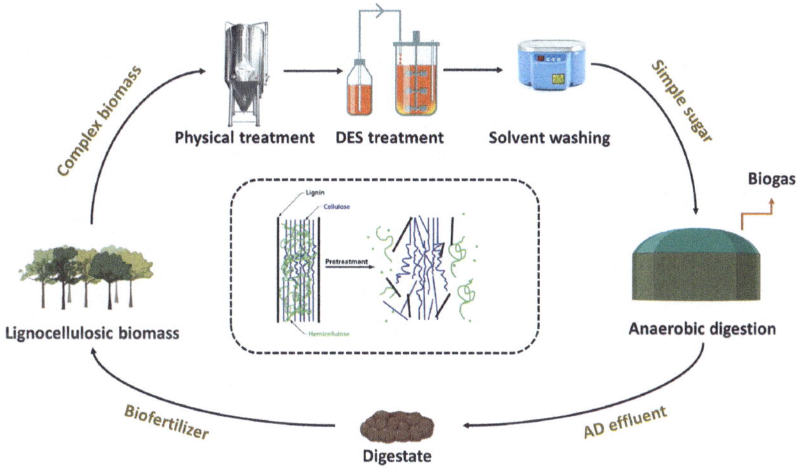


**Fig. 6.2** Schematic representation of biogas and biofertilizer production from DES-treatment lignocellulosic biomass

Yu et al. (2019) in their pioneering work used in situ DES pretreatment, which involved adding choline chloride (HBA) to liquid hot water (LHW) into which leaf sheaths of *Roystones regia* (LSR) and leaves of *Roystones regia* (LR) were separately loaded, to provide HBDs i.e. hemicellulose-derived acids (including formic, acetic, and glucuronic acids) and hydronium ions for the formation of DES. A significant improvement of 309% in biomethane production was reported for DES pretreated LSR compared with the untreated LSR while LR was about 46.7% higher than the control.

Lima et al. (2021) studied the beneficial and detrimental effects of choline chloride (ChCl)-oxalic acid (1:1 molar ratio) based DES on the pretreatment of an organic fraction of municipal solid wastes (OFMSW) for biogas production. Different concentrations (0.3–78 g $L^{-1}$) of choline chloride-oxalic acid were used to pretreat OFMSW, each of the DES components (HBA and HBD) were also used as pretreatment reagents in order to understand the antagonistic or synergistic effects of these solvents on the AD process. The authors reported a direct correlation between lag phase and the DES concentration in the pretreated substrates in the concentration range of 0.3–12.5 g $L^{-1}$. The inhibition concentrations of DES at 10 and 50% ($IC_{10}$ and $IC_{50}$) effects were calculated using lag phase duration, which further established the fact that as the DES concentration increased, the inhibition level also increased. However, this problem can be mitigated by washing the DES-pretreated biomass with water prior to AD to remove inhibitory compounds. In an on-going work by the authors, a lag phase duration of three days was experienced when an unwashed DES-pretreated biomass was used for biogas production. Nonetheless, when the substrate was washed, production commence within the first twenty-four hours of the experiment. The highest biogas production was reported at 12.5 g $L^{-1}$ concentration while in the higher concentration category (19.8–78.1 g $L^{-1}$) biogas production was significantly inhibited.

In a separate investigation, Olugbemide et al. (2021) assessed the viability of utilizing corn stover waste biomass for biomethane production by subjecting it to pretreatment with DES. The researchers conducted a comprehensive analysis of the effects of DES pretreatments using ammonium thiocyanate and urea, considering different conditions such as varying component ratios and temperatures. To understand the impact of pretreatment on the natural biocomposite, the lignocellulosic biomass was characterized in terms of its chemistry and morphology. The study also investigated the implications of DES pretreatment on biomethane production through anaerobic digestion. Different loadings of corn stover biomass (35 and 50 g $L^{-1}$) were tested, and the cumulative biomethane production was measured over a 21-day period. The results showed a significant increase of 48% in biomethane production for the DES-pretreated biomass compared to the untreated sample. This indicated that DES pretreatment positively enhanced the biomethane production process. Furthermore, the authors examined the physicochemical and structural properties of corn stover after pretreatment with ammonium thiocyanate:urea DES under varying conditions. The changes observed in these properties were discussed in relation to their impact on biomethane production. The authors recommended the need for future extensive research to delve deeper into the impacts of DES pretreatment on biomethane production. They also put forward the idea of investigating the composition and usage of digestate, along with its role in resource circulation. It is worth mentioning that this study emphasizes the significance of exploring alternative DESs for pretreating diverse substrates, aiming to assess their biomethane potential and enhance process efficiency for better economic viability in anaerobic digestion.

In another study, Miftah et al. (2022) investigated the optimum conditions for pretreating sugarcane leaves through the utilization of DES. The primary objective of their research was to ascertain the most advantageous parameters for the pretreatment process. Furthermore, they sought to identify the optimal fermentation method that would facilitate the efficient generation of hydrogen and methane from the pretreated sugarcane leaves. The researchers discovered that combining ChCl with monoethanolamine (MEA) proved to be the most effective solvent for removing lignin from the sugarcane leaves. The optimal pretreatment conditions were established with ChCl:MEA at a 1:6 molar ratio, a temperature of 120 °C, a reaction time of 3 h, and a substrate-to-DES solution ratio of 1:12. These conditions resulted in an $86.37 \pm 0.36\%$ removal of lignin and a $73.98 \pm 0.42\%$ removal of hemicellulose, while retaining $84.13 \pm 0.77\%$ of the cellulose content. Two fermentation modes were compared for the subsequent process: simultaneous saccharification and fermentation (SSF) and separate hydrolysis and fermentation (SHF). When the substrate loading was set at 4 g of volatile solids (VS), the SSF process achieved a maximum hydrogen production of $3187 \pm 202$ mL $H_2$ $L^{-1}$, whereas the SHF process yielded a slightly lower hydrogen production of $2135 \pm 315$ mL $H_2$ $L^{-1}$under the same conditions. In the second stage of the study, the researchers utilized the hydrogenic effluent to generate methane. The SSF process demonstrated methane production of $5923 \pm 251$ mL $CH_4$ $L^{-1}$, while the SHF process yielded $3583 \pm 128$ mL $CH_4$ $L^{-1}$. Furthermore, in a one-stage methane production process, the SSF process achieved a maximum methane production of $4067 \pm 320$ mL $CH_4$ $L^{-1}$ with a substrate loading of 4 g of VS. In general, the study revealed that the SSF process was more efficient than the SHF process in producing hydrogen from DES-pretreated sugarcane leaves, both in a two-stage hydrogen and methane production process and a one-stage methane production process.

Basak et al. (2022) employed an integrated pretreatment method of hydrothermal and DES treatments to enhance biogas production from rice straw (RS) under mesophilic (37 °C) and thermophilic (55 °C) conditions. Betaine-lactic acid-based DES was used to pretreat the biomass while pre-produced methanogenic microbial consortium (MMC) was added to digesters to alleviate the toxic effects of inhibitors during AD process. The digesters seeded with anaerobic digester sludge (ADS) served as the control. Results obtained revealed that AD conducted under mesophilic conditions seeded with MMC, achieved 81% increase in comparison to the untreated biomass, while under thermophilic conditions, 54.7% increase was achieved. The authors concluded that integrated hydrothermal-DES pretreatment was effective in making the feedstock amenable to enzymatic attack as confirmed by the changes in morphology and physicochemical properties thereby improving substrate utilization for enhanced biomethane production.

In a more recent study, Nawaz et al. (2023) conducted a study on the production of biogas from DES-pretreated Halophyte *Atriplex crassifolia* under mesophilic conditions. ChCl was used as the HBA while different HBDs under three categories namely carboxylic acid, polyols/glycols and amine/amide were used to prepare various mixtures of DESs. They achieved 89.5% delignification of the substrates by using 15% diluted ChCl:lactic acid (LA) DES mixture at 1:2 molar ratio and a biomass loading of 1:15. Biogas yield was 37.9% higher in comparison to the raw sample which confirms the capability of the ChCl:LA based pretreatment as a viable means of improving biogas production from Halophyte *Atriplex crassifolia.*

Kang et al. (2023) pretreated willow in a microwave oven with ChCl:LA (1:10 molar ratio) at 160 °C for 15 min, ChCl:acetic acid (1:4 molar ratio) at 160 °C for 45 min and ChCl:propionic acid at 160 °C for 32 min (1:4 molar ratio). The ChCl:lactic acid DES recorded the highest biomethane increase of 36.3% in comparison to the untreated willow followed by choline chloride:acetic acid DES with 35.5% while ChCl:propionic acid DES had 17.0% increase.

### 6.1.9 Correlation Between Enhanced Biogas Production Due to DES and Achievement of SDGs

The SDGs, also known as the Global Goals, were adopted by the United Nations in 2015 to serve as a universal call to action for ending poverty, protecting the planet, and ensuring that all people can enjoy peace and prosperity by 2030 (Axon and James 2018). These 17 SDGs are interconnected, recognizing that actions taken in one area can significantly impact outcomes in others, and that development must be balanced to promote social, economic, and environmental sustainability (Olabi et al. 2022). Utilizing DES pretreated lignocellulosic biomass to enhance biogas production shows significant promise in tackling diverse sustainability issues and making notable contributions to multiple SDGs. Below are highlighted some of the SDGs that this study addresses.

#### 6.1.9.1 Environmental Pillar (SDGs 6, 7, 12, 13, 14, 15)

- Sustainable water usage (SDG 6—Clean Water and Sanitation): Traditional pretreatment methods for lignocellulosic biomass, such as acid and alkali treatments, often require large volumes of water. In contrast, the use of DESs may reduce the need for excessive water usage, promoting more sustainable water management practices. SDG 6 aims to ensure access to clean water and sanitation for all. By exploring DES pretreatment as a more water-efficient alternative, the study aligns with the goal of sustainable water use and environmental protection.

- Affordable and clean energy (SDG 7—Affordable and Clean Energy): SDG 7 focuses on ensuring access to affordable, reliable, sustainable, and modern energy for all. Biogas production from lignocellulosic biomass, facilitated by DES pretreatment, can provide a clean and renewable energy source that contributes to achieving this goal. By making biogas production more efficient and environmentally friendly, the study supports the transition to affordable and clean energy sources, especially in areas with limited access to conventional energy.
- Responsible production and consumption (SDG 12—Responsible Consumption and Production): The research on the application of DESs in pretreating feedstocks for biogas production assesses the impacts of this technology on biomass structures. Understanding how DES pretreatment alters lignin and hemicellulose structures can lead to more responsible and sustainable biomass production processes. SDG 12 emphasizes the need for sustainable production and consumption patterns. By investigating the environmental impacts of DES pretreatment, the study contributes to identifying more sustainable practices in the biomass and biogas production industries.
- Reduced greenhouse gas emissions (SDG 13—Climate Action): Biogas production from lignocellulosic biomass is a form of renewable energy that can replace fossil fuels. By converting organic waste and biomass into biogas through pretreatment with DESs, the study contributes to reducing greenhouse gas emissions, particularly methane. Methane emissions from waste and biomass can be significantly higher if left untreated, making this an essential aspect in addressing climate change. SDG 13 specifically targets the urgent need to combat climate change and its impacts. The study's focus on biogas production aligns with the goal of reducing greenhouse gas emissions and promoting low-carbon energy alternatives.
- Conservation of ecosystems and biodiversity (SDGs 14—Life Below Water and 15—Life on Land): The study focuses on utilizing lignocellulosic feedstocks for biogas production. By promoting the use of waste biomass as a renewable energy source, it encourages the conservation of ecosystems and reduces pressure on land use for conventional energy sources. SDGs 14 and 15 aim to protect and restore marine and terrestrial ecosystems, respectively. By diverting biomass waste to biogas production, the study contributes to reducing the environmental impact of conventional energy sources, thereby supporting biodiversity conservation efforts.

#### 6.1.9.2 Economic Pillar (SDGs 8, 9)

- Job creation and economic growth (SDG 8—Decent Work and Economic Growth): The study's emphasis on investigating DES pretreatment for biogas production could lead to increased research and development activities. This, in turn, may create job opportunities in the renewable energy sector and waste management, contributing to economic growth and development.

- Innovation and infrastructure (SDG 9—Industry, Innovation, and Infrastructure): By exploring the application of DESs in pretreatment processes, the study contributes to innovation in biogas production technology and related infrastructure. Identifying challenges and solutions for DES pretreatment may lead to more efficient and cost-effective biogas production systems.

#### 6.1.9.3 Social Pillar (SDGs 1, 2, 3, 4, 10)

- Poverty reduction and food security (SDG 1—No Poverty and SDG 2—Zero Hunger): Biogas production from lignocellulosic biomass provides an alternative energy source that can be harnessed in rural areas. This could lead to increased energy access and job opportunities, potentially contributing to poverty reduction and improving food security in these communities. SDG 1 aims to eradicate poverty in all its forms, and SDG 2 focuses on ending hunger, achieving food security, and improving nutrition. The digestates from biogas production can be used as biofertilizers or soil conditioners that could improve the quality of the soils used for crop production in both rural and urban areas thus increasing crop yields and thereby promoting food security.
- Health and well-being (SDG 3—Good Health and Well-being): The study's evaluation of DES pretreatment may lead to safer and less toxic biomass processing methods. This can help mitigate potential health risks associated with other pretreatment techniques and promote the well-being of workers and communities involved in biogas production. SDG 3 aims to ensure healthy lives and promote well-being for all ages. By exploring safer pretreatment methods, the study contributes to the goal of improving public health and reducing potential health hazards in biomass-related industries. The use of biogas in place of the unclean, traditional fuels used by women in the rural areas that can lead to respiratory diseases in women and the risk women are exposed to in fetching firewood from the forest for cooking can be adequately addressed.
- Quality education (SDG 4—Quality Education): The study's comprehensive review and evaluation of DESs' application in biogas production can contribute to knowledge dissemination and awareness about sustainable energy technologies and their environmental benefits. By providing valuable insights into the potential of DES pretreatment, the study can support educational initiatives and research efforts related to renewable energy and sustainable biomass utilization, aligning with SDG 4's objective of ensuring inclusive and quality education for all.
- Reduced inequalities (SDG 10—Reduced Inequalities): Access to sustainable biogas technology can bridge the energy gap between rural and urban areas, reducing inequalities in energy access and opportunities for development. SDG 10 aims to reduce

inequalities within and among countries. By promoting the use of biogas technology in rural areas, the study contributes to equalizing energy access and promoting sustainable development in marginalized communities.

In summary, the study's aim to review the application of DESs in pretreating feedstocks for biogas production has far-reaching implications for various SDGs. By addressing environmental challenges, contributing to economic growth, and promoting social well-being, such research can play a crucial role in advancing sustainable development and a more inclusive and equitable future.

### 6.1.10 Challenges and Future Prospects

A. **Challenges Associated with DES Pretreatment for Biogas Production**

1. **Residual DES toxicity and its impact on microbial activity**: Residual DES toxicity and its impact on microbial activity is another important challenge associated with DES pretreatment for biogas production. While DESs are known to be less toxic than traditional solvents, there have been reports of their toxic effects on some microorganisms including facultative anaerobes like *Bacillus cereus* (Marchel et al. 2022). Residual DESs in pretreated biomass may inhibit the growth and activity of microorganisms responsible for biogas production, which can ultimately lead to a decrease in biogas yield (Lima et al. 2021). It has been suggested that toxicity of DESs may be species-dependent and the components of the DESs. Therefore, it is essential to understand the effects of residual DES toxicity on microbial activity to optimize the pretreatment conditions and minimize any negative impacts. However, there is a lack of well-designed standard protocols for measuring and assessing the toxicity of residual DESs. Different types of DESs may have different toxicities, and the toxicity may also vary depending on the specific microorganisms present in the anaerobic digestion process (Wen et al. 2015; Marchel et al. 2022). This clearly evidences that more research is needed to develop standardized toxicity testing methods that can be used to assess the impact of residual DESs on different microbial communities.

   Another challenge is the development of effective methods for removing residual DESs from pretreated biomass. While some DESs can be easily removed by washing with water, others may require more complex recovery methods (Al Ameri 2017). Therefore, it is essential to identify efficient and cost-effective methods for removing residual DESs from pretreated biomass to prevent any negative impact on the anaerobic digestion process and that simultaneously allow the reuse of the DESs. Overall, while DES pretreatment shows great promise for enhancing biogas production from lignocellulosic biomass, the potential toxicity of residual DESs (usually associated with their high acidity) and their impact on microbial activity must be carefully considered and addressed to maximize the benefits of DES pretreatment.

2. **Sustainability of DES production and reuse**: DESs are generally considered to be more environmentally friendly than traditional solvents because they are made from renewable resources and have lower toxicity. However, the production of DESs still requires energy and resources, and the sustainability of DES production must be carefully considered to ensure that it does not have a negative impact on the environment. In addition, the reuse of DESs could create a bottleneck. While some DESs can be easily recovered and reused, others may require more complex disposal methods. It is important to develop sustainable methods for reusing of DESs to prevent economic loss and any negative impact on the environment. Furthermore, the sustainability of DES production and disposal can be closely linked to the cost-effectiveness of DES pretreatment. If DES production and disposal methods are not sustainable, it may lead to higher costs for DES pretreatment, which could limit its adoption for large-scale biogas production. Therefore, more research is needed to develop sustainable and cost-effective methods for producing, disposing and/or reusing of DESs, as well as to optimize DES pretreatment conditions to minimize the overall environmental impact. This includes exploring new DES production methods, such as using waste materials as starting materials, and developing efficient and environmentally friendly methods for DES recovery and reuse. By addressing these challenges, DES pretreatment can become a more sustainable and environmentally friendly option for enhancing biogas production from lignocellulosic biomass (Xu et al. 2020).
3. **Variability of biomass composition and properties**: Different types of lignocellulosic biomass, such as agricultural residues, forest residues, and energy crops have different chemical compositions and physical properties, which can affect the effectiveness of DES pretreatment (Al Ameri 2017). For example, some types of biomasses may require more severe pretreatment conditions to achieve the desired level of delignification and increase biogas yields. Furthermore, the properties of the biomass, such as particle size, moisture content, and density, can also affect the effectiveness of DES pretreatment. For example, if the biomass particles are too large, the DES may not be able to penetrate the cell walls effectively, leading to lower biogas yields. On the other hand, if the biomass particles are too small, the DES may react too quickly, leading to over-delignification and reduced biogas yields. To overcome these challenges, researchers need to develop methods for assessing the composition and properties of different types of biomasses and optimizing the pretreatment conditions accordingly. This may involve using statistical models to identify the most important factors affecting pretreatment effectiveness for specific biomass types and developing methods for adjusting the DES composition, temperature, and reaction time to achieve the desired level of delignification. Generally, the variability of biomass composition and properties is an important challenge that needs to be addressed to optimize DES pretreatment

for biogas production. By developing effective strategies for dealing with this variability, researchers can help making DES pretreatment a more reliable and efficient option for enhancing biogas yields from lignocellulosic biomass.

B. **Optimization of DES Pretreatment Conditions**

The optimization of DES pretreatment conditions is a critical factor in the future prospects of DES pretreatment for biogas production. With further research and development, there is significant potential for DES pretreatment to become a more effective and efficient method for enhancing biogas yields from lignocellulosic biomass. One possible avenue for optimizing DES pretreatment conditions is the use of statistical models and experimental design approaches. These methods can help researchers identifying the most important factors affecting the pretreatment effectiveness and optimize the DES composition, temperature, reaction time, and other variables accordingly. Another potential approach is the development of new techniques for DES recovery and reuse. By recovering and reusing DESs, it may be possible to reduce the overall cost of DES pretreatment and improve its sustainability. In addition, there is also the potential for developing new types of DESs that are more effective and cost-efficient than existing ones. For example, researchers may explore the use of novel DESs based on natural or renewable compounds, which could offer a more sustainable and environmentally-friendly alternative to traditional DESs. Overall, the future prospects of DES pretreatment for biogas production are promising, provided that researchers continue to optimize pretreatment conditions and develop new and innovative approaches to overcome the challenges associated with DES pretreatment. As biogas production becomes increasingly important for meeting global energy demands and reducing greenhouse gas emissions, DES pretreatment may become an even more vital technology in the years to come.

1. **Development of new DES formulations**: The development of new DES formulations is an exciting area of research with promising future prospects for DES pretreatment in biogas production. As DESs are still a relatively new technology, there is considerable scope for exploring and developing novel formulations that offer improved performance and cost-effectiveness. One possible approach is to explore new combinations of existing DES components to create more effective formulations. Researchers may also investigate the use of different ratios or concentrations of DES components to achieve optimal results. Another potential area of research is the use of natural and renewable compounds as DES components. For example, choline chloride and urea, which are commonly used in DESs, can be derived from renewable sources such as plants or animal waste products. Using such renewable compounds could help to reduce the environmental impact and cost of DES production. Additionally, there is also the potential for developing DES formulations that are tailored to specific types of lignocellulosic biomass. By considering the unique properties and characteristics of

different types of biomasses, researchers may be able to create DES formulations that are optimized for each specific type, leading to a more efficient and effective pretreatment. The development of new DES formulations has significant potential to improve the effectiveness and cost-effectiveness of DES pretreatment in biogas production. As research in this area continues to progress, it may be possible to create novel DES formulations that offer even greater benefits over existing formulations, paving the way for a more sustainable and environmentally-friendly future for biogas production.

2. **Integration of DES pretreatment with other technologies**: The integration of DES pretreatment with other technologies is an exciting area of research with promising future prospects for biogas production. By combining DES pretreatment with other technologies, it may be possible to improve the overall efficiency of the biogas production process and reduce its environmental impact. One possible approach is to integrate DES pretreatment with enzymatic hydrolysis. Enzymatic hydrolysis involves using enzymes to break down the lignocellulosic biomass into its constituent sugars, which can then be fermented into biogas. Combining DES pretreatment with enzymatic hydrolysis may help to improve the efficiency of the hydrolysis process and increase the yield of fermentable sugars, leading to higher biogas yields, especially considering DES can act simultaneously as a catalyst and reaction media, and in some cases even promote the enzyme super-activity (Toledo et al. 2019). Overall, the integration of DES pretreatment with other technologies has significant potential to improve the efficiency and sustainability of biogas production. As research in this area continues to progress, it may be possible to create novel integrated systems that offer even greater benefits over existing technologies, paving the way for a more sustainable and environmentally-friendly future for biogas production.
3. **Implementation of DES pretreatment in industrial settings**: The implementation of DES pretreatment in industrial settings for biogas production is an important future prospect for this technology. To achieve this, several challenges associated with DES pretreatment, such as cost-effectiveness, residual DES toxicity, and sustainability of DES production and disposal, need to be addressed. However, there are several potential ways to overcome these challenges and facilitate the implementation of DES pretreatment in industrial settings. One effective approach is the optimization of DES pretreatment conditions for specific lignocellulosic biomass types and processing requirements (Sharma et al. 2023). For instance, researchers can employ Artificial Neural Networks (ANNs) to model the complex relationship between DES pretreatment parameters and biogas production performance (Arias et al. 2023). By training the neural network on historical data from various pretreatment experiments, the ANN can learn patterns and correlations, enabling it to predict optimal DES conditions with high accuracy, thus reducing the cost of DES pretreatment while maximizing biogas yields (Beltramo et al. 2016; Cruz et al. 2022). Moreover, Genetic Algorithms (GAs) can play a crucial role in the optimization process by searching through a vast parameter space to identify the optimal combination of DES compositions and process conditions

for maximum biogas yield (Li et al. 2021). This evolutionary approach can efficiently identify near-optimal solutions and reduce the need for extensive experimental trials, thereby accelerating the implementation of DES pretreatment in industrial biogas production (Li et al. 2021). To address concerns related to residual DES toxicity and its adverse environmental impact, intelligent systems like Support Vector Machines (SVMs) can aid in classifying biomass types based on their properties and characteristics (Akinpelu et al. 2023). This classification can guide the selection of the most suitable DES formulations and pretreatment conditions for each biomass type, ensuring a more sustainable and tailored approach to biogas production (Akinpelu et al. 2023). Machine learning can also contribute to the development of new DES formulations that are more efficient and cost-effective for biogas production (Faria et al. 2022). Techniques such as Reinforcement Learning (RL) can be utilized to develop intelligent controllers for DES pretreatment systems. RL algorithms learn by interacting with the DES pretreatment process, adjusting parameters, and observing the outcomes (Faria et al. 2022). Over time, RL can optimize the process to achieve desired biogas production targets while minimizing the use of DES and its associated costs (Faria et al. 2022). Additionally, Random Forests (RF), an ensemble learning method, can analyze various factors that affect biogas production, such as feedstock characteristics, DES formulations, and pretreatment conditions (Xing et al. 2019). RF can provide valuable insights for process optimization, guiding researchers and engineers in designing more efficient and sustainable DES pretreatment strategies (Xing et al. 2019). To address the challenge of cost-effectiveness, researchers can employ Principal Component Analysis (PCA) to identify the most influential factors affecting biogas production efficiency during DES pretreatment (Senthilkumar et al. 2016). By focusing on critical parameters, PCA can guide targeted efforts for process improvement (Senthilkumar et al. 2016). Finally, the implementation of Deep Reinforcement Learning (DRL) can address complex and high-dimensional data scenarios, making it suitable for optimizing DES pretreatment with numerous variables and interactions (Viquerat et al. 2021). DRL adapts to changing conditions and improves the DES pretreatment process through continuous learning, ensuring the most efficient biogas production (Jeong et al. 2021). By incorporating these machine learning and intelligent systems tools into the optimization and implementation of DES pretreatment in industrial settings, researchers and engineers can accelerate progress, overcome challenges, and make DES pretreatment a practical and viable option for large-scale biogas production. Embracing these technological advancements will not only enhance biogas production but also contribute to the sustainable and eco-friendly utilization of lignocellulosic biomass resources.

## 6.2 Conclusions

The processing of lignocellulosic biomass with DESs holds considerable promise for increasing biogas output. Studies have shown that DESs are successful at breaking down lignocellulosic biomass and improving its digestibility for biogas production, despite some difficulties related to DES preparation and reuse. The efficiency and sustainability of biogas production systems could be improved through further optimization of DES pretreatment conditions, creation of innovative DESs (which are task-specific in design), and integration with other technologies. The advantages of DES pretreatment in the production of renewable energy and value-added products from biomass emphasize the need for more studies and advancement in this research area.

**Acknowledgements** The authors would like to express their deepest gratitude to Tertiary Education Trust Fund (TETfund) for the financial support and the Slovenian Research Agency under research core funding P2-0152 and research project NC-0024. The authors would also like to appreciate Dr. Damilare Olugbemide for his technical support.

**Declarations (Conflict of Interest)** Authors declare no conflict of interest.

## References

Abbott AP, Capper G, Davies DL, Rasheed RK, Tambyrajah V (2003) Novel solvent properties of choline chloride/urea mixtures. Chem Commun 1:70–71

Abbott AP, Boothby D, Capper G, Davies DL, Rasheed RK (2004) Deep eutectic solvents formed between choline chloride and carboxylic acids: versatile alternatives to ionic liquids. J Am Chem Soc 126:9142–9147

Abraham A, Mathew AK, Park H, Choi O, Sindhu R, Parameswaran B, Pandey A, Park JH, Sang BI (2020) Pretreatment strategies for enhanced biogas production from lignocellulosic biomass. Bioresour Technol 301:122725

Abranches DO, Coutinho JAP (2023) Everything you wanted to know about deep eutectic solvents but were afraid to be told. Annu Rev Chem Biomol Eng 14:141–163

Abu-Reesh IM (2014) Kinetics of anaerobic digestion of labaneh whey in a batch reactor. Afr J Biotechnol 13(16):1745–1755

Ai B, Li W, Woomer J, Li M, Pu Y, Sheng Z, Zheng L, Adedeji A, Ragauskas AJ, Shi J (2020) Natural deep eutectic solvent mediated extrusion for continuous highsolid pretreatment of lignocellulosic biomass. Green Chem 22:6372–6383. https://doi.org/10.1039/D0GC01560A

Akinpelu DA, Adekoya OA, Oladoye PO, Ogbaga CC, Okolie JA (2023) Machine learning applications in biomass pyrolysis: from biorefinery to end-of-life product management. Digit Chem Eng 8:100103

Al Ameri M (2017) Deep eutectic solvent pretreatment for enhancing biochemical conversion of switchgrass. University of Missouri-Columbia

Allison BJ, Cádiz JC, Karuna N, Jeoh T, Simmons CW (2016) The effect of ionic liquid pretreatment on the bioconversion of tomato processing waste to fermentable sugars and biogas. Appl Biochem Biotechnol 179:1227–1247

Arias A, Feijoo G, Moreira MT (2023) How could artificial intelligence be used to increase the potential of biorefineries in the near future? A review. Environ Technol Innov 32:103277. https://doi.org/10.1016/j.eti.2023.103277
Arumugam V, Redhi G, Gengan RM (2018) The application of ionic liquids in nanotechnology. In: Fundamentals of nanoparticles. Elsevier, pp 371–400
Axon S, James D (2018) The UN Sustainable Development Goals: how can sustainable chemistry contribute? A view from the chemical industry. Curr Opin Green Sustain Chem 13:140–145
Badiei M, Asim N, Jahim JM, Sopian K (2014) Comparison of chemical pretreatment methods for cellulosic biomass. APCBEE Procedia 9:170–174
Bai Y, Zhang XF, Wang Z, Zheng T, Yao J (2022) Deep eutectic solvent with bifunctional Brønsted-Lewis acids for highly efficient lignocellulose fractionation. Bioresour Technol 347:126723. https://doi.org/10.1016/J.BIORTECH.2022.126723
Basak B, Patil S, Kumar R, Ha GS, Park YK, Khan MA, Yadav KK, Fallatah AM, Jeon BH (2022) Integrated hydrothermal and deep eutectic solvent-mediated fractionation of lignocellulosic biocomponents for enhanced accessibility and efficient conversion in anaerobic digestion. Bioresour Technol 351:127034
Beltramo T, Ranzan C, Hinrichs J, Hitzmann B (2016) Artificial neural network prediction of the biogas flow rate optimized with an ant colony algorithm. Biosys Eng 143:68–78. https://doi.org/10.1016/j.biosystemseng.2016.01.006
Chen Z, Bai X, Wan C (2018) High-solid lignocellulose processing enabled by natural deep eutectic solvent for lignin extraction and industrially relevant production of renewable chemicals. ACS Sustain Chem Eng 6(9):12205–12216
Chen Z, Sun Y, Wan C (2019a) Effects of alkaline hydrogen peroxide treatment on cellulose accessibility of switchgrass pretreated by acidic deep eutectic solvent. Cellulose 26:9439–9446. https://doi.org/10.1007/S10570-019-02759-5/FIGURES/4
Chen Z, Jacoby WA, Wan C (2019b) Ternary deep eutectic solvents for effective biomass deconstruction at high solids and low enzyme loadings. Bioresour Technol 279:281–286
Chu Q, Tong W, Chen J, Wu S, Jin Y, Hu J, Song K (2021) Organosolv pretreatment assisted by carbocation scavenger to mitigate surface barrier effect of lignin for improving biomass saccharification and utilization. Biotechnol Biofuels 14(1):136
Ci YH, Yu F, Zhou CX, Mo HE, Li ZY, Ma YQ, Zang LH (2020) New ternary deep eutectic solvents for effective wheat straw deconstruction into its high-value utilization under near-neutral conditions. Green Chem 22:8713–8720. https://doi.org/10.1039/D0GC03240A
Cruz IA, Nascimento VRS, Felisardo RJA, Santos AMGD, Jesus AAD, Vasconcelos BRD, Bilal M, Souza RLD, Ferreira LFR (2022) Evaluation of artificial neural network models for predictive monitoring of biogas production from cassava wastewater: comparison of training algorithms. SSRN Electron J. https://doi.org/10.2139/ssrn.4241643
Dahmen N, Lewandowski I, Zibek S, Weidtmann A (2019) Integrated lignocellulosic value chains in a growing bioeconomy: status quo and perspectives. GCB Bioenergy 11(1):107–117
Dheyab AS, Abu Bakar MF, AlOmar M, Sabran SF, Muhamad Hanafi AF, Mohamad A (2021) Deep eutectic solvents (DESs) as green extraction media of beneficial bioactive phytochemicals. Separations 8(10):176
El Asri O (2023) Anaerobic biodegradation: the anaerobic digestion process. Handbook of biodegradable materials. Springer International Publishing, Cham, pp 85–110
Faria RdR, Capron BDO, Secchi AR, de Souza MB Jr (2022) Where reinforcement learning meets process control: review and guidelines. Processes 10:2311. https://doi.org/10.3390/pr10112311
Fernandes C, Melro E, Magalhães S, Alves L, Craveiro R, Filipe A, Valente AJM, Martins G, Antunes FE, Romano A, Medronho B (2021) New deep eutectic solvent assisted extraction

of highly pure lignin from maritime pine sawdust (Pinus pinaster Ait.). Int J Biol Macromol 177:294–305. https://doi.org/10.1016/J.IJBIOMAC.2021.02.088
Gomes MG, de Oliveira Paranhos AG, Camargos AB, Baêta BEL, Baffi MA, Gurgel LVA, Pasquini D (2022) Pretreatment of sugarcane bagasse with dilute citric acid and enzymatic hydrolysis: use of black liquor and solid fraction for biogas production. Renewable Energy 191:428–438
Gönen Ç, Deveci EÜ, Akter Önal N (2021) Evaluation of biomass pretreatment to optimize process factors for different organic acids via Box-Behnken RSM method. J Mater Cycles Waste Manage 23(5):2016–2027
Gu Y, Zhang Y, Zhou X (2015) Effect of Ca (OH)2 pretreatment on extruded rice straw anaerobic digestion. Bioresour Technol 196:116–122
Jankovičová B, Hutňan M, Czölderová MN, Hencelová K, Imreová Z (2022) Comparison of acid and alkaline pre-treatment of lignocellulosic materials for biogas production. Plant Soil Environ 68(4):195–204
Jeong K, Abbas A, Shin J, Son M, Kim YM, Cho KH (2021) Prediction of biogas production in anaerobic co-digestion of organic wastes using deep learning models. Water Res 205:117697. https://doi.org/10.1016/j.watres.2021.117697
Kang X, Deng C, Shinde R, Lin R, Murphy JD (2023) Renewable deep eutectic solvents pretreatment improved the efficiency of anaerobic digestion by lignin extraction from willow. Energy Convers Manage 288:117115
Khan MU, Ahring BK (2019) Lignin degradation under anaerobic digestion: influence of lignin modifications-a review. Biomass Bioenergy 128:105325
Kohli K, Katuwal S, Biswas A, Sharma BK (2020) Effective delignification of lignocellulosic biomass by microwave assisted deep eutectic solvents. Bioresour Technol 303:122897
Kumar R, Strezov V, Weldekidan H, He J, Singh S, Kan T, Dastjerdi B (2020) Lignocellulose biomass pyrolysis for bio-oil production: a review of biomass pre-treatment methods for production of drop-in fuels. Renew Sustain Energy Rev 123:109763
Li P, Zhang Q, Zhang X, Zhang X, Pan X, Xu F (2019) Subcellular dissolution of xylan and lignin for enhancing enzymatic hydrolysis of microwave assisted deep eutectic solvent pretreated Pinus bungeana Zucc. Bioresour Technol 288:121475. https://doi.org/10.1016/J.BIORTECH.2019.121475
Li B, Tuo A, Kong H, Liu S, Chen J (2021) Application of multilayer perceptron genetic algorithm neural network in Chinese-English parallel corpus noise processing. Comput Intell Neurosci 2021:7144635. https://doi.org/10.1155/2021/7144635
Li N, Meng F, Yang H, Shi Z, Zhao P, Yang J (2022) Enhancing enzymatic digestibility of bamboo residues using a three-constituent deep eutectic solvent pretreatment. Bioresour Technol 346:126639
Lima F, Branco LC, Lapa N, Marrucho IM (2021) Beneficial and detrimental effects of choline chloride–oxalic acid deep eutectic solvent on biogas production. Waste Manage 131:368–375
Maibam PD, Goyal A (2022) Approach to an efficient pretreatment method for rice straw by deep eutectic solvent for high saccharification efficiency. Bioresour Technol 351:127057. https://doi.org/10.1016/J.BIORTECH.2022.127057
Mamilla JL, Novak U, Grilc M, Likozar B (2019) Natural deep eutectic solvents (DES) for fractionation of waste lignocellulosic biomass and its cascade conversion to value-added bio-based chemicals. Biomass Bioenergy 120:417–425
Mankar AR, Pandey A, Pant KK (2022) Microwave-assisted extraction of lignin from coconut coir using deep eutectic solvents and its valorization to aromatics. Bioresour Technol 345:126528
Marchel M, Cieśliński H, Boczkaj G (2022) Deep eutectic solvents microbial toxicity: current state of art and critical evaluation of testing methods. J Hazard Mater 425:127963

Miftah AK, Sittijunda S, Imai T, Salakkam A, Reungsang A (2022) Biohydrogen and methane production from sugarcane leaves pretreated by deep eutectic solvents and enzymatic hydrolysis by cellulolytic consortia. Fermentation 8(8):396. https://doi.org/10.3390/fermentation8080396

Mirmohamadsadeghi S, Karimi K, Horváth IS (2016) Improvement of solid state biogas production from wood by concentrated phosphoric acid pretreatment. Bioresources 11(2):3230–3243

Mirmohamadsadeghi S, Karimi K, Azarbaijani R, Yeganeh LP, Angelidaki I, Nizami AS, Bhat R, Dashora K, Vijay VK, Aghbashlo M, Gupta VK, Tabatabaei M (2021) Pretreatment of lignocelluloses for enhanced biogas production: a review on influencing mechanisms and the importance of microbial diversity. Renew Sustain Energy Rev 135:110173

Morone A, Pandey RA (2014) Lignocellulosic biobutanol production: gridlocks and potential remedies. Renew Sustain Energy Rev 37:21–35

Muthu P, Muniappan G, Jeyakumar RB (2023) Efficacious utilization of food waste for bioenergy generation through the anaerobic digestion method. Processes 11(3):702

Nawaz A, Chaudhary R, Haq IU, Fu X, Huang R, Mukhtar H, Jiang K (2023) Delignification of halophyte Atriplex crassifolia by green recyclable deep eutectic solvents for enhanced production of biogas. Fermentation 9(3):314

New EK, Wu TY, Lee CBTL, Poon ZY, Loow YL, Foo LYW, Procentese A, Siow LF, Teoh WH, Daud NNN, Jahim JM (2019) Potential use of pure and diluted choline chloride-based deep eutectic solvent in delignification of oil palm fronds. Process Saf Environ Prot 123:190–198

Ngan NVC, Chan FMS, Nam TS, Van Thao H, Maguyon-Detras MC, Hung DV, Cuong DM, Van Hung N (2020) Anaerobic digestion of rice straw for biogas production. Sustain Rice Straw Manag 65–92

Nugraha WD, Keumala CF, Matin HHA (2018) The effect of acid pre-treatment using acetic acid and nitric acid in the production of biogas from rice husk during solid state anaerobic digestion (SS-AD). In: E3S web of conferences, vol 31. EDP Sciences, p 01006

Obaideen K, Abdelkareem MA, Wilberforce T, Elsaid K, Sayed ET, Maghrabie HM, Olabi AG (2022) Biogas role in achievement of the sustainable development goals: evaluation, challenges, and guidelines. J Taiwan Inst Chem Eng 131:104207

Olabi AG, Obaideen K, Elsaid K, Wilberforce T, Sayed ET, Maghrabie HM, Abdelkareem MA (2022) Assessment of the pre-combustion carbon capture contribution into sustainable development goals SDGs using novel indicators. Renew Sustain Energy Rev 153:111710

Oliva A, Tan LC, Papirio S, Esposito G, Lens PN (2021) Effect of methanol-organosolv pretreatment on anaerobic digestion of lignocellulosic materials. Renewable Energy 169:1000–1012

Olugbemide AD, Lajide L, Adebayo A, Owolabi BJ (2016) Kinetic study of biogas production from raw and solid-state organosolv pretreated rice husk 7(2):110–118. https://doi.org/10.5958/0976-4763.2016.00014.3

Olugbemide AD, Lajide L, Adebayo A, Owolabi BJ (2020) Enhanced biogas production from rice husk through solid-state chemical pretreatments. Waste Biomass Valorization 11:2397–2407

Olugbemide AD, Oberlintner A, Novak U, Likozar B (2021) Lignocellulosic corn stover biomass pre-treatment by deep eutectic solvents (DES) for biomethane production process by bioresource anaerobic digestion. Sustainability 13:10504. https://doi.org/10.3390/su131910504

Orlando MQ, Borja VM (2020) Pretreatment of animal manure biomass to improve biogas production: a review. Energies 13(14):3573

Padrino B, Lara-Serrano M, Morales-delaRosa S, Campos-Martin JM, Fierro JLG, Martinez F, Melero JA, Puyol D (2018) Resource recovery potential from lignocellulosic feedstock upon lysis with ionic liquids. Front Bioeng Biotechnol 6(119):1–13

Pérez-Pimienta JA, Icaza-Herrera JP, Méndez-Acosta HO, González-Álvarez V, Méndoza-Pérez JA, Arreola-Vargas J (2020) Bioderived ionic liquid-based pretreatment enhances methane production from Agave tequilana bagasse. RSC Adv 10(24):14025–14032

Petravić-Tominac V, Nastav N, Buljubašić M, Šantek B (2020) Current state of biogas production in Croatia. Energy, Sustain Soc 10(1):1–10

Radošević K, Čanak I, Panić M, Markov K, Bubalo MC, Frece J, Srček VG, Redovniković IR (2018) Antimicrobial, cytotoxic and antioxidative evaluation of natural deep eutectic solvents. Environ Sci Pollut Res 25:14188–14196

Sarto S, Hildayati R, Syaichurrozi I (2019) Effect of chemical pretreatment using sulfuric acid on biogas production from water hyacinth and kinetics. Renewable Energy 132:335–350

Satlewal A, Agrawal R, Bhagia S, Sangoro J, Ragauskas AJ (2018) Natural deep eutectic solvents for lignocellulosic biomass pretreatment: recent developments, challenges and novel opportunities. Biotechnol Adv 36(8):2032–2050

Senthilkumar N, Deepanraj B, Vasantharaj K, Sivasubramanian V (2016) Optimization and performance analysis of process parameters during anaerobic digestion of food waste using hybrid GRA-PCA technique. J Renew Sustain Energy. https://doi.org/10.1063/1.4972884

Shafiei M, Karimi K, Zilouei H, Taherzadeh MJ (2014) Enhanced ethanol and biogas production from pinewood by NMMO pretreatment and detailed biomass analysis. BioMed Res Int 1–10

Sharma V, Nargotra P, Sharma S, Bajaj BK (2021) Efficacy and functional mechanisms of a novel combinatorial pretreatment approach based on deep eutectic solvent and ultrasonic waves for bioconversion of sugarcane bagasse. Renewable Energy 163:1910–1922. https://doi.org/10.1016/J.RENENE.2020.10.101

Sharma V, Tsai ML, Chen CW, Sun PP, Nargotra P, Dong CD (2023) Advances in machine learning technology for sustainable biofuel production systems in lignocellulosic biorefineries. Sci Total Environ 886:163972. https://doi.org/10.1016/j.scitotenv.2023.163972

Smith EL, Abbott AP, Ryder KS (2014) Deep eutectic solvents (DESs) and their applications. Chem Rev 114(21):11060–11082

Sumiati T, Suryadi H (2021) Potency of deep euteutic solvent as an alternative solvent on pretreatment process of lignocellulosic biomass. In: Journal of physics: conference series, vol 1764, no 1. IOP Publishing, p 012014

Suopajärvi T, Ricci P, Karvonen V, Ottolina G, Liimatainen H (2020) Acidic and alkaline deep eutectic solvents in delignification and nanofibrillation of corn stalk, wheat straw, and rapeseed stem residues. Ind Crops Prod 145:111956

Svigelj R, Dossi N, Grazioli C, Toniolo R (2021) Deep eutectic solvents (DESs) and their application in biosensor development. Sensors 21(13):4263

Taherzadeh MJ, Karimi K (2008) Pretreatment of lignocellulosic wastes to improve ethanol and biogas production: a review. Int J Mol Sci 9(9):1621–1651

Tantayotai P, Mutrakulchareon P, Tawai A, Roddecha S, Sriariyanun M (2019) Effect of organic acid pretreatment of water hyacinth on enzymatic hydrolysis and biogas and bioethanol production. In: IOP conference series: earth and environmental science, vol 346, no 1. IOP Publishing, p 012004

Toledo ML, Pereira MM, Freire MG, Silva JP, Coutinho JA, Tavares AP (2019) Laccase activation in deep eutectic solvents. ACS Sustain Chem Eng 7(13):11806–11814

Ubando AT, Felix CB, Chen WH (2020) Biorefineries in circular bioeconomy: a comprehensive review. Bioresour Technol 299:122585

Uddin MN, Siddiki SYA, Mofijur M, Djavanroodi F, Hazrat MA, Show PL, Ahmed SF, Chu YM (2021) Prospects of bioenergy production from organic waste using anaerobic digestion technology: a mini review. Front Energy Res 9:627093

Viquerat J, Rabault J, Kuhnle A, Ghraieb H, Larcher A, Hachem E (2021) Direct shape optimization through deep reinforcement learning. J Comput Phys 428:110080. https://doi.org/10.1016/j.jcp.2020.110080

Wang S, Li H, Xiao LP, Song G (2020) Unraveling the structural transformation of wood lignin during deep eutectic solvent treatment. Front Energy Res 8:48

Wen Q, Chen JX, Tang YL, Wang J, Yang Z (2015) Assessing the toxicity and biodegradability of deep eutectic solvents. Chemosphere 132:63–69

World Biogas Association (2018) The contribution of anaerobic digestion and biogas towards achieving the UN Sustainable Development Goals. World Biogas Association, London, UK

Xing J, Luo K, Wang H, Fan J (2019) Estimating biomass major chemical constituents from ultimate analysis using a random forest model. Bioresour Technol 288:121541. https://doi.org/10.1016/j.biortech.2019.121541

Xu GC, Ding JC, Han RZ, Dong JJ, Ni Y (2016) Enhancing cellulose accessibility of corn stover by deep eutectic solvent pretreatment for butanol fermentation. Bioresour Technol 203:364–369

Xu H, Peng J, Kong Y, Liu Y, Su Z, Li B, Song X, Liu S, Tian W (2020) Key process parameters for deep eutectic solvents pretreatment of lignocellulosic biomass materials: a review. Bioresour Technol 310(2020):123416

Yang GH, Wu TY, Loow YL, Lin YA (2016) Application of deep eutectic solvents in biomass valorization. In: The 4th international conference on sustanable solid waste management (CYPRUS), Malaysia

Yogalakshmi KN, Mohamed Usman TM, Sachdeva S, Thakur S (2023) Lignocellulosic biorefinery technologies: a perception into recent advances in biomass fractionation, biorefineries, economic hurdles and market outlook. Fermentation 9(3):238

Yu Q, Qin L, Liu Y, Sun Y, Xu H, Wang Z, Yuan Z (2019) In situ deep eutectic solvent pretreatment to improve lignin removal from garden wastes and enhance production of bio-methane and microbial lipids. Bioresour Technol 271:210–217

# 7 Development of Novel Sustainable Waste-Based Adsorbents for Water Treatments

Oscar E. Medina, Yurany Villada Villada, Farid B. Corteś, and Camilo A. Franco

**Abstract**

This chapter focuses on utilizing waste materials as potential adsorbents for water treatment, thereby offering a twofold solution—effective removal of contaminants from water and reducing the waste burden on the environment. The research involves exploring and characterizing different waste-derived materials, such as agricultural residues, industrial by-products, and discarded materials, to evaluate their adsorption capacities for various water pollutants. Through a comprehensive series of laboratory experiments and analyses, the adsorption efficiency of these novel waste-based adsorbents is evaluated concerning different pollutants like heavy metals, organic compounds, and emerging contaminants. The study also investigates the influence of various factors, including pH, temperature, contact time, and adsorbent dosage, on the adsorption process, enabling the optimization of treatment parameters for maximum efficiency. Additionally, the work delves into computational models and artificial intelligence for removing pollutants in waste-derived adsorbents to provide valuable insights for enhancing the adsorbent's performance and designing future water treatment systems. The environmental sustainability and economic viability of waste-based adsorbents are assessed, considering factors like availability, cost-effectiveness, life cycle assessment, and ease of regeneration. Furthermore, the study evaluates the potential for large-scale implementation of waste-based adsorbents in practical water treatment scenarios, considering their compatibility with existing treatment processes. In conclusion, this

O. E. Medina (✉) · Y. V. Villada · F. B. Corteś · C. A. Franco
Grupo de Investigación en Fenómenos de Superficie Michael Polanyi, Facultad de Minas, Universidad Nacional de Colombia Sede Medellín, Kra 80 No. 65-223, 050034 Medellín, Colombia
e-mail: oemedinae@unal.edu.co



M. Nasr and B. Ravindran (eds.), *Solid Waste Management in Delta Region for SDGs Fulfillment*, Deltas of the World, https://doi.org/10.1007/978-3-031-58253-0_7

research showcases the significant promise of waste-based adsorbents as efficient and eco-friendly solutions for water treatment.

## 7.1 Introduction

In the face of escalating environmental concerns and mounting challenges in water quality management, the search for innovative and sustainable solutions has intensified (Zhou et al. 2019a). The treatment of water resources to ensure their safety and purity is paramount for human health and environmental preservation. The ever-increasing challenges posed by population growth, industrialization, and urbanization have propelled the need for innovative and sustainable solutions in water treatment technologies (Dutta et al. 2012). Among these solutions, developing novel waste-based adsorbents has emerged as a promising avenue to address the complex and diverse array of contaminants in water sources (Medina et al. 2021a, 2021b; Giraldo et al. 2023).

Using waste materials as adsorbents addresses the pressing issue of waste management and provides a unique opportunity to create value from materials that would otherwise contribute to environmental degradation (Kooh et al. 2016). Their inherent abundance, renewability, low cost, and eco-friendliness make them an ideal alternative to conventional adsorbents. The concept of waste-based adsorbents capitalizes on the intrinsic properties of various waste materials, such as agricultural residues, industrial by-products, and discarded biomasses, to adsorb and remove pollutants from water. This approach aligns with the principles of green chemistry and circular economy by transforming waste into a valuable resource for water treatment applications. Researchers have made substantial strides in harnessing various biomass sources, such as agricultural residues, forest waste, and industrial by-products, to fabricate adsorbents with tailored physicochemical properties (Ma et al. 2020; Rodriguez Acevedo et al. 2020; Sherugar et al. 2022; Olugbenga et al. 2023). The advancement of biomass adsorbents is tightly interwoven with material science and engineering principles. Researchers have delved into modifying and functionalizing biomass materials to enhance their surface area, porosity, and adsorption capacities. Techniques such as carbonization, activation, and chemical modification have been explored to create adsorbents with well-defined structures and improved adsorption performance (Mudhoo et al. 2019). This has led to the development of materials like activated carbon, biochar, and hydrothermal carbon as effective adsorbents for a wide range of contaminants, including heavy metals, organic pollutants, and emerging micropollutants (Li et al. 2019; Wang et al. 2019).

However, biomass adsorbents extend beyond material synthesis. Researchers are dedicated to unraveling the intricate mechanisms governing adsorption processes, with a keen eye on kinetic and equilibrium modeling (Desai et al. 2019; Kumari et al. 2019). Understanding these mechanisms is crucial for optimizing adsorbent performance and predicting their behavior in complex water matrices. Moreover, the advent of computational techniques has enabled the exploration of adsorption phenomena at the molecular level, providing insights into surface interactions and driving forces (Afolabi et al. 2020a; Coruh et al. 2014).

Integrating biomass adsorbents into practical water treatment systems is another pivotal aspect of the landscape. Researchers are actively engaged in translating laboratory-scale successes to pilot and full-scale applications. Challenges related to adsorbent regeneration, scalability, and compatibility with existing treatment processes are being addressed to ensure the feasibility of large-scale implementation (Li et al. 2019; Wang et al. 2019; Desai et al. 2019; Kumari et al. 2019).

Developing effective waste-based adsorbents necessitates a multidisciplinary approach involving materials science, chemistry, engineering, and environmental science. This approach encompasses the design, synthesis, characterization, and application of adsorbents with high affinity for target contaminants, robust adsorption capacities, and economic viability (Zhou et al. 2019a; Chakraborty et al. 2013). This scientific exploration embarks on an investigation into the development of novel waste-based adsorbents for water treatments. By delving into the intricate interplay between waste material properties, adsorption mechanisms, and water treatment efficiency, this research seeks to advance our fundamental understanding of waste-based adsorption processes and pave the way for translating this knowledge into practical and scalable solutions.

The main objective of this study is to explore the recent developments in waste-based adsorbents for water purification based on an exhaustive review. With growing concerns about water scarcity and pollution, the quest for innovative materials that can effectively and efficiently remove contaminants from water sources has gained paramount importance. The following information was considered to accomplish the objective.

Section 7.2 synthesis and characterization of waste-derived adsorbents; Sect. 7.3 Water treatment with waste-derived adsorbents. Here, the adsorption efficiency of these novel waste-based adsorbents is analyzed concerning different pollutants like heavy metals, organic compounds, and emerging contaminants and external factors that influence the adsorption mechanism like pH, temperature, contact time, among others. Section 7.4 Regeneration studies of the adsorbents, Sect. 7.5 Computational models and artificial intelligence for removal of pollutants in waste-derived adsorbents to provide valuable insights for enhancing the adsorbent's performance and designing future water treatment systems. Section 7.6 Discussion of the environmental sustainability and economic viability of waste-derived adsorbent application. Embark on a voyage of scientific exploration as we unveil the intricate processes behind these novel adsorbents' design, synthesis, and application, offering a glimmer of hope for a cleaner and more sustainable future.

## 7.2 Synthesis and Characterization of Waste-Derived Adsorbents

Figure 7.1 presents the main waste-based adsorbents applied in water treatments. Agricultural waste offers an available, accessible and renewable source for the obtaining adsorbent precursors, using as strategies for the water treatment (Solangi et al. 2021). Agricultural wastes are solid residues derived from agricultural activities. They are composed of lignocellulosic materials (cellulose, hemicellulose, pectin, and lignin), proteins, lipids, starch, and simple sugars. The application of these materials in water treatment has demonstrated a high remotion percentage of organic and inorganic pollutants due to their porosity and the presence of several functional groups on their surface (Ambaye et al. 2021a). Specifically, lignin contains aliphatic and phenolic hydroxyl, methoxyl groups, and carbonyl groups contributing to the increases in the adsorptive properties. The investigations reported using agro-wastes to produce adsorbents from raw biomass, or chemically and/or physically modified biomass (Singh and Nagpal 2018; Jain et al. 2021). In general, biomass is transformed into activated carbon through the carbonization process and an activation stage (physical or chemical).

Actually, the use and requirement of food and fruit have reached billions of tons per year for human consumption. This promotes the generation of high amounts of fruit waste including peels, leaves, stalks, seeds, and stones, among others. Several research studies have shown that large quantities of waste promote environmental cost pollution for waste management (Bhatnagar et al. 2015). The fruit wastes are composed of lignocellulosic materials. Particularly their percentage of carbon is around 50%, which can be converted to activated carbons through thermal treatment processes for water purification (Nyika and Dinka 2023).

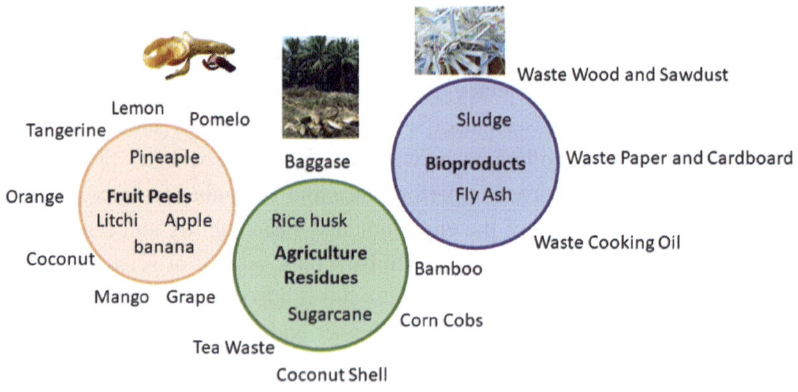


**Fig. 7.1** Waste-based adsorbents types applied in water treatments

Several methods have been used for the synthesis of waste-based adsorbents, including *pyrolysis, hydrothermal activation, precipitation, and carbonization*. The activation of the carbonized product involves a chemical process with KOH, $H_3PO_4$, and $ZnCl_2$, or physical/thermal activation at high temperatures (800–1000 °C) utilizing hot gases such as $CO_2$ or water vapor steams (Solangi et al. 2021; Hamad and Idrus 2022; Mo et al. 2018; Ahmad and Danish 2018). The activation stage alters the adsorbent surface adding new functional groups such as carboxyl, lactone, phenol, and carbonyl. The selection of the activation method and the activating agent is based on the type of waste and its physiochemical characteristics.

Fly ash, metal hydroxide sludge, biosolids, red mud, and waste slurry are industrial products used as dye adsorbents due to their high performance and low cost. Particularly, fly ash is an industrial waste generated during the combustion process. Likewise, the red mud is an industrial byproduct obtained from the production of alumina (Aigbe et al. 2021).

### 7.2.1 Characterization of Waste Derived Adsorbents and Byproducts

Characterization of synthesized adsorbents is performed using several characterization techniques such as scanning electron microscopy (SEM), energy dispersive X-ray spectroscopy (EDX), Brunauer–Emmett–Teller (BET), Fourier transmission infrared spectrometry (FTIR), X-ray diffraction (XRD), Elemental analysis, ultraviolet–visible (UV–Vis) spectrophotometer (Lv et al. 2023). A brief description of each technique is presented as follows.

*Scanning electron microscopy (SEM)*. To observe the morphology of the absorbents and the surface characteristics (swelling, agglomeration, surface porosity, and molecular arrangement) SEM technique is employed. Likewise, SEM can be coupled with other with energy dispersive X-ray analysis to evaluate the formation or adsorption of compounds. It is important to note that the morphology of bio-adsorbents changes with synthesis processes and the employed activating agents. For adsorption evaluation of the adsorbent materials, the surface morphology and cavities at the surface of the adsorbents fulfill a significant role in adsorption. However, after the adsorption process, the surface of the adsorbent exhibits some changes only appreciable with high-resolution FESEM images (Pellenz et al. 2022).

Farias et al. (2023) investigated the morphological properties of the banana peel powder (BPP) biosorbents for efficient sorption of methylene blue (MB), atrazine, and glyphosate pollutants. The micrographs showed an irregular morphology evidencing the presence of agglomerated and microsized particles with smooth surfaces as well as microporous structures. The BPP showed more agglomerated particles.

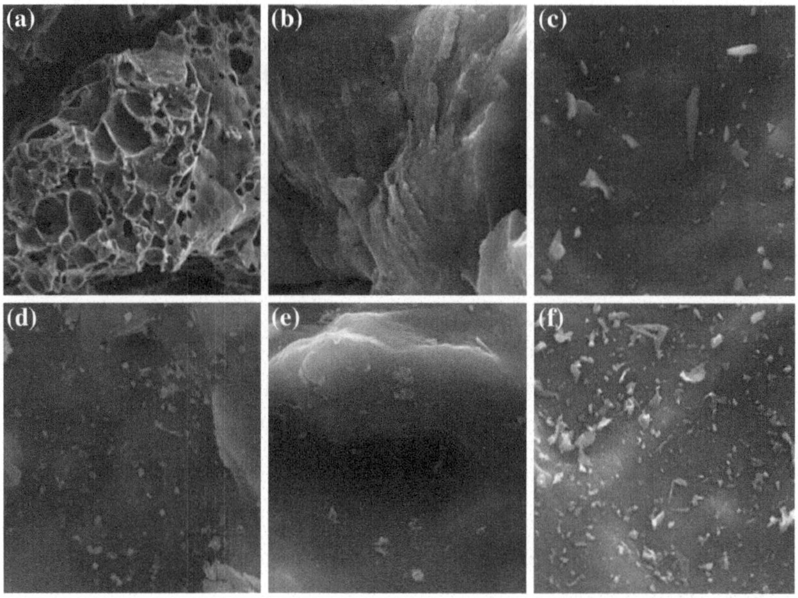


**Fig. 7.2** SEM images of the adsorbent surface before and after heavy metal adsorption. The micrographs were adapted from Gomravi et al. (2021)

Gomravi et al. (2021) studied the use of apple waste as an adsorbent of heavy metal ions. The authors evaluated the morphologic changes of the adsorbent surface before and after heavy metal adsorption. The results showed no presence of the heavy metal on the adsorbent surface before the adsorption process. Contrary, the presence of heavy metal small particles was observed on the adsorbent surface and confirmed that the adsorption process by adsorbent has been successful (see Fig. 7.2).

*Brunauer–Emmett–Teller (BET).* The surface area, pore size, and pore volumes are the main parameters to estimate the adsorption characteristics of the synthesized adsorbents. An adsorbent with a high surface area, and pore volume promotes to increase the adsorption of organic impurities. Likewise, the analysis of the internal structure mainly concerns the pore structure of the adsorbent. The variety of pore sizes such as micropores (pore size $\leq 2$ nm), mesopores (2 nm $\leq$ pore size $\leq$ 50 nm), and macropores (pore size > 50 nm) have a significant impact on the adsorption process (Kali et al. 2022a).

Tummala et al. (2023) investigated the feasibility of using iron-coated nutshells as adsorbents for phosphate removal and their potential for fertilizer application after adsorption. The authors assessed the surface area of adsorbents on the unmodified and iron-coated nutshells through BET. The surface area of the iron-coated shells increased by an order of magnitude, from 0.3–0.6 $m^2\ g^{-1}$ to 3.3–4.0 $m^2\ g^{-1}$. Likewise, the pore volume decreased from 0.0004 $cm^3\ g^{-1}$ for walnut shells to 0.004–0.005 $cm^3\ g^{-1}$ for the iron-coated walnut and pistachio shells. The average pore sizes were 4.9 and 4.8 nm for walnut shells and iron-coated walnut shells, respectively.

*Fourier transmission infrared spectrometry.* Fourier transmission infrared spectrometry (FTIR) is used to identify the functional groups associated with the adsorption bands of bio-adsorbents. FTIR is used for a wide variety of biomass adsorbents. Additionally, several bi-adsorbents are modified physicochemically to improve their adsorptive capacity. In this sense, the FTIR technique permits the identification of these functional groups to guarantee that the chemical or physical modification is appropriate. Besides, FTIR contributes to the determination of functional groups of the adsorbent that participate in chemical or physical interactions with the adsorbate molecules. Researchers have applied this technique before and after the adsorption process to evaluate the changes in the transmittance peaks as a result of the adsorption of pollutants (Rajesh and Jeeru 2022).

*X-ray diffraction.* X-rays diffraction technique is employed to identify the lattice structure of a material. To this end, X-rays penetrate at different angles on the sample and provide the patterns of diffracted peaks related to material crystallinity and its unit cell structure (Njewa et al. 2022). Sahu et al. (2020) synthesized activated carbons from Kendufruit peel for the removal of methylene blue dye. The authors employed XRD to determine the crystalline structure of kendu fruit peel (KFP), ammonium carbonate-treated kendu fruit peel (MKFP), and methylene blue dye adsorbed. On the other hand, several works include other analytical techniques for the characterizations of the adsorbents such us thermogravimetric analyses (TGA), X-ray photoelectron spectroscopy (XPS), Ultraviolet–visible (UV–Vis) spectroscopy, and elemental analyzer. Specifically, TGA contributes to the determination of thermal stability adsorbents and XPS determines surface composition and elemental chemistry (Singh et al. 2017).

Wang et al. (2022) evaluated the efficient removal of Pb(II) and Cd(II) from aqueous solutions by mango seed biosorbent. The authors studied the crystalline properties of adsorbents through DRX before and after the adsorption process (see Fig. 7.3). The results showed that the broad peak at $2\theta$ of 7–30° is characteristic of the amorphous phase of lignin, hemicellulose, and cellulose in the biosorbents. On the other hand, the distinct reflections at $2\theta$ of 15.3° and 23.0° and corresponded to mango kernel starch. The weaker and broader reflections in the XRD diffractograms after Pb(II)/Cd(II) adsorption revealed the reduced crystallinity of bioadsorbent. Finally, the author argued that the adsorption process mainly occurs in the amorphous region of material.

Kali et al. (2022b) studied the adsorption performance of orange peel and magnetized orange peel as adsorbents of crystal violet dye from polluted water. The authors characterized the adsorbent materials by DRX. The XRD patterns indicated the presence of amorphous materials due to the hemicellulose and lignin. The crystalline peaks at 15.3 and 22.3° proposed the existence of cellulosic material. The XRD pattern of the magnetized orange peel exhibited the same peaks orange peel and small intermittent peaks at 30.8, 36.9, 41.2, 49.56, 53.26, 55.4, 61.82, and 75.12° associated with the iron oxide of the magnetization.

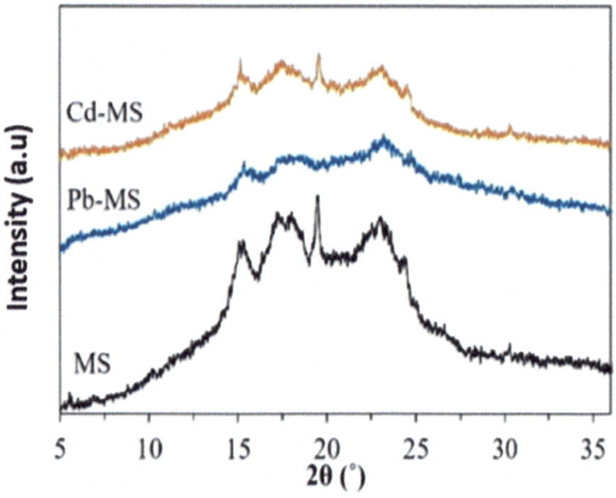


**Fig. 7.3** XRD diffractograms of mango adsorbent before and after Pb(II)/Cd(II) adsorption. The figure was adapted from Wang et al. (2022)

Reddy et al. (2016) evaluated the thermal stability of Corn cobs adsorbent to be applied in the methylene blue adsorption. The TGA analyses (Fig. 7.4) showed that the decomposition of material was performed in three stages. The first stage indicates a weight loss of 11.23% between 596–656°K related to the loss of moisture. The second stage represented a weight loss of 46.5% in the range of 656–946 °K. This result can be attributed to the decomposition of chemically bound water, cellulose, hemicelluloses, and lignin. The third stage showed a weight loss of 35.2% for temperatures greater than 946 °K, and it could correspond to the formation of volatile products.

Akköz and Coşkun (2023) evaluated cellulose fibers as a bioadsorbent for the remotion of anionic dyes. The TGA/DTA thermograms bioadsorbents indicated that the decomposition processes occur in a single step. The thermal decomposition process started at 330 °C

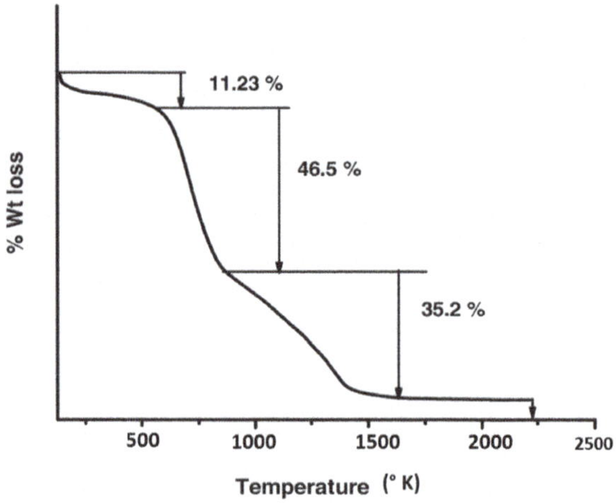


**Fig. 7.4** Thermogravimetric analysis of corn cob adsorbent. The figure was adapted from Reddy et al. (2016)

and ended at 400 °C. Likewise, the maximum decomposition temperature was determined at 360 °C.

XPS is widely used to characterize the chemical composition of the surface of adsorbent materials, evaluate the several synthesis and functionalization processes, as well as the characterization of an adsorbent saturated with the adsorbate. Studies have reported binding energies of 283.8–289.0 eV for $C_{1s}$ (C–OH/C = O/ O–C = O/C–C/C = N/ = C–NH), 531.1–533.8 eV for $O_{1s}$ (C–O–C/C = O/C–O–H); 397.9–401.6 eV for $N_{1s}$ (–NH–/ = N–/–$NH^{+-}$/–$NH^{3+}$/C–N/amine/amide/pyrrolic N/graphitic N); 709.8–711.4 eV for $Fe_{2p3/2}$, 724.3–725.2 eV for $Fe_{2p1/2}$ (Pellenz et al. 2022). In general, XPS analysis can be complementary to FTIR analysis in the chemical characterization of adsorbents.

The FTIR absorption peaks related to bending and stretching of a specific bond occur in different wavenumbers and different shapes (strong/weak; broad/sharp) depending on the functional group and structure. Several studies reported FTIR wavenumbers related to specific bonds (Pellenz et al. 2022). The main groups are: O–H stretching (2500–3750 $cm^{-1}$), O–H bending (1160–1440 $cm^{-1}$), C–H stretching (2600–3000 $cm^{-1}$), C–H bending (1280–1465; 1650–2000 $cm^{-1}$), C = O stretching (1560–1815 $cm^{-1}$), C–O stretching (900–1400 $cm^{-1}$), N–H stretching (2800–3500 $cm^{-1}$), N–H bending (1400–1660 $cm^{-1}$), S–H stretching (2550–2600 $cm^{-1}$). Particularly, the activated carbon showed the presence of carboxylic acids, quinones, phenolic groups, lactones, ketones, ethers, pyridine-like groups, amides, and amines, among others. Furthermore, the functionalization of adsorbents to increase their adsorption sites is performed with oxygen, nitrogen, and sulfur-containing groups. The comparison between signal strengths of different groups can be used to quantify the formation of the functional groups through FTIR.

Uddin et al. (2017) studied the use of mango leaf powder as a potential adsorbent for the removal of methylene blue from aqueous solutions. The author employed FTIR as an characterization technique for the adsorbent (see Fig. 7.5). The spectra displayed a number of absorption peaks, indicating the complex nature of the material examined. The broad band observed at 3435 $cm^{-1}$ corresponded to the stretching vibrations of O–H bond in hydroxyl groups. The O–H stretching vibrations occurred within a broad range of frequencies that indicate the presence of phenols and alcohols, which confirms the presence of cellulose and lignin in the adsorbent. The bands observed at 2918 $cm^{-1}$ were assigned to asymmetric C–H bands present in alkyl groups. The peak at 1707 $cm^{-1}$ was characteristic of carbonyl group stretching of carboxylic acid. The peak at 1623 $cm^{-1}$ was attributed to bending vibration of molecular water. The band at 1383 $cm^{-1}$ was related to COO- symmetric stretching vibration group. The peaks that appeared in the region 1320–1210 $cm^{-1}$ were assigned to the stretching vibration of C–O in carboxylic acid. Thus, the FTIR analysis indicated that the adsorbent contained functional groups such as OH and COOH that contribute to the adsorption sites for interaction with the cationic dye.

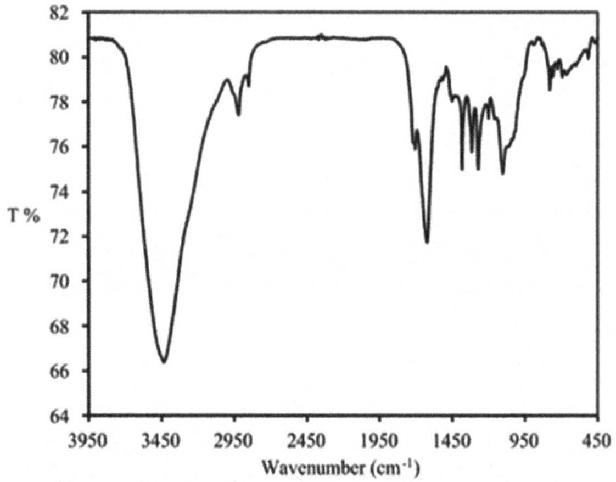


**Fig. 7.5** FTIR spectra of mango leaf powder adsorbent. The figure was adapted from Uddin et al. (2017)

## 7.3 Water Treatment with Waste-Derived Adsorbents

Water pollution due to agricultural, industrial, and domestic activities has caused significant human and environment issues. Several water treatment technologies have been employment. However, adsorption still remains the most efficient process in pollutants removal from water due to its simplicity in design and implementation and low cost (Mekouar 2018).

### 7.3.1 Type of Waste-Derived Adsorbents

The use of agricultural wastes and plants to adsorb organic and inorganic polluants is considered an attractive alternative to conventional wastewater treatment procedures (Amalina et al. 2022). In general, dead or living agro-waste, algae, fungi, fruit peels, seeds, leaves, straw, sawdust, bark, sludge, and ash, cashew nutshell, dalbergia sissoo, ficus benghalensis, cow hooves, lentil husk, pomegranate peel, anadara inaequivalvis shells, chitosan-based sorbent, eggshells and eggshell membrane, papaya peel, mangifera indica have been used as adsorbents (Nayagam 2022; Kosheleva et al. 2018; Rajan and Anish 2022). On the other hand, by-products such as sugar beet pulp (SBP) and sugarcane bagasse (SCB) obtained from sugar have been applied in the pollutant removal from water (Muthanna 2017).

Ahmad et al. (2023) investigated the application of agro-waste-based adsorbents as sustainable alternatives for the removal of ibuprofen and carbamazepine from contaminated waters. The author highlights the mechanisms implicated in the adsorption of ibuprofen and carbamazepine, as well as the operational parameters that contribute to the efficient design of the adsorption process. The results showed that the obtained data not allwe the selction of the best adsorbent due to the comparison could include the production steps, the adsorbent yield, and the main experimental conditions (pH and the temperature). Likewise, it is important to note that high adsorption capacity depends on acidic/basic pH and/or temperatures. For this reason, the studied parameters needs to be considered prior to scaling up process.

Anastopoulos et al. (2017) studied the use of sugar waste-based adsorbents in water treatment. To this end, they studied the isotherms, kinetics, desorption, and thermodynamic information of the adsorption process. Besides, the effect of initial concentration, pH, contact time, temperature, and adsorbent dosage, on the adsorption efficiency are also evaluated. The authors concluded that the performance of the adsorbents slightly depended on the type of pollutants and experimental conditions. Likewise, the higher removal of dyes was at pH 7–9, 1 h, and 45 °C. On the other hand, they found that surface modification with chemicals promotes the removal efficiency favoring the kinetics and adsorption mechanism.

Hamad and Idrus (2022) evaluated the use of the bio-waste-derived adsorbents for methylene blue removal, as well as the parameter for the performance of process. The work highlighted the regeneration processes, economic challenges, and the valorization of post-sorption materials. The autors argued that bio-waste has emerged as a low cost, effective, and renewable source of activated carbon for the removal of methylene blue with efficiency around 90%. Low-cost bio-waste-derived adsorbents are defined in terms of the initial costs, local availability, stability, eco-friendliness, transportation, applied treatment processes, recycling, lifespan concerns, regeneration potential, and pore volume after deactivation. On the other hand, the authors concluded that the pH level (5–10), the initial dye concentration (65–100 mg $L^{-1}$), temperature (25–45), adsorbent dose (2–6 g $L^{-1}$), type, and contact time (60–300 min) had a significance effect in the the adsorption of cationic dyes due to the chemical nature. Likewise, steam activation was the most cost-effective methodology for the adsorption process, whereas chemical activation promote the high porosity and surface area.

Bilal et al. (2021) recopiled the results of several studies in a paper type review. In particular, the modern developments in applications of bio-sorbents for water purification with an emphasis on adsorption were the topic. To this end, the heavy metals adsorption on different bio-waste derived adsorbents was evaluated through an adsorption mechanism and their regeneration abilities. The results were compared with conventional adsorbents such as carbon nanotubes, metal organic frameworks, graphene, boron nitride, carbon nitride and dendrimers and biadsorbents. Despite the fact that conventional adsorbents exhibit excellent performance in the adsorption of heavy metals for the water purification, the biadsorbents have an advantage of low cost, easy availability, and synthesis.

### 7.3.2 Factors that Influence the Adsorption Process

Several authors have concluded that the adsorption procees is influenced by the dosage of adsorbent, the initial solution pH, the contact time, and the temperature (Bello et al. 2021; Jia et al. 2013; Ambaye et al. 2021b; Wang et al. 2015; Fagbohungbe et al. 2017; Qiao et al. 2018). These parameters are discussed in detailed.

*The effect of adsorbent dosage.* The optimization of adsorbent dosage is an essential parameter in the design of an efficient adsorption system. The adequate adsorbent dosage determines the equilibrium relationship between adsorbent-adsorbate and predicts the associated cost of the process (Bello et al. 2021).

*Effct of pH solution.* The pH of solution has a significant effect in the adsorption mechanis due of change in surface properties of material adsorbent (Jia et al. 2013).

*Effect of particle size.* The physical modification of adsorbents can modify the sorption capacity of adsorbents. It was reported that smaller particle size adsorbents exhibit higher surface area that promotes high active vacant sites to enhance sorption capacity (Ambaye et al. 2021b).

*The effect of contact time.* The sorption capacity of an adsorbent increases with the contact time increase. During the adsorption process, the maximum sorption occurs at the initial stage, when the vacant adsorbent sites are abundant. However, the rate of adsorption decreases with the saturation level of the adsorbent. The appropriate selection of the time for the adsorption process has a significant influence on the operation cost (Wang et al. 2015).

*Effect of temperature.* Temperature is an essential parameter in the adsorption mechanism, which determines the nature of the adsorption process (endothermic or exothermic). In general, the adsorption processes are endothermic. As the temperature increases the sorption capacity of bio-adsorbent enhances. The increase in sorption capacity could be the fact the to enlargement of active sites on the surface of the adsorbent or due to mobility of pollutant molecules (Fagbohungbe et al. 2017).

The banana peel has been the main adsorbent used in the water treatment. Figure 7.6 shows several parameters that should be considered in the adsorption process for the remotion of metals dyes, pesticides, pollutants, and radionuclides from water. The studies include the evaluation of different parameters such as adsorbent dosage, temperatures, contact times, and pH solution. The results indicated that the efficiency of the adsorbent is in the range of 89–99% for evaluated parameters. Likewise high dosage of banana peel are requerid for the removal of pesticides. On the contrary, low dosages of adsorbents are used in the radionuclides removal. The temperature varies from 25 to 55 °C and the pH adequate for achieving the maximum removal is neutral or basic pH (Qiao et al. 2018).

The parameters involved in the water treatment using orange, kiwi, and tangerine peel for the removal of heavy metals are presented in Fig. 7.7. All adsorbents exhibit a removal

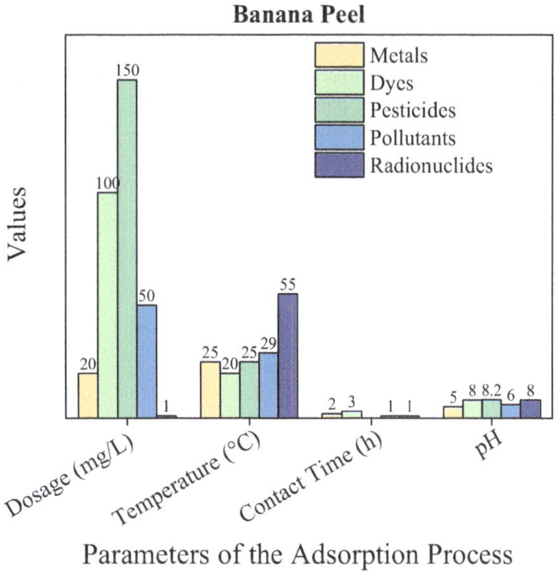


**Fig. 7.6** Effect of dosage, temperature, contact time and pH in the removal of metals, dyes, pesticides, pollutants, radionuclides using banana peels (Farias et al. 2023; Akpomie and Conradie 2020; Gupta and Gupta 2016)

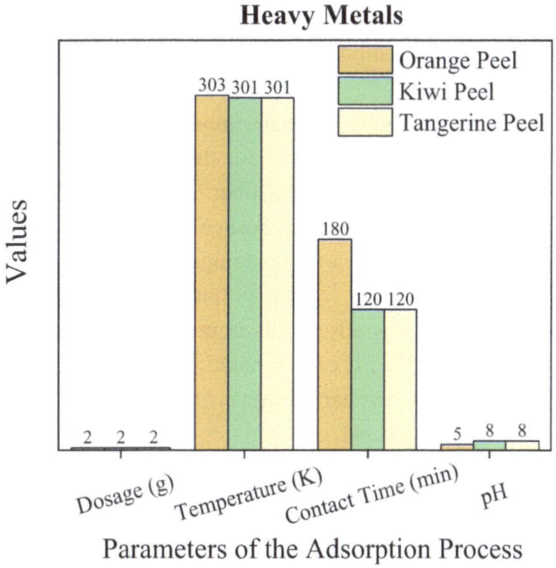


**Fig. 7.7** Effect of dosage, temperature, contact time and pH in the removal of heavy metals using orange, kiwi, and tangerine peels (Feng et al. 2020; Al-Qahtani 2016)

efficiency between 90 and 95%. The values of dosage, temperature, contact time and pH using kiwi and tangerine peel are the same for the heavy metal removal. However, the parameters using orange peel varies for the contact time and pH.

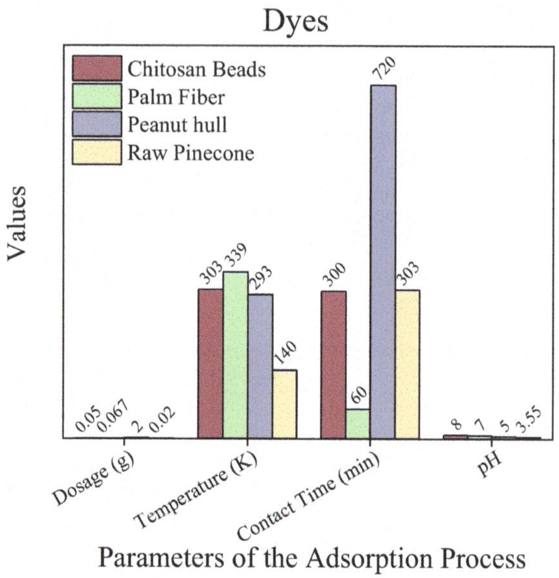


**Fig. 7.8** Effect of dosage, temperature, contact time, and pH in the removal of dyes using chitosan beads, palm fiber, peanut hull, and raw pinecone adsorbents (Bekçi et al. 2008; Laskar and Kumar 2022; Gong et al. 2005; Dawood and Sen 2012)

The authors found that the values of the dosage, temperature, contact time, and pH used to achieve higher efficiency in the adsorption process are 2 g, 301 K, 120 min, and 8, respectively.

Figure 7.8 shows the water treatment with chitosan beads, palm fiber, peanut hull, and raw pinecone as adsorbents for the removal of dyes. The removal efficiency percentage using chitosan beads, palm fiber, peanut hull, raw pinecone is 99, 90, 96.32, and 60%, respectively. Note that the dosage of adsorbent in all cases is low and the pH is around of 7.0. Likewise the average temperature is 303 K.

The removal of organic pollutants from water using palm, paper mill sludge, and coconut shells as adsorbents is presented in Fig. 7.9. The studies showed removal procent around 90%. It can be noted that the values of dosage, and temperature, using these adsorbents are similar. However, there are significant differences between the values of contact time and pH.

Overall, the removal of heavy metals required a higher dosage of adsorbents than the removal of dyes and organic pollutants. However, the contact time, pH, and temperature employed in the adsorption process are similar.

## 7.4 Regeneration Studies

The source, method, and dosage of the adsorbent are all factors that affect how the adsorbent is regenerated. The desorption and decomposition principles can be used in a regeneration study to recover the adsorbates (Bello et al. 2021; Jia et al. 2013;

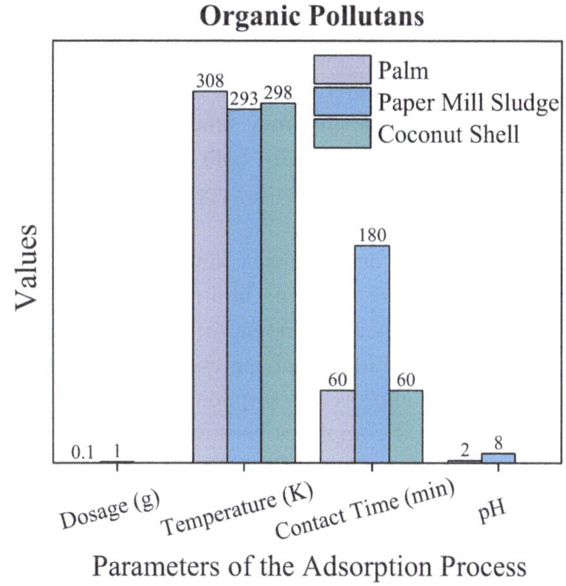


**Fig. 7.9** Effect of dosage, temperature, contact time, and pH in the removal of organic pollutants using palm, paper mill sludge and raw coconut shell (Bekçi et al. 2008; Sathishkumar et al. 2007)

Ambaye et al. 2021b; Wang et al. 2015). Desorption is the process of removing adsorbent molecules or ions from the surface of a material, such as an adsorbent, through various methods. It is the reverse of adsorption, where substances are captured or retained on the surface of a solid or liquid. Desorption can occur through techniques like heating, flushing with a desorbing agent, or changing environmental conditions to release the adsorbent species from adsrobent's surface. Decompositon on the other hand, refers to the breakdown or disintegration of a substance into its constituent parts or simpler compounds. This process can occur through various mechanisms, such as thermal decomposition (breaking down due to heat), chemical decomposition (involving chemical reactions), or biological decomposition (by microorganisms). Decomposition can lead to the formation of different products or byproducts depending on the nature of the original substance and the conditions under which it decomposes.

For regeneration (removal of pollutants) of spent bio-adsorbents, using a low concentration (0.1–0.2 M) of acids or bases is suggested. Acids and bases that have been used as regenerative solvents are HCl, $HNO_3$, $H_2SO_4$, EDTA (ethylenediaminetetraacetic acid), $Ca(NO_3)_2$, NaOH, and $NaNO_3$. Regenerative solvents such as HCl, $HNO_4$, $H_2SO_3$, EDTA (ethylenediaminetetraacetic acid), $Ca(NO_3)_2$, NaOH, and $NaNO_3$ have all been employed (Gupta et al. 2020). It was demonstrated by Hassan et al. (2020) and Baig et al. (2019) that As could be removed from magnetic sorbents and that 0.5 M NaOH could be used to regenerate the adsorbents. When used as regenerative solvents, HCl, $HNO_3$, and $H_2SO_4$ showed a notable desorption efficiency. Adsorbent regeneration is enhanced by a low pH, which also causes metals to desorb from the adsorbent surface. Strong acids cause the hydronium ion ($H_3O^+$), the hydrogen ion ($H^+$), and heavy-metal (HM) ions to compete for active sites. For instance, Kołodyńska et al. (2017) found that 95% Cu desorption efficiency was achieved when 3.5 M $HNO_3$ was used as an eluent. Conversely, environments with higher acidity may simultaneously deform the adsorbent structure, leading to a decrease in the capacity for both adsorption and desorption. For example, Kołodyńska et al. (2017) discovered that the application of 3.5 M $HNO_3$ as an eluent resulted in 95% Cu desorption efficiency.

Higher acidity environments, however, have the potential to simultaneously deform the adsorbent structure, resulting in a reduction in both adsorption and desorption capacity. Therefore, acid treatment ought to be reserved for sorbents with respectable mechanical qualities and stability. Applying chelators such as ethylenediaminetetraacetic acid could also result in improved desorption rates (Yang et al. 2020a). Chelators can form stable chelator-HM complexes because they have a variety of electron-donating functional groups, such as carboxy (COOH) and amine, which exhibit a strong attraction towards HM ions. Chelators are used to remove the adsorbed HM ions from the adsorbents, and these chelators and the adsorbed HM ions subsequently form complexes.

Hu and Shipley (2013) assessed the ability of nano-$TiO_2$ to desorb $Pb^{2+}$, $Zn^{2+}$, and $Cu^{2+}$ using EDTA and a mixed solution (NaCl, $NaNO_4$, $CaCl_2$, $NaHCO_3$, $MgSO_4$, and $NaHPO_4$) of common ions. While the EDTA chelator demonstrated 92% desorption efficiency, the application of a solution of common ions had only nominal desorption efficiency. Better regeneration cycles transpired, most likely as a result of EDTA's strong chelating properties. The creation and activation of new adsorption sites in the adsorbent by EDTA-4Na may have contributed to this increased adsorption capacity.

The thermal regeneration method has recently been acknowledged as the best method for recovering biomass because, especially when thermal heating is finished at high temperatures, it allows for the development of tiny pore sizes compared to the initial biomass's first pores. The scientists found that raising the temperature enhanced regeneration effectiveness. In a study on removing organic compounds like pyrene and benzopyrene using biomass adsorbent derived from *Enteromorpha prolifera* and regeneration studies conducted at various temperatures of 80, 150, and 200 °C, an increase in temperature improved the regeneration efficiency. According to research by Qiao et al. pyrene's adsorption efficiency increased by 35.0%, 45.0%, and 48.0%, while benzopyrene's increased by 31.0, 41.0, and 40.0%) (Anastopoulos et al. 2017). Additionally, it was discovered that biomass regeneration using the thermal method could lower dissolved organic carbon. The use of microwave irradiation has also been used to regenerate biomass. The results obtained demonstrate how quickly the biomass was renewed. Sun et al. (2017) used granular activated carbon for chloramphenicol removal usin microwave radiation at 2450 MHz for 10 min. The addition of an electrodeless lamp to the regeneration reactor increased the percentage of chloramphenicol mineralization from 5 to 37%. Furthermore, 83% of the chloride present in chloramphenicol has the ability to change into inorganic chloride. The oxidizability of the microwave regeneration process can be enhanced by adding UV radiation to the reactor. Furthermore, GAC's adsorption capacity can continue to function well even after five cycles of absorption and regeneration. Zhang et al. (2014) used magnesium hydroxide-coated pyrolytic biochar for anionic dye removal at 320, 480, and 640 W (microwave radiation) for 5 min. When used for 5 min at 320 W, 480 W, or 640 W, respectively, microwave irradiation of used magnesium hydroxide-coated pyrolytic biochar produced regeneration yields of 98.5%, 89.0%, and 85.5%. Foo and Hameed (2012) removed methylene blue dye using durian shell and jackfruit peel ACs by microwave radiation operated at 2.45 GHz for 3 and 4 min. Even after five adsorption–regeneration cycles, the regenerated activated carbons' adsorption uptake and carbon yield could remain between 181.43–207.57 mg $g^{-1}$ and 80.51–81.63%, respectively. The spent ACs' porous structure was effectively preserved by microwave treatment, allowing the original active sites and adsorption capacity to be restored.

In order to study biomass regeneration, the operating pressure was recently changed using supercritical fluids (SCF). Nowadays, the use of a SCF to regenerate used adsorbents is widespread and is being considered as a replacement for chemical-solvent and incineration processes. The SCF acts as a typical solvent in the soil matrix, aiding in the process of pollutant desorption. By lowering the pressure, the contamination is further condensed until it can be collected into a small volume container. The most preferred and widely used SCF is carbon dioxide ($CO_2$), which is inexpensive, non-hazardous, and incombustible. This approach has benefits like a small amount of biomass loss, a quick operating cycle, and low operating temperatures. Zhang et al. (2019) used alkali metal catalyst and supercritical water regeneration to regenerate activated carbon while utilizing $H_2O_2$. Remarkably, the samples that underwent supercritical water regeneration demonstrated a 107% regeneration efficiency and a greater specific surface area (813 $m^2$ $g^{-1}$) when compared to the fresh sample (765 $m^2$ $g^{-1}$). Furthermore, it has been discovered that the regeneration temperatures (385, 405, and 425 degrees Celsius), the $H_2O_2$ concentration, and the alkali metal concentration act as catalysts to alter the phenol's ability to adsorb contaminants. Supercritical $CO_2$ is used in another study by Carmona et al. (2014) to regenerate granular activated carbon. In this case, the temperature (45 °C, 60 °C) and pressure (6, 15, 20, 31 MPa) had an impact on the desorption yield of the contaminants. The highest desorption yields of phenol, 2-chlorophenol, 4-chlorophenol, and 2,4-dichlorophenol were found to be 97.9%, 68.3%, 71.5%, and 64.5% at 31 MPa and 45 °C, respectively. Applying a co-solvent in addition to an SCF was another method used to enhance solvent polarity and subsequently remove impurities from a spent sorbent. Both with and without ethanol as a co-solvent, the desorption of 4-nitrophenol (4-NP) and phenol from organically functionalized smectite has been accomplished successfully (Park and Yeo 1999). According to the study's findings, when 2.5% (v/v) ethanol was added to the reaction mixture at 70 °C and 413.6 bar of pressure, the percentage recovery increased to 90.8%. However, when no co-solvent was added, the contaminant desorption rate was only 73.6%. Similar to this, Salgn et al. (2004) used ethanol to remove salicylic acid from bentonite that had been organically functionalized. When no ethanol was present, the desorption efficiency was 76 weight percent, and when 10% (v/v) ethanol was applied, it was 98 weight percent.

The method is still in the experimental stage and needs to be scaled up due to the method's complexity, expensive materials, equipment, and high-pressure resistance. Additionally, due to its low operating costs and broad economic applicability, the thermal regeneration process appears to be the most effective of all the techniques used for the regeneration and management of biomass in earlier studies. When compared to other methods, carbon loss is extremely high. It is necessary to conduct additional research to concentrate on the economic worth of biomass used for adsorption.

Multiple studies on the adsorption of different organic pollutants were conducted because chemical modification has gained considerable popularity for enhancing the uptake of organic pollutants by adsorbents. These studies confirmed that this approach is feasible for converting the applicability of modified biomass adsorbents into industrial applications. This is based on effective adsorbent characteristics, such as the affordable or free availability of biomass material, improved porosities, and larger surface area. However, it is crucial to assess the risks these modified adsorbents pose to guarantee the safety of the environment worldwide. Following a successful adsorption process, the life cycle experiment through desorption studies is required to guarantee the number of cycles these adsorbents can complete before exhausting their maximum adsorption capacities. In order to eliminate the various imbalances that the spent (used) adsorbent poses to the nearby environment, multiple strategies, including catalytic technology, biological treatment, ion exchange, and advanced biophysical methods, have been developed for waste disposal (Xu et al. 2017; Ogunlalu et al. 2021).

## 7.5 Computational Models and Artificial Intelligence for Removal of Pollutants in Waste-Derived Adsorbents

The central composite design (CCD) and the Box-Behnken design (BBD) are the two most frequently used designs in the response surface methodology (RSM). Malachite green, methylene blue, azo, and anthraquinone dyes have been reported to be adsorptively removed from effluents using RSM (Song et al. 2015; Chowdhury et al. 2019; Ahmad et al. 2020). Additionally, research findings on the employment of RSM for the adsorptive removal of organic contaminants from aqueous systems as well as the adsorptive removal of acid dye using derived activated carbon for antibiotic (ciprofloxacin) removal from wastewater using rice husk have established the effective application of RSM as an efficient computational model in wastewater treatment (Zhou et al. 2019b; Amini et al. 2008; Niad et al. 2014). The effectiveness of RSM in examining the removal of methyl orange using modified sugar beet bagasse has been confirmed, and the results obtained demonstrate agreement between experimental and predicted values. The ideal pH, initial methyl orange concentration, and adsorbent dose determined by RSM for the removal of methyl orange from aqueous solutions were 2.51, 100 mg $L^{-1}$, and 0.37 g $L^{-1}$, with predicted adsorption capacity and adsorptive removal efficiency of 221.5 mg $g^{-1}$ and 51.8%, respectively (Ghorbani and Kamari 2017). The preparation of functionalized pomegranate fruit peels has been optimized using RSM using two-factor interaction models and quadratic models. When used to remove Remazol brilliant blue R dye, it produced 31.2% of activated carbon from functionalized pomegranate fruit peels and removed Remazol brilliant blue R dye 81.35% of the time. Analysis of variance was used to determine whether the model was appropriate. A strong correlation between experimental and predicted results was inferred from the percentage error between the actual and predicted results, which was calculated to be 1.92% (Ahmad et al. 2020).

In a related study, KOH impregnation and $CO_2$ gasification were used to prepare pomegranate fruit peel activated carbon (PFPAC), which was then used to model the adsorption of methylene blue (MB) from an aqueous system. According to the results, the ideal radiation power, activation time, and impregnation ratio for removing MB dye is 375 W, 4.5 min, and 0.9, respectively. 83.4% of the MB dye was removed under these ideal adsorption conditions, and 30.8% of the PFPAC was produced (Ahmad et al. 2021). To investigate the effects of operating conditions on the adsorptive removal of the dye, using eggshell in the adsorptive removal of Malachite green (MG) dye was studied using CCD. At an $R^2$ value of 0.9388, there was a significant degree of agreement between the predicted values and the experimental findings. At operating conditions of 1.25 g adsorbent dosage, pH 6, and contact time of 40 min, the best adsorptive removal was 90.66%. According to the study's findings, eggshells may work well as an adsorbent to remove MG from aqueous solutions (Hoo et al. 2022). According to reports, the second-order polynomial equation of CCD has been used to optimize the adsorption parameters for the adsorptive removal of Congo red, brilliant green, and methylene blue with their variable levels. The predicted values for temperature, pH, adsorbent dose, and the percentage of adsorptive removal of dye using CCD were 30 °C, 9.8, 2.5 g $L^{-1}$, and 99%. The $R^2$ of 0.9966, which was higher than the $R^2$ of 0.88 suggested as the best correlation from previous studies (Fegousse et al. 2019) provided support for the model's qualitative fit. To remove MB from the aqueous solution, activated carbon made from crushed grape leaves was used. Utilizing RSM, the experiment's design and data modeling were completed. The obtained results indicate that at the adsorption conditions of adsorbent dose 12.5 g $L^{-1}$, MB concentration 100 mg $L^{-1}$, pH 11, and contact time 90 min, the optimal percentage removal of MB was 97.4%. Additionally, the results show that initial dye concentration and adsorbent dosage significantly boost MB adsorption efficiency (Mousavi et al. 2022). Using a Box-Behnken design, the low-cost adsorbent green pea was examined for the adsorptive removal of the dye reactive blue 19 to determine the effects of independent and combined influences of process variables, including temperature, pH, and adsorbent dosage. The RSM experiment results showed that at the ideal temperature (35 °C), pH level 2, and adsorbent dosage (1.5 g/100 mL), the maximum removal was 99.42%. In addition, the kinetic analysis demonstrated that the Freundlich isotherm, which followed a pseudo-second-order model, best explained the removal of the dye (Demirhan 2020). By using sugarcane bagasse biomass and various physical and chemical processes, Direct Violet 51 dye was adsorptively removed from aqueous solution by biosorption (Sadaf et al. 2015). The effects of three independent variables (biosorbent dose, pH, and initial dye concentration) were examined using the Box-Behnken experimental design. The best removal was achieved at pH 2, 0.05 g of biosorbent, and 63.0 mg/g of Direct Violet 51 dye. Using a 1 M NaOH solution, a desorption study on the loaded biosorbent removed 61.58% of the dye (Sadaf et al. 2015). Additionally, the statistical optimization of Gracilariaedulis for the adsorption removal of dye from textile effluents has used Plackett–Burman design and RSM (Venkataraghavan et al. 2020).

The use of artificial intelligence (AI) models for various applications in various fields, such as water purification, catalysis, and medicine, has increased due to recent technological advancements. Artificial neurons or nodes are interconnected assemblies of units used in the design of AI, and this idea is based on the biological nervous system. This idea is currently gaining attention on a global scale as a highly effective computational-based modeling tool for predicting the pollutant removal rate using data sets. This gives the correlation coefficient obtained from experimental and anticipated data a comparative study opportunity. As a result, an experiment takes less money and time to conduct. Along with demonstrating correlations of simulated output and the interactive influence among variables, AI has also been demonstrated to be capable of identifying complex and complicated nonlinear relationships among a number of parameters and variables (Nasr et al. 2017; Mazloom et al. 2020; Liao and Yao 2021; Martini and Roni 2021).

Afolabi et al. (2020b), for instance, reported the use of an artificial neural network (ANN) to simulate a pseudo-second-order kinetic model for the adsorptive removal of paracetamol using modified orange peels from aqueous media under various adsorption conditions, including initial concentration (10–50 mg $L^{-1}$), contact time (0–330 min), and temperature (30–50 °C).

It was discovered that the ANN with hyperbolic tangent sigmoid transfer functions at the input and output layers, eighteen (18) hidden neurons, and Levenberg–Marquardt as its backpropagation algorithm demonstrated the best prediction performance. It also demonstrates the usefulness of using modified orange peels to model a pseudo-second-order kinetic model for the removal of paracetamol (Afolabi et al. 2020b). In the most recent studies, numerous reports on the use of AI for modeling the effectiveness of the organic adsorption process have been published. Table 7.1 compares different adsorbents' adsorption performance using artificial intelligence.

**Table 7.1** Comparison of the various adsorbents performances using AI for the adsorptive removal process of organic pollutants

| Adsorbent | Pollutants | Optimum number of neurons | Evaluation indexes | | | | | | References |
|---|---|---|---|---|---|---|---|---|---|
| | | | $R^2$ | MRE | AE | MSE | RMSE | SSE | |
| AC from Acacia *auriculiformis* | Direct blue 86 | 5–8–1 | 0.982 | | | | | | Dutta et al. (2012) |
| AC from coconut shell | Indigo dye | 6–5–5–1 | | | | | 0.07283 | | Deshmukh (2012) |
| AC from orange wood tree | Sunset yellow | 3–13–1 | 0.997 | | | 0.0001 | | | Ghaedi et al. (2015) |
| Activated carbon | Methylene blue | – | | | 0.0014 | | | | Xu and Hu (2010) |
| Activated spent tea | Methylene blue | 5–10–1 | 0.999 | | | 0.00082 | | | Babaei et al. (2016) |
| Au NP AC Tamarisk | Methyl Owange | 3–11–1 | 0.958 | | | 0.0006 | | | Ghaedi et al. (2014) |
| Bamboo biochar | Acid black 172 | 5–5–1 | 0.996 | | | 0.00001 | | | Yang et al. (2014) |
| *Chara contratia* | Lanaset Red G | 4–23–1 | 0.999 | | | | | | Çelekli et al. (2012) |
| *Eucalyptus camaldulensis* | Basic blue 41 | 3–5–1 | | | | 0.006206 | | | Balci et al. (2011) |
| Iron-coated pumice | Ethidium bromide | 4–1–2–1 | 0.999 | | | | | | Heibati et al. (2016) |
| Modified rice straw | Crystal violet | 3–17–1 | 0.997 | | | | | | Chakraborty et al. (2013) |

(continued)

**Table 7.1** (continued)

| Adsorbent | Pollutants | Optimum number of neurons | Evaluation indexes | | | | | | References |
|---|---|---|---|---|---|---|---|---|---|
| | | | $R^2$ | MRE | AE | MSE | RMSE | SSE | |
| NaOH-modified rice husk | Methylene blue | 3–13–1 | 0.995 | | | | | | Chowdhury et al. (2019) |
| Orange peel | Paracetamol | – | | | | | | | Afolabi et al. (2020a) |
| *Posidonia oceanica* L | Methylene blue | 5–15–1 | 0.998 | | | | | | Cavas et al. (2011) |
| Potato peel powder | Reactive Red 198 | 5–7–1 | 0.98 | | | | 4.3 | | Maleki et al. (2013) |
| Powdered activated carbon | Acid orange 7 | 3–2–1 | | 5.81% | | | | | Dutta et al. (2012) |
| Rice husk | Safranin onto | 3–10–1 | 0.988 | | | | | | Saha and Das Mishra (2012) |
| Rice husk carbon | Bromocresol red | 3–5–5–2 | | | | | | 0.025 | Khonde and Pandharipande (2012) |
| Sawdust | Dye and metal ion | – | 0.99 | | | | | | Dolatabadi et al. (2018) |
| Soya bean waste | Methyl violet 2B | 6–4–1 | 0.995 | | | | | | Kooh et al. (2016) |
| Walnut husk | Lanaset red G | 4–20–1 | 0.995 | | 0.499–5.006 | | | | Çelekli et al. (2012) |

(continued)

**Table 7.1** (continued)

| Adsorbent | Pollutants | Optimum number of neurons | Evaluation indexes | | | | | | References |
|---|---|---|---|---|---|---|---|---|---|
| | | | $R^2$ | MRE | AE | MSE | RMSE | SSE | |
| Walnut husk | Basic red (BR) 46 | 5–25–1 | 0.999 | | | | | | Çelekli et al. (2016) |
| Walnut husk | Azo dye | – | | | | | | | Çelekli et al. (2016) |
| Waste marble dust | Malachite green | 4–12–1 | >0.89 | | <0.01 | | | | Coruh et al. (2014) |

## 7.6 Environmental Sustainability and Economic Viability

Different adsorbents can be created, altered, and used on various environmental contaminants. The method used to prepare adsorbents for activation and modification affects their behavior. Surface area, void active sites, and reproducibility in the activation processes are generally considered to be the most important properties of a good adsorbent (Zhang et al. 2021). No matter the activation methods used, the biomass-derived adsorbent is preferred in comparison because it has the highest surface area, which increases the adsorption of contaminants depending on the source types. Due to their adaptability and affordability, bioadsorbents are the most widely used adsorbents for the adsorption of dyes and other contaminants from wastewater. Their operations are also simple, and their by-products are not hazardous. Additionally, biomass-derived adsorbents are frequently composed of cellulose-based structures and can generate high-quality porous carbon that can absorb a variety of contaminants in the environment. Because of its enormous availability, one of the key benefits of using biomass-derived adsorbent for the removal of dyes is that it can be applied on a large scale. Additionally, adsorbents made from biomass have good regeneration capacities as well as high affinities and binding capacities for various types of contaminants. Additionally, the removal of dyes from wastewater is aided by their favorable surface chemistry, which includes pore distribution, specific surface area, and functional groups (Sherugar et al. 2022).

However, environmental variables like temperature, pH, ionic strength, nutrient availability, and photodegradation from natural light, which can occur in some seasonal and temporal variations, may affect the biomass's structural characteristics and chemical components. The actual adsorption capacity of biomass-based adsorbents may, therefore, be significantly impacted by this. As a result, to comprehend the adsorption mechanisms for specific contaminants, the biomass-derived adsorbents should be thoroughly characterized. Additionally, research needs to be done in a multi-pollutant system to satisfy the needs of actual wastewater treatment, which includes a variety of contaminants that can ensure stability, leading to the suitability of biomass for the real-world biosorption process (Zhou et al. 2019a)

The scaling up and cost analysis are the other major barriers to using biomass-derived adsorbents for dye removal. Even though there are many inexpensive adsorbents available, their processing (activation, cross-linking, drying, autoclaving, etc.), modification, and transportation cost.

Despite not including a cost–benefit analysis in their research, several authors claimed that the cost of bioadsorbents made from microbes and agricultural and forest waste is lower than that of conventional treatment technologies. The properties of the adsorbent material used in dye adsorption processes are related to the benefits of biosorption. Along with the principles of green chemistry, the availability of a large quantity of adsorbent material, the simplicity of preparation or processing, and the requirements of activation are critically important for the economy (Bulgariu et al. 2019). According to Adegoke and Bello et al. (2015) the price of forest wastes only includes the cost of transportation from their storage location to the location where they will be used. According to some authors, the term "low-cost adsorbents" only refers to their original prices, so it is important to critically examine their local availability, transportation, treatment process, recycling, and lifetime issues, as well as the processes of regeneration and treatment (Zhou et al. 2019a). The cost analysis of biomass-based adsorbents is unfortunately limited, especially for microbial biomass, as the majority of studies on biomass-based adsorption have been carried out on the laboratory scale using simulated wastewater. The Basic Red 09 dye removal performance of biochars made from agricultural waste (coconut shell, groundnut shell, and rice husk) from wastewater was thoroughly cost-analyzed. By taking into account operational costs such as manufacturing costs, maintenance costs, feedstock costs, transportation costs, labor costs, and distribution costs, it was estimated that the price of 1 g of adsorbent for coconut shell, groundnut shell, and rice husk is ₹4.54, ₹0.91, and ₹0.97, respectively (Praveen et al. 2021). The biochar made from groundnut shells had the highest adsorption capacity (46.3 mg $g^{-1}$) and the lowest cost (0.91) per gram of Basic Red 09 dye removal. Bello et al. (2019) prepared locust bean pod AC functionalized with phosphoric acid for removing the dye RhB and investigated the initial cost analysis of this adsorbent. They came to the conclusion that the AC made from this plant material was roughly six times less expensive than traditional AC. The main cost drivers are phosphoric acid and deionized water (Bello et al. 2019).

## 7.7 Conclusions and Future Remarks

While most studies on biomass-based adsorption focused on removing just one pollutant, actual wastewater containing dye typically contains a variety of dyes. To meet the needs of wastewater treatment for practical applications, additional research in the mixed and multi-pollutant systems should be done. The stability of the adsorbent and the price of adsorbents are also crucial factors that can determine the adsorbent's suitability for real-world applications. Most reported studies only evaluate adsorption capacity in the lab; cost analyses are ignored, and other factors like local availability, transportation, the treatment process, and recycling and lifetime issues must be considered. Adsorbent regeneration research is also crucial in lowering process costs, recovering adsorbed compounds, and reducing waste.

Another top priority is to assess the biosorbents' removal abilities on a pilot scale and against actual wastewater. Cost estimates should also be made for using chemical solutions to remove pollutants from surfaces. Future research should concentrate on the engrave strategy regarding the handling of the chemical solution wastes obtained from adsorption/desorption studies as well as the disposal of the polluted loaded wastes.

**Acknowledgements** We acknowledge to the research group "Fenómenos de superficie Michael Polanyi" and Universidad Nacional de Colombia Sede Medellín.

**Declaration of Competing Interest** The authors declare that they have no known competing financial interests or personal relationships that could have appeared to influence the work reported in this paper. Likewise, the figures included have not been published previously.

## References

Adegoke KA, Bello OS (2015) Dye sequestration using agricultural wastes as adsorbents. Water Resour Ind 12:8–24

Afolabi IC, Popoola SI, Bello OS (2020a) Machine learning approach for prediction of paracetamol adsorption efficiency on chemically modified orange peel. Spectrochim Acta Part A Mol Biomol Spectrosc 243:118769

Afolabi IC, Popoola SI, Bello OS (2020b) Modeling pseudo-second-order kinetics of orange peel-paracetamol adsorption process using artificial neural network. Chemom Intell Lab Syst 203:104053

Ahmad T, Danish M (2018) Prospects of banana waste utilization in wastewater treatment: a review. J Environ Manage 206:330–348

Ahmad MA, Eusoff MA, Oladoye PO, Adegoke KA, Bello OS (2020) Statistical optimization of Remazol Brilliant Blue R dye adsorption onto activated carbon prepared from pomegranate fruit peel. Chem Data Collect 28:100426

Ahmad MA, Eusoff MA, Oladoye PO, Adegoke KA, Bello OS (2021) Optimization and batch studies on adsorption of Methylene blue dye using pomegranate fruit peel based adsorbent. Chem Data Collect 32:100676

Ahmed MJ (2017) Adsorption of non-steroidal anti-inflammatory drugs from aqueous solution using activated carbons: review. J Environ Manag:274–282

Aigbe UO, Ukhurebor KE, Onyancha RB, Osibote OA, Darmokoesoemo H, Kusuma HS (2021) Fly ash-based adsorbent for adsorption of heavy metals and dyes from aqueous solution: a review. J Market Res 14:2751–2774

Akköz Y, Coşkun R (2023) Cellulose-supported bioadsorbent from natural hemp fiber for removal of anionic dyes from aqueous solution. Int J Biol Macromol 252:126447

Akpomie KG, Conradie J (2020) Banana peel as a biosorbent for the decontamination of water pollutants: a review. Environ Chem Lett 18:1085–1112

Al-Qahtani KM (2016) Water purification using different waste fruit cortexes for the removal of heavy metals. J Taibah Univ Sci 10:700–708

Amalina F, Abd Razak AS, Krishnan S, Zularisam AW, Nasrullah M (2022) Dyes removal from textile wastewater by agricultural waste as an absorbent–a review. Clean Waste Syst:100051

Ambaye TG, Vaccari M, van Hullebusch ED, Amrane A, Rtimi SJ (2021) Mechanisms and adsorption capacities of biochar for the removal of organic and inorganic pollutants from industrial wastewater. Int J Environ Sci Technol:1–22

Ambaye T, Vaccari M, van Hullebusch ED, Amrane A, Rtimi S (2021) Mechanisms and adsorption capacities of biochar for the removal of organic and inorganic pollutants from industrial wastewater. Int J Environ Sci Technol:1–22

Amini M, Younesi H, Bahramifar N, Lorestani AAZ, Ghorbani F, Daneshi A, Sharifzadeh M (2008) Application of response surface methodology for optimization of lead biosorption in an aqueous solution by Aspergillus niger. J Hazard Mater 154(1–3):694–702

Anastopoulos I, Bhatnagar A, Hameed BH, Ok YS, Omirou M (2017) A review on waste-derived adsorbents from sugar industry for pollutant removal in water and wastewater. J Mol Liq 240:179–188

Babaei AA, Khataee A, Ahmadpour E, Sheydaei M, Kakavandi B, Alaee Z (2016) Optimization of cationic dye adsorption on activated spent tea: equilibrium, kinetics, thermodynamic and artificial neural network modeling. Korean J Chem Eng 33:1352–1361

Baig N, Ihsanullah I, Sajid M, Saleh TA (2019) Graphene-based adsorbents for the removal of toxic organic pollutants: a review. J Environ Manag 244:370–382

Balci B, Keskinkan O, Avci M (2011) Use of BDST and an ANN model for prediction of dye adsorption efficiency of Eucalyptus camaldulensis barks in fixed-bed system. Expert Syst Appl 38(1):949–956

Bekçi Z, Özveri C, Seki Y, Yurdakoç K (2008) Sorption of malachite green on chitosan bead. J Hazard Mater 154:254–261

Bello OS, Adegoke KA, Inyinbor AA, Dada AO (2021) Trapping Rhodamine B dye using functionalized mango (Mangifera indica) pod. Water Environ Res 93(10):2308–2328

Bello OS, Adegoke KA, Sarumi OO, Lameed OS (2019) Functionalized locust bean pod (Parkia biglobosa) activated carbon for Rhodamine B dye removal. Heliyon 5(8)

Bhatnagar A, Sillanpää M, Witek-Krowiak A (2015) Agricultural waste peels as versatile biomass for water purification–a review. Chem Eng J 270:244–271

Bilal M, Ihsanullah I, Younas M, Shah MUH (2021) Recent advances in applications of low-cost adsorbents for the removal of heavy metals from water: a critical review. Sep Purif Technol 278:119510

Bulgariu L, Escudero LB, Bello OS, Iqbal M, Nisar J, Adegoke KA, Alakhras F, Kornaros M, Anastopoulos I (2019) The utilization of leaf-based adsorbents for dyes removal: a review. J Mol Liq 276:728–747

Carmona M, Garcia MT, Carnicer Á, Madrid M, Rodríguez JF (2014) Adsorption of phenol and chlorophenols onto granular activated carbon and their desorption by supercritical $CO_2$. J Chem Technol Biotechnol 89:1660–1667

Cavas L, Karabay Z, Alyuruk H, Doğan H, Demir GK (2011) Thomas and artificial neural network models for the fixed-bed adsorption of methylene blue by a beach waste Posidonia oceanica (L.) dead leaves. Chem Eng J 171(2):557–562

Çelekli A, Birecikligil SS, Geyik F, Bozkurt H (2012) Prediction of removal efficiency of Lanaset Red G on walnut husk using artificial neural network model. Biores Technol 103(1):64–70

Çelekli A, Bozkurt H, Geyik F (2016) Artificial neural network and genetic algorithms for modeling of removal of an azo dye on walnut husk. Desalin Water Treat 57(33):15580–15591

Chakraborty S, Chowdhury S, Saha PD (2013) Artificial neural network (ANN) modeling of dynamic adsorption of crystal violet from aqueous solution using citric-acid-modified rice (Oryza sativa) straw as adsorbent. Clean Technol Environ Policy 15:255–264

Chowdhury ZZ, Chandran RRR, Jahan A, Khalid K, Rahman MM, Al-Amin M, Akbarzadeh O, Badruddin IA, Khan TY, Kamangar S (2019) Extraction of cellulose nano-whiskers using ionic liquid-assisted ultra-sonication: optimization and mathematical modelling using Box–Behnken design. Symmetry 11(9):1148

Coruh S, Gürkan H, Kilic E, Geyikci F (2014) Prediction of adsorption efficiency for the removal malachite green and acid blue 161 dyes by waste marble dust using ANN. Global Nest J 16(4):676–689

Dawood S, Sen TK (2012) Removal of anionic dye Congo red from aqueous solution by raw pine and acid-treated pine cone powder as adsorbent: equilibrium, thermodynamic, kinetics, mechanism and process design. Water Res 46:1933–1946

Demirhan E (2020) Response surface methodology approach for adsorptive removal of Reactive Blue 19 onto green pea pod. Water Sci Technol 81(6):1137–1147

Desai AV, Sharma S, Let S, Ghosh SK (2019) N-donor linker based metal-organic frameworks (MOFs): advancement and prospects as functional materials. Coord Chem Rev 395:146–192

Deshmukh S (2012) Indigo dye removal by using coconut shell adsorbent and performance evaluation by artificial neural network. In: Emerging trends in science, engineering and technology: proceedings of international conference (INCOSET 2012). Springer, pp 655–663

Dolatabadi M, Mehrabpour M, Esfandyari M, Alidadi H, Davoudi M (2018) Modeling of simultaneous adsorption of dye and metal ion by sawdust from aqueous solution using of ANN and ANFIS. Chemom Intell Lab Syst 181:72–78

Dutta M, Ghosh P, Basu JK (2012) Application of artificial neural network for the decolorization of direct blue 86 by using microwave assisted activated carbon. J Taiwan Inst Chem Eng 43(6):879–888

Fagbohungbe MO, Herbert BM, Hurst L, Ibeto CN, Li H, Usmani SQ, Semple KT (2017) The challenges of anaerobic digestion and the role of biochar in optimizing anaerobic digestion. Waste Manage 61:236–249

Farah AA (2023) The use of agro-waste-based adsorbents as sustainable, renewable, and low-cost alternatives for the removal of ibuprofen and carbamazepine from water. Heliyon

Farias KC, Guimarães RC, Oliveira KR, Nazário CE, Ferencz JA, Wender H (2023) Banana peel powder biosorbent for removal of hazardous organic pollutants from wastewater. Toxics 11(8):664

Fegousse A, El Gaidoumi A, Miyah Y, El Mountassir R, Lahrichi A (2019) Pineapple bark performance in dyes adsorption: optimization by the central composite design. J Chem 2019

Feng Z, Chen H, Li H, Yuan R, Wang F, Chen Z, Zhou B (2020) Preparation, characterization, and application of magnetic activated carbon for treatment of biologically treated papermaking wastewater. Sci Total Environ 713:136423

Foo KY, Hameed BH (2012) A cost effective method for regeneration of durian shell and jackfruit peel activated carbons by microwave irradiation. Chem Eng J 193:404–409

Ghaedi M, Ghaedi A, Ansari A, Mohammadi F, Vafaei A (2014) Artificial neural network and particle swarm optimization for removal of methyl orange by gold nanoparticles loaded on activated carbon and Tamarisk. Spectrochim Acta Part A Mol Biomol Spectrosc 132:639–654

Ghaedi A, Ghaedi M, Karami P (2015) Comparison of ultrasonic with stirrer performance for removal of sunset yellow (SY) by activated carbon prepared from wood of orange tree: artificial neural network modeling. Spectrochim Acta Part A Mol Biomol Spectrosc 138:789–799

Ghorbani F, Kamari S (2017) Application of response surface methodology for optimization of methyl orange adsorption by Fe-grafting sugar beet bagasse. Adsorpt Sci Technol 35(3–4):317–338

Giraldo LJ, Medina OE, Ortiz-Pérez V, Franco CA, Cortés FB (2023) Enhanced carbon storage process from flue gas streams using rice husk silica nanoparticles: an approach in shallow coal bed methane reservoirs. Energy Fuels 37(4):2945–2959

Gomravi Y, Karimi A, Azimi H (2021) Adsorption of heavy metal ions via apple waste low-cost adsorbent: characterization and performance. Korean J Chem Eng 38:1843–1858

Gong R, Li M, Yang C, Sun Y, Chen J (2005) Removal of cationic dyes from aqueous solution by adsorption on peanut hull. J Hazard Mater 121:247–250

Gupta H, Gupta B (2016) Adsorption of polycyclic aromatic hydrocarbons on banana peel activated carbon. Desalin Water Treat 57:9498–9509

Gupta S, Sireesha S, Sreedhar I, Patel CM, Anitha KL (2020) Latest trends in heavy metal removal from wastewater by biochar based sorbents. J Water Process Eng 38:101561

Hamad HN, Idrus S (2022) Recent developments in the application of bio-waste-derived adsorbents for the removal of methylene blue from wastewater: a review. Polymers 14(4):783

Hassan M, Naidu R, Du J, Liu Y, Qi F (2020) Critical review of magnetic biosorbents: their preparation, application, and regeneration for wastewater treatment. Sci Total Environ 702:134893

Heibati B, Rodriguez-Couto S, Ozgonenel O, Turan NG, Aluigi A, Zazouli MA, Ghozikali MG, Mohammadyan M, Albadarin AB (2016) A modeling study by artificial neural network on ethidium bromide adsorption optimization using natural pumice and iron-coated pumice. Desalin Water Treat 57(29):13472–13483

Hoo HM, Mohamad M, Wannahari R, Masri MN, Ter TP, Mohidem NA (2022) Optimization study of malachite green dye adsorption by eggshell using response surface methodology. Key Eng Mater 908:367–372

Hu J, Shipley HJ (2013) Regeneration of spent $TiO_2$ nanoparticles for Pb (II), Cu (II), and Zn (II) removal. Environ Sci Pollut Res 20(8):5125–5137

Jain A, Kumari S, Agarwal S, Khan S (2021) Water purification via novel nano-adsorbents and their regeneration strategies. Process Saf Environ Prot 152:441–454

Jia M, Wang F, Bian Y, Jin X, Song Y, Kengara FO, Xu R, Jiang X (2013) Effects of pH and metal ions on oxytetracycline sorption to maize-straw-derived biochar. Biores Technol 136:87–93

Kali A, Amar A, Loulidi I, Jabri M, Hadey C, Lgaz H, Boukhlifi F (2022a) Characterization and adsorption capacity of four low-cost adsorbents based on coconut, almond, walnut, and peanut shells for copper removal. Biomass Convers Biorefin:1–12

Kali A, Amar A, Loulidi I, Jabri M, Hadey C, Lgaz H, Boukhlifi F (2022b) Characterization and adsorption capacity of four low-cost adsorbents based on coconut, almond, walnut, and peanut shells for copper removal. Biomass Convers Biorefin:1–12

Khonde R, Pandharipande S (2012) Artificial neural network modeling for adsorption of dyes from aqueous solution using rice husk carbon. Int J Comput Appl 41(4)

Kołodyńska D, Krukowska JA, Thomas P (2017) Comparison of sorption and desorption studies of heavy metal ions from biochar and commercial active carbon. Chem Eng J 307:353–363

Kooh MRR, Dahri MK, Lim LB, Lim LH, Malik OA (2016) Batch adsorption studies of the removal of methyl violet 2B by soya bean waste: isotherm, kinetics and artificial neural network modelling. Environ Earth Sci 75:1–14

Kosheleva R, Mitropoulos C, Kyzas GZ (2018) Activated carbon from food waste. In: Crini G, Lichtfouse E (eds) Green adsorbents for pollutant removal. Springer, Cham

Kumari P, Alam M, Siddiqi WA (2019) Usage of nanoparticles as adsorbents for waste water treatment: an emerging trend. Sustain Mater Technol 22:e00128

Laskar N, Kumar U (2022) Application of low-cost, eco-friendly adsorbents for the removal of dye contaminants from wastewater: current developments and adsorption technology. Environ Qual Manage 32(1):209–221

Li Y, Tsend N, Li T, Liu H, Yang R, Gai X, Wang H, Shan S (2019) Microwave assisted hydrothermal preparation of rice straw hydrochars for adsorption of organics and heavy metals. Biores Technol 273:136–143

Liao M, Yao Y (2021) Applications of artificial intelligence-based modeling for bioenergy systems: a review. GCB Bioenergy 13(5):774–802

Lv H, Zhang W, Hosseini M, Samani MR, Toghraie D (2023) Characterization and synthesis of new adsorbents with some natural waste materials for the purification of aqueous solutions. J Environ Manage 336:117660

Ma X, Chen R, Zhou K, Wu Q, Li H, Zeng Z, Li L (2020) Activated porous carbon with an ultra-high surface area derived from waste biomass for acetone adsorption, $CO_2$ capture, and light hydrocarbon separation. ACS Sustain Chem Eng 8(31):11721–11728

Maleki A, Daraei H, Khodaei F, Bayazid-Aghdam K, Rezaee R, Naghizadeh A (2013) Investigation of potato peel-based bio-sorbent efficiency in reactive dye removal: artificial neural network modeling and genetic algorithms optimization. J Adv Environ Health Res 1(1):21–28

Martini S, Roni KA (2021) The existing technology and the application of digital artificial intelligent in the wastewater treatment area: a review paper. J Phys Conf Ser, p 012013. IOP Publishing

Mazloom MS, Hemmati-Sarapardeh A, Husein MM, Behbahani HS, Zendehboudi S (2020) Application of nanoparticles for asphaltenes adsorption and oxidation: a critical review of challenges and recent progress. Fuel 279:117763

Medina OE, Galeano-Caro D, Ocampo-Pérez R, Perez-Cadenas AF, Carrasco-Marín F, Franco CA, Corteś FB (2021a) Development of a monolithic carbon xerogel-metal composite for crude oil removal from oil in-saltwater emulsions: evaluation of reuse cycles. Microporous Mesoporous Mater 327:111424

Medina OE, Galeano-Caro D, Castelo-Quibén J, Ocampo-Pérez R, Perez-Cadenas AF, Carrasco-Marín F, Franco CA, Corteś FB (2021b) Monolithic carbon xerogels-metal composites for crude oil removal from oil in-saltwater emulsions and subsequent regeneration through oxidation process: composites synthesis, adsorption studies, and oil decomposition experiments. Microporous Mesoporous Mater 319:111039

Mekouar MA (2018) Food and Agriculture Organization of the United Nations (FAO), Yearbook of International Environmental Law

Mo J, Yang Q, Zhang N, Zhang W, Zheng Y, Zhang Z (2018) A review on agro-industrial waste (AIW) derived adsorbents for water and wastewater treatment. J Environ Manage 227:395–405

Mousavi SA, Mahmoudi A, Amiri S, Darvishi P, Noori E (2022) Methylene blue removal using grape leaves waste: optimization and modeling. Appl Water Sci 12(5):112

Mudhoo A, Gautam RK, Ncibi MC, Zhao F, Garg VK, Sillanpää M (2019) Green synthesis, activation and functionalization of adsorbents for dye sequestration. Environ Chem Lett 17:157–193

Nasr M, Mahmoud AED, Fawzy M, Radwan A (2017) Artificial intelligence modeling of cadmium (II) biosorption using rice straw. Appl Water Sci 7:823–831

Nayagam JOP, Prasanna K (2022) Utilization of shell-based agricultural waste adsorbents for removing dyes: a review. Chemosphere 291:132737

Niad M, Rasoolzadeh L, Zarei F (2014) Biosorption of copper (II) on Sargassum angostifolium C. Agardh phaeophyceae biomass. Chem Speciat Bioavailab 26(3):176–183

Njewa JB, Vunain E, Biswick T (2022) Synthesis and characterization of activated carbons prepared from agro-wastes by chemical activation. J Chem:1–13

Nyika J, Dinka MO (2023) The potential of reusing fruit bio-adsorbents for water purification: a minireview. Mater Today Proc

Ogunlalu O, Oyekunle IP, Iwuozor KO, Aderibigbe AD, Emenike EC (2021) Trends in the mitigation of heavy metal ions from aqueous solutions using unmodified and chemically-modified agricultural waste adsorbents. Curr Res Green Sustain Chem 4:100188

Olugbenga OS, Adeleye PG, Oladipupo SB, Adeleye AT, John KI (2023) Biomass-derived biochar in wastewater treatment-a circular economy approach. Waste Manage Bull

Park SJ, Yeo SD (1999) Supercritical extraction of phenols from organically modified smectite. Sep Sci Technol 34(1):101–113

Pellenz L, de Oliveira CRS, da Silva Júnior AH, da Silva LJS, da Silva L, de Souza AAU, da Silva A (2022) A comprehensive guide for characterization of adsorbent materials. Sep Purif Technol:122435

Praveen S, Gokulan R, Pushpa TB, Jegan J (2021) Techno-economic feasibility of biochar as biosorbent for basic dye sequestration. J Indian Chem Soc 98(8):100107
Qiao K, Tian W, Bai J, Dong J, Zhao J, Gong X, Liu S (2018) Preparation of biochar from Enteromorpha prolifera and its use for the removal of polycyclic aromatic hydrocarbons (PAHs) from aqueous solution. Ecotoxicol Environ Saf 149:80–87
Rajan MJ, Anish CI (2022) Role of activated carbon in water treatment. In: Dincer S, Mercimek HA, Sumengen M (eds) Water quality—new perspectives. IntechOpen, Rijeka
Rajesh Y, Jeeru LR (2022) Synthesis and characterization of low-cost wood based biosorbent. Mater Today Proc 57:34–37
Reddy PMK, Verma P, Subrahmanyam C (2016) Bio-waste derived adsorbent material for methylene blue adsorption. J Taiwan Inst Chem Eng 58:500–508
Rodriguez Acevedo E, Franco CA, Carrasco-Marín F, Pérez-Cadenas AF, Cortés FB (2020) Biomass-derived carbon molecular sieves applied to an enhanced carbon capture and storage process (e-CCS) for flue gas streams in shallow reservoirs. Nanomaterials 10(5):980
Sadaf S, Bhatti HN, Arif M, Amin M, Nazar F, Sultan M (2015) Box-Behnken design optimization for the removal of Direct Violet 51 dye from aqueous solution using lignocellulosic waste. Desalin Water Treat 56(9):2425–2437
Saha P, Das Mishra R (2012) Adsorption of safranin onto chemically modified rice husk in a upward flow packed bed reactor: artificial neural network modeling. Biotechnol Adv 44(1):7579–7583
Sahu S, Pahi S, Sahu JK, Sahu UK, Patel RK (2020) Kendu (Diospyros melanoxylon Roxb) fruit peel activated carbon—an efficient bioadsorbent for methylene blue dye: equilibrium, kinetic, and thermodynamic study. Environ Sci Pollut Res 27:22579–22592
Salgın U, Yıldız N, Çalımlı A (2004) Desorption of salicylic acid from modified bentonite by using supercritical fluids in packed bed column. Sep Sci Technol 39(11):2677–2694
Sathishkumar M, Binupriya A, Kavitha D, Yun S (2007) Kinetic and isothermal studies on liquid-phase adsorption of 2, 4-dichlorophenol by palm pith carbon. Bioresour Technol 98:866–873
Sherugar P, Padaki M, Naik NS, George SD, Murthy DH (2022) Biomass-derived versatile activated carbon removes both heavy metals and dye molecules from wastewater with near-unity efficiency: mechanism and kinetics. Chemosphere 287:132085
Singh H, Chauhan G, Jain AK, Sharma SK (2017) Adsorptive potential of agricultural wastes for removal of dyes from aqueous solutions. J Environ Chem Eng 5(1):122–135
Singh NB, Nagpal G, Agrawal S (2018) Water purification by using adsorbents: a review. Environ Technol Innov 11:187–240
Solangi NH, Kumar J, Mazari SA, Ahmed S, Fatima N, Mubarak NM (2021) Development of fruit waste derived bio-adsorbents for wastewater treatment: a review. J Hazard Mater 416:125848
Song Y, Ding S, Chen S, Xu H, Mei Y, Ren J (2015) Removal of malachite green in aqueous solution by adsorption on sawdust. Korean J Chem Eng 32:2443–2448
Sun Y, Zhang B, Zheng T, Wang P (2017) Regeneration of activated carbon saturated with chloramphenicol by microwave and ultraviolet irradiation. Chem Eng J 320:264–270
Tummala CM, Dardona M, Praneeth S, Mohanty SK, Dittrich TM (2023) Iron-coated nutshell waste bioadsorbents: synthesis, phosphate remediation, and subsequent fertilizer application. Environ Res:117468
Uddin MT, Rahman MA, Rukanuzzaman M, Islam MA (2017) A potential low cost adsorbent for the removal of cationic dyes from aqueous solutions. Appl Water Sci 7:2831–2842
Venkataraghavan R, Thiruchelvi R, Sharmila D (2020) Statistical optimization of textile dye effluent adsorption by Gracilaria edulis using Plackett-Burman design and response surface methodology. Heliyon 6(10)
Wang S, Gao B, Zimmerman AR, Li Y, Ma L, Harris WG, Migliaccio KW (2015) Removal of arsenic by magnetic biochar prepared from pinewood and natural hematite. Biores Technol 175:391–395

Wang L, Wang Y, Ma F, Tankpa V, Bai S, Guo X, Wang X (2019) Mechanisms and reutilization of modified biochar used for removal of heavy metals from wastewater: a review. Sci Total Environ 668:1298–1309

Wang Q, Wang Y, Yang Z, Han W, Yuan L, Zhang L, Huang X (2022) Efficient removal of Pb (II) and Cd (II) from aqueous solutions by mango seed biosorbent. Chem Eng J Adv 11:100295

Xu Y, Hu X (2010) Activated carbon adsorption of methylene blue based on artificial neural network. Mater Rev 24:23

Xu X, Hu X, Ding Z, Chen Y, Gao B (2017) Waste-art-paper biochar as an effective sorbent for recovery of aqueous Pb (II) into value-added PbO nanoparticles. Chem Eng J 308:863–871

Yang Y, Lin X, Wei B, Zhao Y, Wang J (2014) Evaluation of adsorption potential of bamboo biochar for metal-complex dye: equilibrium, kinetics and artificial neural network modeling. Int J Environ Sci Technol 11:1093–1100

Yang H, Ye S, Zeng Z, Zeng G, Tan X, Xiao R, Wang J, Song B, Du L, Qin M, Yang Y (2020) Utilization of biochar for resource recovery from water: a review. Chem Eng J 125502

Zhang XN, Mao GY, Jiao YB, Shang Y, Han RP (2014) Adsorption of anionic dye on magnesium hydroxide-coated pyrolytic bio-char and reuse by microwave irradiation. Int J Environ Sci Technol 11(5):1439–1448

Zhang Y, Yang D, Ning P, Li Y, Tian S, Gu J (2019) Regeneration of phenol-saturated activated carbon by supercritical water: effect of $H_2O_2$ and alkali metal catalysts. J Environ Eng 145:04019083

Zhang X, Navarathna CM, Leng W, Karunaratne T, Thirumalai RV, Kim Y, Pittman CU Jr, Mlsna T, Cai Z, Zhang J (2021) Lignin-based few-layered graphene-encapsulated iron nanoparticles for water remediation. Chem Eng J 417:129199

Zhou Y, Lu J, Zhou Y, Liu Y (2019a) Recent advances for dyes removal using novel adsorbents: a review. Environ Pollut 252:352–365

Zhou Z, Kearnes S, Li L, Zare RN, Riley P (2019b) Optimization of molecules via deep reinforcement learning. Sci Rep 9(1):10752

# Integrated Waste Management System for Environmental Protection and Sustainability of Resources

8

Yahya Jani and Monica Odlare

**Abstract**

Modern civilization is defined by the unsustainable consumption of most of the Earth's natural resources and the contribution with different kinds of hazards to the air, water and soil. More than two centuries since the industrial revolution have caused the accumulation of different contaminants in different forms in most of the environmental compartments, therefore, there is an urgent need to shift this trend to sustainable solutions where protecting our environment needs to be in the core of new strategies. Waste management is one of the major areas where development towards sustainable development has a potential by implementing new strategies to protect the environment and also providing societies with secondary resources for the continuation of living on this Earth. In this chapter, an integrated waste management system based on the European hierarchy is introduced as an input towards reducing the impact of losing virgin materials as wastes. In addition, international and European regulations and policies are also presented as a crucial tool to save the environment.

## 8.1 Introduction

Waste is generated due to human daily activities regardless of the activity type simple like drinking water, eating food or more advance activities like production processes of goods, transportations and others. In any production process converting raw materials to products is connected with the generation of waste because it is challenging to reach

Y. Jani (✉) · M. Odlare
Department of Sustainable Environment and Construction, School of Business, Society and Engineering, Mälardalen University, Västerås, Sweden
e-mail: yahya.jani@mdu.se



M. Nasr and B. Ravindran (eds.), *Solid Waste Management in Delta Region for SDGs Fulfillment*, Deltas of the World, https://doi.org/10.1007/978-3-031-58253-0_8

to 100% of yield. There are different kinds of losses due to mechanical, chemical and physical mechanisms like as an example friction or losing energy to the surrounding due to temperature differences. These losses are regarded as wastes and can take the gas, liquid or a solid form or a combination of the three forms. For instance, the production process of each mobile phone relates to the generation of about 86 kg of different kinds of wastes from the first stage of mining metals to the final product which is the mobile phone (Swedish Waste Management 2023). These wastes are invisible to the consumer who usually deal with the final device of a maximum average weight of around 200 g. Invisible wastes have considerable impact on the environment due to that it is a waste completely hidden for consumers. Consumers can deal with the daily wastes generated at homes and can understand why waste management systems are needed to handle such wastes while hidden waste will be an undisclosed that in needs to be clarified for society. During our lectures in the different environmental engineering courses, we realized that students as part of the society did not have any clue about invisible waste as an example the waste generated from the production of mobile phone or even other products such as jeans and cloths. In general, society is missing the environmental and health impact of municipal solid wastes generated from anthropogenic activities and more concern about household wastes. Therefore, there is an urgent necessity to explore the effect of such wastes and why waste management systems and regulations are important tools to mitigate the shortage of resources, demands of raw materials, environmental impacts and most importantly a tool to reach to the sustainable development goals in waste management sector.

Living in modern societies has initiated new activities where people are engaged with the consumption of limited life items such as mobile phones and electronic devices, cloths, furniture and others. For instance, the lifespan of electronic devices has been shortened to roughly 2–3 years due to rapid scientific expansion (Vermeșan et al. 2020a). This expansion results in the generation of more than 44 million tons of e-waste in 2016 with a global share of about 6 kg per person (Kaya 2019; Vermeșan et al. 2020b) and it is estimated that the e-waste will raise to approximately 55 million tons by 2022 (Balde et al. 2017). It is well known that metals are essential elements in the production of electronic devices with a content of more than 60% by weight for some devices (Kaya 2019; Vermeșan et al. 2020b; Balde et al. 2017). Metals such as rare earth elements, platinum group and others are crucial for the future green technology due to their chemical and physical properties (Gómez et al. 2023; Burlakovs et al. 2018). These elements, on the other hand, are negatively impacting human health and the environment when accumulating in considerable concentrations in living organisms (Tan et al. 2017). Metals with their persistence nature cannot be degraded and are able to cause carcinogenic effects in living organisms (Tan et al. 2017). Therefore, an integrated waste management system (IWMS) is crucial to handle generated waste to protect human and the environmental health (Jani 2018).

The objective of this chapter is to analyze the integrated waste management system (IWMS) as a tool to handle municipal solid waste and to protect the environment from the toxicities associated with the generation of wastes. Managing solid waste including a series of actions following the European waste management hierarchy will be provided by this chapter and the focus will be on the connection of these steps to protect the environment and human health. This chapter is acting to bridge the missing knowledge between the integrated waste management system (IWMS) and the need for resources to guarantee the sustainability of the Earth natural reserves for future needs.

## 8.2 Background

According to the World Bank Group (2023), about 2.01 billion tons of municipal solid waste are generated annually on a global level. This sum of waste is estimated to jump to around 3.4 billion tons within 2050. On average each person is generating 270 kg of municipal solid waste yearly, ranging between 40.15 and 1657.1 kg/year based on socioeconomic perspectives. The economic factor is proportionally correlated to the waste generation as increasing the income can be a driving force towards increasing the consumption intensity (Kaza et al. 2018).

On environmental perception, the annually generated waste is responsible for the emission of 1.6 billion tons of carbon dioxide equivalent to the atmosphere in 2016 and it is expected to reach to 2.6 billion tons of carbon dioxide equivalent in 2050 (Kaza et al. 2018). This can be due to that 33% of the world's total generated wastes are still disposed of at open dumpsites and 30% of all countries are missing a proper waste management system (Jani 2018). The recycling of municipal solid waste is barely 19% and energy recovery by modern incineration is covering 11% globally (World Bank Group 2023).

On the European level, the total waste generation was around 2.2 billion tons in 2020 where the construction industry dominated by 37.5%, as shown in Fig. 8.1 (Eurostat 2023). However, major minerals were dominating the sort of waste from all industrial and other activities with 64% (Eurostat 2023). Comparing the member states waste generation statistics by excluding the major minerals reveals that Estonia is having the higher rate of waste generation per capita with 6.3 tons, see Fig. 8.2. The EU recycling of materials and recovery of energy (composting and incineration) rates reached to 49.6% in 2021 (European parliament, Waste management in the EU 2023).

All the above facts indicate that society is facing a major challenge named waste generation which needs solutions to stand for the growing rate of generation due to population growth, consumption and urbanization. The European Commission, as an example, specify a plan (*European Parliament resolution of 10 February 2021 on the New Circular Economy Action Plan (2020/2077(INI))* (European Parliament 2021) to mitigate the impact of the solid waste generation by specifying milestones for all member states starts with reusing and recycling 60% of the generated waste within 2030 and reducing landfilling to 10% by 2035. However, statistical results from the European Parliament show that only 4 countries out of 27 (Austria, Germany, Slovenia and Sweden) have already reached to

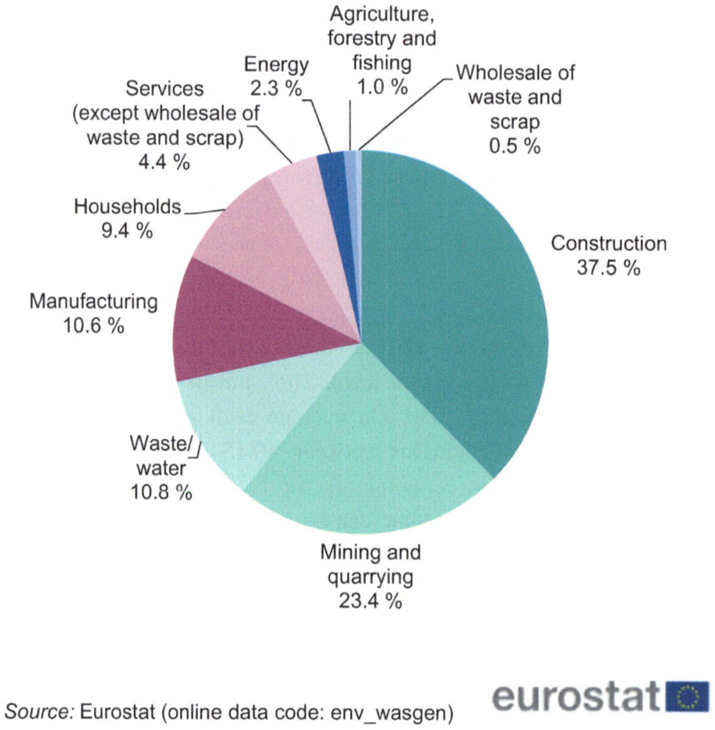


**Fig. 8.1** The share of generated waste per activity within EU (Eurostat 2023). For a high-resolution interactive image, please consult this figure at (https://ec.europa.eu/eurostat/statistics-explained/index.php?title=Waste_statistics)

recycle and reuse 60% of generated waste, while 19 countries out of 27 are missing to reach near the milestone of landfilling less than 10% of waste and some countries like Greece, Malta and Estonia are still landfilling 60%, 73% and 83% of their generated total waste respectively (European parliament, Waste management in the EU 2023).

The carbon-neutrality, sustainability of resources and circular economy are the current available approaches to mitigate the climate related changes and protecting the environment. In Sect. 8.5, the European waste management hierarchy as an example of the IWMS will be discussed with focus on how the different steps contribute to protect the environmental and human health.

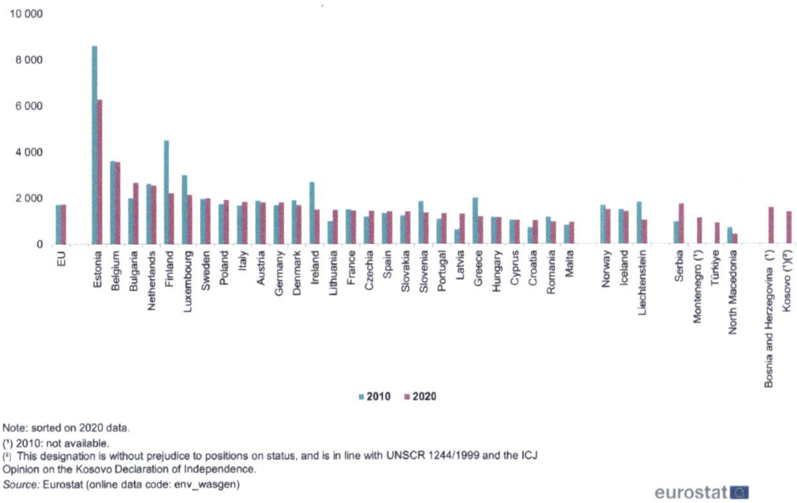


**Fig. 8.2** The share of generated waste per capita in the European member states in the years 2010 and 2020 (Eurostat 2023). For a high-resolution interactive image, please consult this figure at (https://ec.europa.eu/eurostat/statistics-explained/index.php?title=Waste_statistics)

## 8.3 History of Solid Waste Management

Historically, solid waste management is linked to the rise of urban areas and big cities. Before the industrial revolution, the generated wastes were small in amount and the management was unsatisfactory and based on collecting the wastes from cities and discarded nearby the city borders leading to contaminate the surroundings. This management developed with the concept of open dumpsites where wastes from sewers, homes, demolition and buildings were collected and discarded at open holes in the mark at the city gates (Barles 2014).

After the industrial revolution, concerned regarding the fertilizers and resources grew to support industries and growing population with food. Food wastes, blood, mud from streets and all kinds of organic wastes were collected by scavengers from streets to be used as fertilizers which could be regarded as the first big reusing practice in the waste management systems in Europe. However, health concerns linked to the increase of mortality in cities due to the contamination of drinking water and soils led to the inhibition of discarding wastes in streets as the case of Sweden where it was forbidden by the law to discard wastes to streets in 1868 (Rambo and Avfallshistoria 2023). Scavengers were an important part of the solid waste management systems to collect wastes, sorting the wastes to supply it to different industries and finally to dump the residues. These activities were run by the municipalities based on specific fees (Barles 2014). At that time the economic factor related to the recycling of wastes as raw materials was the main driving force for collecting and reusing/recycling wastes as materials or lately to produce energy through incineration. The purpose of incineration first was to reduce the volume of wastes and the cost related to transport the wastes to long distances outside urban areas as these incinerators were located nearby cities. An economic benefit raised from the recycling of ashes in construction industry such as mortars and the energy production to run, as an example, tramways (Melosi et al. 1988).

Sanitary landfill was first introduced in England in 1912 with the idea of separating 1.5–2 m of layers of wastes by inert materials (Barles 2014). Landfills located near cities were cost effective solution for accumulated wastes in urban areas. However, landfills are still considered as the most attractive economic solution for dumping wastes in different parts of the world (Kaza et al. 2021).

During the 1960s, there was a paradigm change in the thinking of international public concerning environmental awareness which initiated by the published of different important books in this regard such as *Silent Spring* by Carson in (1962). In 1972, the United Nations Conference on the Human Environment was held in Stockholm/Sweden with the attendance of more than 100 nations and many organizations and NGOs. The main outcome from Stockholm conference was the declaration which focused on human rights to develop within the frame of healthy environment. The first principle in the Stockholm declaration emphasized on placing the efforts towards environmental protection. Stockholm declaration was a milestone in specifying preservation of the earth natural resources through the implementation of international environmental policies.

In the USA, the Resource Recovery Act was approved by the Congress in 1970 which led to the shift from the linear economy model of resource consumption and disposal to introduce materials recycling and energy recovery through technology development (Pavoni et al. 1975). The fact that the earth's natural resources are finite was getting more attention and driving the needs to implement an integrated waste management systems where wastes are regarded as secondary resources to stand to the demanding society and industries requests. It was also apparent that landfills and open dumpsites are responsible for spreading of contaminated leachates to the ground and surface water, occupying land and contaminating soils and polluting air by diffusing toxic compounds such as carbon dioxide and methane. All these issues are related to human and environmental health and led the United Nations to adopt the 2030 Agenda for Sustainable Development in 2015 (more discussion regarding the SDG and IWMS is given in Sect. 8.7) (United Nations 2015). This act was developed further by the European Union where circular economy and sustainability of resources through well-defined goals was implemented with clear milestones that need to be fulfilled by the member states within different time frames (European Commission 2016). As an example, the European Waste Framework Directive defined clearly that the waste management systems within the European boundaries should "*by 2025, the preparing for re-use and the recycling of municipal waste shall be increased to a minimum of 55%, 60% and 65% by weight by 2025, 2030 and 2035 respectively*" (European Commission 2023), this among other actions were taken by the European Commission to protect our environment as well as guarantee the continuous future development of mankind on the earth.

## 8.4 Integrated Solid Waste Management System (IWMS)

The objective of IWMS is to optimize the management of solid waste to maximize the environmental protection and recovery of valuables from wastes at the lowest possible cost (McDougall et al. 2001). To achieve the sustainability of the earth natural resources the three R's concept (Reduce, Reuse and Recycle) should be key steps in the implementation of IWMS. Therefore, the basic definition of IWMS is the system which can handle solid wastes in any municipality with a circularity of resources to sustain sustainability, in this regard the European waste management hierarchy can be an example of such an IWMS (Fig. 8.3).

However, IWMS is a function of the society properties because such a system is dealing with all the wastes streams at a municipal level (households, industrial, agricultural, etc.), thus IWMS must be designed according to the municipality needs, resources, population, regulations and other factors connected to the characterization of waste materials and recycling/recovery potentials (Jani et al. 2016). In addition, the final destination of the generated wastes at municipality level should also be specified based on the share of hazardous and non-hazardous wastes as well as specify methods to neutralize the hazards.

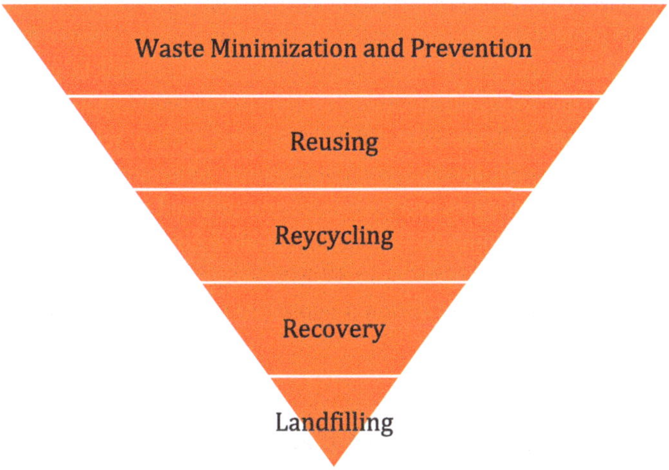


**Fig. 8.3** European waste management hierarchy as an example of current IWMS, the most preferred step from an environmental and sustainability perspectives is in the top and ranging accordingly to the less preferred action in the bottom which is landfilling

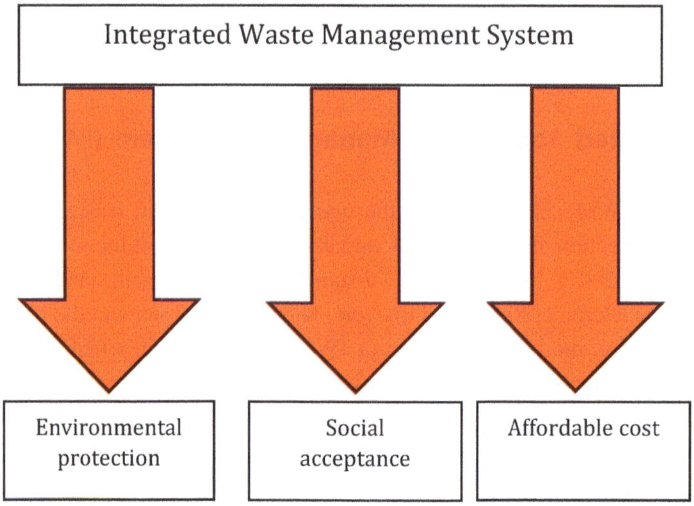


**Fig. 8.4** The main three pillars of the Integrated waste management system which are reciprocally, equally and continuously affecting each other and the final performance of IWMS designed for any municipality

A sustainable waste management system involves the interaction between society, industry and decision makers to fulfill the requirements of a healthy environment at an economic affordable system. As shown in Fig. 8.4, the main pillars of ISWM system are: regulations to protect the environment, society needs and acceptance and economy to run the system (Xiao et al. 2017). These three pillars are essential to guarantee sustainable development of the waste management system in any municipality, hence a balance should be assured between these three pillars. Engaging society requires clear plan with specific goals to reach to all levels from kindergarten to seniors. Industry is the main force for the development of cities and their infrastructure and understanding the needs of industrial partners and describing the regulations with a continuous dialogue are important factors to enhance the sustainability of IWMS. While decision makers have the tools to link between the different actors within a municipal (or a country) level to reach to the already defined goals of IWMS.

## 8.5 Components of IWMS and Environmental Protection

Figure 8.3 shows the component of an integrated waste management system and in this section, we will briefly describe these components and their effects on the final performance of IWMS.

### 8.5.1 Waste Minimization and Prevention

Prevention of waste can be either qualitative through reducing the amount of hazards within the waste or quantitative by minimizing the total amount of the generated wastes. There are many environmental benefits from minimizing or prevention of waste including:

1. Reduce the landfill amount of wastes and hence the spaces occupied by landfills and open dumpsites.
2. Minimize the emission of hazards to the environment through landfill leachate, air pollutants from the decomposition of organic wastes such as methane or carbon dioxide and contamination of soil.
3. Reduce the cost connected to treat the waste and landfilling.

The main goal of waste minimization is to reduce the waste at the source of generation while prevention is to avoid hazards to minimize the environmental impacts. There are different techniques used to minimize and prevent the generation of waste such as changing the raw materials with other materials with less impact on the environment. As an example, in the production of modern crystal glasses barium, potassium or zinc oxides are used instead of lead oxides to reduce the hazardous impact connected with the use of lead (Jani and Hogland 2014).

Waste minimization also includes the reduction of the volume of waste such as incineration process, however incineration is also leading to recover energy and can take an action in two steps in the IWMS. For instance, plastic waste is identified as one of the waste streams responsible for negatively impacting biodiversity and human health (Mumtaz et al. 2023). About 6.3 billion tons of plastic waste generated in the period between 1950 to 2018 with an incineration rate of 12% in the whole world (Cudjoe and Wang 2022). Incineration process can reduce the volume of the plastic waste to 20% also able to handle mixed plastic wastes to recover energy (Cudjoe and Wang 2022). Another way to prevent and minimize waste generation is by using renewable raw materials or by developing the production process. Optimization can also be a key factor to increase the production as well as reducing the amount of generated wastes from any process.

Regulations and guidelines (more details in Sect. 8.7) are important tools on national as well as international levels to control waste minimization and prevention. Local environmental agencies like Swedish EPA, USEPA and others are responsible for protecting the environment through implementing programs guiding local industries and citizens to reduce the waste generated or to prevent specific hazards in the production of products. These agencies are also responsible to apply research and provide information to develop methods to minimize waste generation. As an example, the European Single Use Plastic Directive (Directive 2019/904) (2019) of the European Parliament and of the Council on the reduction of the consumption of plastic bags due to their environmental impact.

### 8.5.2 Reusing and Recycling

Shortage of materials, crises, restrictions and regulations and environmental protection are among the many reasons which have been a driving force for recycling/reusing of different wastes including the hazardous ones. For the sustainability of the earth natural resources reuse and recycle are essential steps. Many natural resources like copper and iron, as examples, are suffering from reduction in their primary sources and reusing/recycling can be essential for saving these resources (Jani et al. 2017). It has been reported that about 50% of copper and iron have been lost as waste (Krook and Baas 2013) while there is an industrial demand for most of the natural resources due to the continuous growth of population and technology development. Recycling/reusing is an approach where waste can be regarded as a source to gain economic benefits. Recycling has covered many waste streams like metals scrabs, glasses, paper, cardboard and others like food waste. Recycling/reusing is done based on four different paths:

1. Direct recycling/reusing in the same industry as a raw material.
2. Indirect recycling/reusing as a raw material in different industries.
3. Extraction of valuables from a waste stream or using as an agent to treat a contamination.
4. As a source to recover energy (will be covered in Sect. 8.5.3).

One of the major problems of landfilling is the occupation of land space that can be used for other purposes like agriculture and urbanization. Many landfills at different part of the world have already reached their capacity and there are no available spaces for erecting new landfills, this is apparent problem in Belgium and Hong Kong as examples (Hernández Parrodi et al. 2019). Reusing and recycling will also enhance the minimization of the waste streams to landfills and avoid environmental impact associated with dumping sites, however, reuse and recycle will not eliminate landfilling but they will extend the life and capacity of the current landfills.

One of the key factors for recycling/reusing is the characterization of the waste material (Jani et al. 2016, 2017; García López et al. 2019). Physicochemical properties like organic and moisture contents, particle size distribution, metals content, density and volume of wastes are crucial to identify the possibility of recycling/reusing of any waste materials (Hull et al. 2005; Quaghebeur et al. 2013; Mönkäre et al. 2016). The value of the waste stream depends on the quality of the material including the content of impurities.

#### 8.5.2.1 Kalundborg Symbiosis in Denmark

One of the most successful examples of recycling/reusing projects is the case of Kalundborg Symbiosis in Denmark where 16 different partners from industries, decision makers and society are working together to reduce the environmental degradation by sharing waste materials, water and energy from one partner to other partners as sources instead of discharging as wastes (Valentine 2016a). Industrial symbiosis as defined by Chopra and Khanna (2014) is:

> a mutually beneficial relationship between industries that achieves productive use of waste and byproducts- promotes sustainable development by providing economic benefits while minimizing environmental degradation caused by the participating industries.

Sustainability and circularity of natural resources as a base for protecting the environment is the core of the industrial symbiosis and the whole industrial ecology concept (Ehrenfeld and Ni 1997; Domenech and Davies 2011; Valentine 2016b). Closing the loop of energy and materials have been followed since the establishment of the Kalundborg symbiosis in 1961 (Ehrenfeld and Ni 1997). The created linkages between the partners at this symbiosis enhance sharing of the surplus energy and byproducts as raw materials instead of losing the value by landfilling. Table 8.1 shows the savings of energy, materials and economic benefits reported by two studies during two different years at Kalundborg Symbiosis.

**Table 8.1** Yearly environmental and economic savings at Kalundborg Symbiosis during two different years (Ehrenfeld and Ni 1997; Domenech and Davies 2011)

| Material and energy | Savings reported 1997 (Ehrenfeld and Ni 1997) | Saving reported 2011 (Domenech and Davies 2011) |
|---|---|---|
| Water | 1.2 million $M^3$ | 3.9 million $M^3$ |
| Sulfur | 2800 tons | 20 000 tons |
| Gypsum | 80 000 tons | 170 000 tons |
| Nitrogen | 800 tons | |
| Phosphorus | 400 tons | |
| Coal | 30,000 tons | |
| Oil | 19 000 tons | |
| Minimized waste | | |
| Fly ash and clinker | 200 000 tons | |
| Sludge | 80 000 tons | |
| Sulfur | 2800 tons | |
| Sulfur dioxide | 1500–2500 tons | 53 tons |
| Carbon dioxide | 130 000 tons | 64 460 tons |
| Biomass | | 319 000 $M^3$ |
| Capital annual savings | | 15 million US dollars |

### 8.5.3 Energy Recovery

Retrieving of energy contained in wastes (also called Waste to Energy) is an important step in the IWMS. Studies have shown that about 50% of the solid waste is biodegradable fractions which usually are renewable materials rich of energy (Kayhanian 1995; Ramasamy and Abbasi 2000; Pangallo et al. 2023). When the (i) recycling/reusing available techniques are costly, (ii) the wastes are a mixture of unsorted materials, or (iii) the wastes are organic materials then energy recovery can be an alternative option. In addition, waste incineration, as part of energy recovery, is also regarded as one of the treatment options of hazardous wastes such as medical wastes (Zhao et al. 2209).

On the other hand, with the current demand of energy as a cause of the growing consumption of industrial, social and economic sectors the need to establish an alternative fuel for the fossil ones is a priority. According to the BP Statistical review of world energy (BP Statistical Review of World Energy 2018) human consumed about 13.5 billion tons of oil equivalent during 2017, see Fig. 8.5 for details. Fossil fuels are well known for the generation of greenhouse gases the methane and carbon dioxide among others when burning. Furthermore, landfilling organic wastes is also responsible for the emission of greenhouse gases. Therefore, using solid wastes as fuel can save the environment through reduce the emission of greenhouse gases from both sectors as well as saving the finite fossil fuels, minimize landfilling and provide cost effective energy. Our focus in the energy recovery step will be given for the incineration of waste as it is the second oldest method of managing municipal solid wastes.

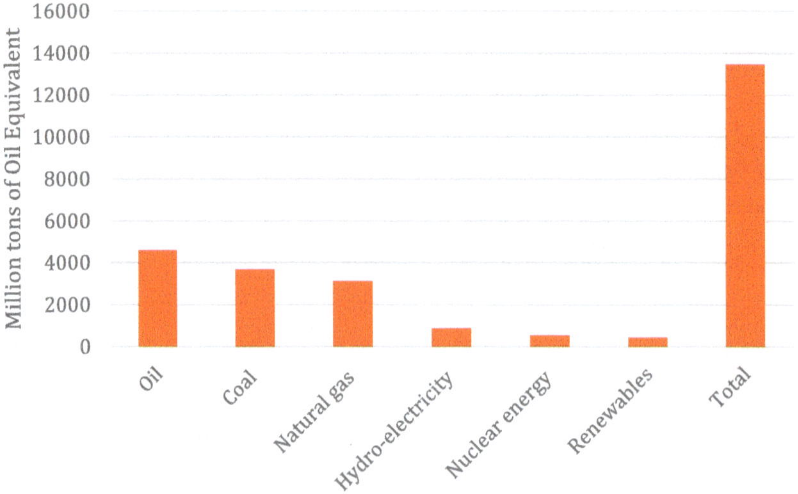


**Fig. 8.5** The world's energy consumption based on fuel type in 2017 (data extracted from BP Statistical review of world energy (BP Statistical Review of World Energy 2018))

#### 8.5.3.1 Waste Incineration

Waste incineration (combustion) is the second oldest method, behind dumping of waste, as a management practice. It was first used to reduce the volume of the generated wastes and to maintain hygiene. The oldest incinerator (called destructor) was built by Alfred Fryer for Nottingham municipality in England year 1874 (Hering and Greeley 1921). Theses incinerators were based on a batch process of loading and cleaning and were common in Europe. However, many countries still depend on burning of wastes in open air even with the environmental consequences connected to this practice (ISWA 2016).

Currently incinerators are appreciated for reducing the waste volume and generating energy for commercial uses. To guarantee a maximum efficiency of an incinerator, the combustion process must be regulated to generate minimum residue and emissions are controlled to minimize hazards to the atmosphere. Crucial factors to design and operate an incinerator are the characterization of the waste stream as well as the generated waste amount by the specific municipality. These factors are affecting the performance of the incinerator and the total heat produce. The type of waste can, as an example, reduce or increase the produced heat due to the differences in the heat value of each waste. Generally, to ensure economical viability of an incinerator, the average heat value of the mixture of waste should be about 0.55 kcal/kg (Pavoni et al. 1975). Another important factor is the moisture content of waste which has different value based on the country of choice, for instance, the average moisture content of waste in Europe and USA range between 20 and 30% while in Asian countries like China it is between 30 and 60% (Jani et al. 2016). Moisture content is affecting the economy and the operation of the incinerator due to increase the need for auxiliary fuel with increasing the moisture content of waste. The auxiliary fuel is essential to guarantee efficient burning with high moisture content or low heat value of waste. Tables 8.2 and 8.3 show the effect of moisture content on the heat value of different waste materials. It clear that moisture content can affect the performance of incineration and therefore it is crucial to control the moisture content in the waste stream before implementing incineration as a method for recovering energy.

In the European Union (EU), waste incineration is playing a major role to reduce the volume of generated waste and hygiene purposes especially hazardous wastes like medical. On the second hand, incineration provides European countries with energy. In 2019, the EU countries incinerated about 100 million tons of residual waste in 500 incinerator

**Table 8.2** Heat value of different types of waste materials, dry matter of almost zero moisture (data was extracted from Pavoni et al. (1975) and Bagchi (2004) with recalculation to SI units)

| Type of waste | Heat value in KJ/kg |
|---|---|
| Rubber | 29.075 |
| Wood | 19.538 |
| Paper | 18.375 |
| Garden (leaves and grass) | 20.003 |
| Mixed | 16.980 |

**Table 8.3** Average heat value of different types of waste at (31% w/w) moisture content (data was extracted from Pavoni et al. (1975) and Bagchi (2004) with recalculation to SI units)

| Type of waste | Heat value in KJ/kg |
|---|---|
| Plastic | 32.000 |
| Paper and cardboard | 13.000 |
| Vegetable | 1.500 |
| Mixed | 10.000 |

plants to generate 43 billion kWh electricity and 99 billion kWh heat, this sum of energy is equal to 13.8 billion $m^3$ of natural gas or 10–54 billion tons of fossil fuel (CEWEP 2019). Incinerators contribute also to recycling materials such as metals content in bottom ash or reusing the ash in construction industry as aggregate or in the production of cement. According to CEWEP (2019), considering only the energy production and recycling the bottom ash can contribute to save 20 equivalent kilogram of carbon dioxide for each ton of waste incinerated at the EU. In addition, residues from other facilities to recover energy like the composting and anaerobic digestion of biowaste are also in need for treatment and therefore incinerators are the available solution for that.

**Sweden's Waste to Energy Scenario**

Sweden has built one of the most successful scenarios of recovering energy from waste through incineration with its total 38 incineration plants. The kingdom reached to the maximum energy recovery per ton of waste in Europe with 3 MWh in 2019 (Swedish Waste Management 2023). Sweden is also in the global lead regarding energy recovery from waste through the incineration of 6.7 million tons of local and imported waste burned during 2019, the recovered energy was 17.8 TWh and 2.7 TWh used for district heating and electricity respectively (Swedish Waste Management 2022).

Figure 8.6 shows the share of energy recovery (55.4%) compared to material recycling (27.5%) and landfilling (1.6%) in Sweden in 2022 compared to results in years 2021, 2020 and 2019 (Swedish Waste Management 2022). The results indicate that Sweden witnessed a decrease in the energy incineration during the last four years connected to decrease the collected wastes on municipal levels. However, it is obvious that Sweden's incineration is dominating the waste management sector which can be a reason for the stable and slow development of the recycling performance.

### 8.5.4 Disposing/Landfilling of Wastes

Disposing waste in a place is the oldest set of managing wastes and is one of the most used and still the cheapest method of handling waste globally. As discussed in Sect. 8.3, disposing has witnessed actual development from just throwing waste outside nearby a house or a city border till the modern sanitary landfills with all technical development to

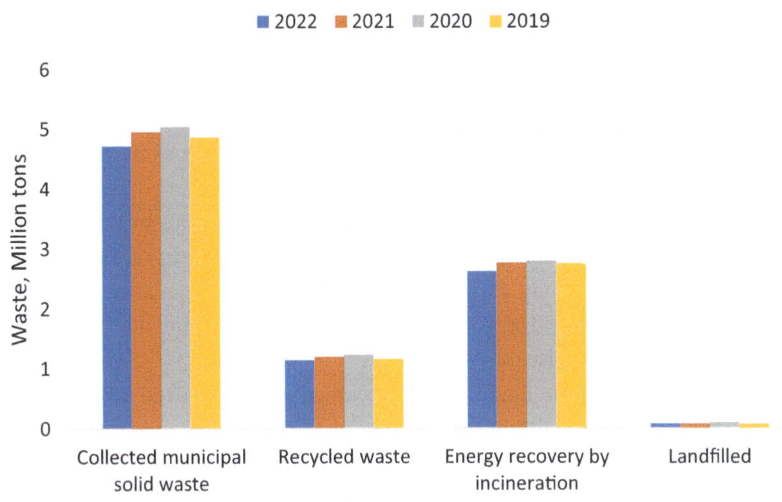


**Fig. 8.6** The collected Swedish municipal solid waste in the last four years with share of each step following the EU waste management hierarchy (data extracted from the Swedish Waste Management (2022)

protect the environment from the contaminated leachate and spreading gases like methane to the atmosphere. The concept sanitary landfilling means a cut and cover operation which involving three steps:

1. Daily covering the waste materials in a layer of soil or composite material and final covering of the waste cell when capacity is reached.
2. Prohibit the incineration of waste in open air and dumpsites.
3. Prevent the emissions of hazardous with both leachate and gases to the atmosphere to protect surface and ground water and air.

Despite the efforts of the United Nations through the 2030 Agenda for Sustainable Development (United Nations 2015) and other regional and local organizations to reduce and prevent landfilling still 37% of the total generated waste globally is disposed at different forms of dumping sites (Kaza et al. 2018) and about 31% is disposed at open dumpsites and only 9% at sanitary landfills. For instance, member states involved in the European Union have ambitious target to reduce landfilling to less than 10%, though, 18% of the municipal solid waste is landfilling at EU annually (European parliament, Waste management in the EU 2023).

According to EURELCO, around half million landfills of different forms is found in EU which contain about 30–50 billion $m^3$ of wastes and about 80% of the waste content is municipal solid waste. About 10% of these landfills sites are sanitary landfills while the rest are missing in large extent any proper environmental protection facilities (EURELCO

2023). However, open dumpsites (and in small extent sanitary landfills) are responsible for the creation of many environmental and health hazards such as:

1. Water leachate with considerable concentrations of both organic and inorganic contaminants like nutrients and metals that can transfer to the surrounding environment and reaching to soil and surface and ground water.
2. Leaking of hazardous gases to the atmosphere and small particles of wastes which can be blown by wind. Gases such as methane and carbon dioxide have direct impacts on the climate change while small particles like PM2.5 or PM10 can affect human health.
3. Pests and animals looking for food and scavengers collecting at landfills can be responsible for spreading diseases.
4. Waste fires specially when landfills are occupied with a mixed of wastes. These fires have dual effect through emitting hazardous gases to the atmosphere or through contaminating water when extinguish.
5. Occupying areas of high economic value which can be used for other purposes.

An important factor concerning landfilling, is that when landfilling any material (organic or inorganic) the economic value of this material will be lost and opportunities of reusing/recycling or even recovery as energy will be eliminated. Mixing materials in the same cell at landfills can create, with time, complex chemical compounds that will be economically unfeasible to separate. Therefore, landfilling should be managing only sorted waste materials and be regarded as a saving place for resources when there will be a development in techniques able to reuse these materials as resources again.

## 8.6 Regulations and Environmental Protection Within the Frame of IWMS

Regulations are the governmental tools to control the performance of a waste management system to ensure the environmental and human health protection. Local, regional and global environmental protection authorities are responsible for performing monitoring programs, guidelines, recommendations and codes to reduce the current and future environmental impacts. However, local regulations are based on country targets and industrial activities, among other factors, but in general harmonized within the international environmental law as a frame. Therefore, the international environmental law main four principles (no-harm, prevention, precaution and polluter-pay) can be found as a background in many of the local and regional environmental regulations (Dupuy and Viñuales 2020). Although, countries need always to find a balance between economic growth and environmental protection policies, thus, countries with powerful economy have established more restricted environmental regulations compared to poor countries. This lead companies from USA and European developed countries to migrate to Asia and Africa where environmental legislations are fragile. Unfortunately, such action was and still responsible for the contamination of the environment in many developing countries where the environment was sacrificed for faster economic benefits.

Waste management sector as one of the main emitters of pollutants to the environment has witnessed paradigm shift from the open loop system where the last destination of wastes is dumpsites to a close loop management system regarding wastes as a secondary resource to guarantee sustainable development and environmental protection. Table 8.4 shows some of the international environmental conventions and treaties connected to the waste management sector.

The function of the international environmental law is to set regulations, norms and practices connected to wastes to protect the environment and human health on a global scale. The international environmental law has succeeded to deal with the generation, transportation and trade of hazardous wastes; however, this law has failed to deal with the waste itself as a material and is still struggling to deal with how to minimize the generation of wastes and transportation of different types of wastes globally (Barsalou and Picard 2018). In addition, the international environmental law is missing the legal definition of waste (Sadeleer 2006). For instance, going through the Basel convention (1989) reveal that the main goal behind this convention is to protect the industrial trade and circulation of wastes instead of minimizing the generation and circulation of hazardous wastes (Lipman 2015). The flow of hazardous wastes between countries under the Basel convention has increased from five million tons in 1945 to about 400 million tons in 2000 which can be direct evidence for the lack of the international environmental law as a tool to tackle environmental impacts. Currently, there is an urgent need to specify accurately the role of the international environmental law regarding waste generation and actions to implement IWMS as a tool to protect the environment.

In contrast, the European Union (EU), as a regional example in this area, has worked since the establishment in 1993 with focus on the minimization of generated waste based on specific targets with an obligate time frame for all member states to follow. This is an important step to secure circularity of resources and sustainability. The European Union (EU) has established a mechanism (i.e. eligible directives for member states) to hunt all the environmental threats within the EU boundaries and beyond to take the lead in the mitigation of all environmental impacts. The European Commission is the responsible legal body in this mechanism to initiate and generate regulations in the form of directives to all member states based on the Environmental Action Programs. While the European Environmental Agency (EEA) is acting as a channel of information and provider of description of actions and environmental work between member states and society. Table 8.5 shows few of the current EU directives and targets specified for all member states involved within the waste management sector.

December 2019 the European Commission displayed the ***European Green Deal*** *a roadmap for making the EU's economy sustainable by turning climate and environmental challenges into opportunities across all policy areas and making the transition just and*

**Table 8.4** International environmental conventions and treaties connected to the waste management sector

| Convention or treaty | Year of adoption | Summary | Reference |
|---|---|---|---|
| London convention convention on the prevention of marine pollution by dumping of wastes and other matter | 1972 | The goal of this convention is to encourage the effective control of dumping of wastes and other hazardous at sea to prevent marine pollution. This convention contains a list of the materials and wastes that are prohibited from dumping at sea | London Convention (1972) |
| London protocol (replacing London convention) | 1996 | This protocol is the modernization version of the London Convention. This protocol is more restrictive and taking in account the precautionary principle. It also includes a list of wastes and materials prohibited from dumping at sea such as incineration at sea and export of wastes for dumping and incineration at sea | London Protocol (1996) |
| United Nations Framework Convention on Climate Change (UNFCCC) | 1992 | This convention aiming to "*stabilisation of greenhouse gas concentrations in the atmosphere at a level that would prevent dangerous anthropogenic interference with the climate system*" (United Nations Framework Convention on Climate Change 1992)<br>And it contains actions to control emissions of greenhouse gases from waste management sector | United Nations Framework Convention on Climate Change (1992) |
| Kyoto protocol (COP3) | 1997 | This protocol binding developed and industrial countries to reduce and limit the greenhouse gases emissions by following agreed targets in the UNFCCC. Actions to reduce emissions of greenhouse gases from waste management were also addressed in this protocol in the frame of the UNFCCC | Kyoto Protocol (1997) |

(continued)

**Table 8.4** (continued)

| Convention or treaty | Year of adoption | Summary | Reference |
|---|---|---|---|
| Basel convention | 1989 | This convention is aiming to manage the disposal and transfer of hazardous and other wastes between countries to protect human health and the environment. Countries with minimum advance storage and dumping facilities are prohibited from importing such wastes unless the purpose and information is well documented and consent with the Basel convention secretariat | Basel Convention (1989) |
| Minamata convention on mercury | 2013 | Managing the life cycle of mercury from production to safe disposal which including actions to mitigate and hinder the use and the release of mercury from industrial activities like gold mining as well as initiating control programs for the emission of mercury to the atmosphere from industrial activities like coal power plants | Minamata Convention on Mercury (2013) |

*inclusive for all* (WECOOP 2023). The goals behind the European Green Deal are to make EU: (a) neutral regarding greenhouse gases emissions in 2050; (b) protect, sustain and restore the natural capital; (c) protecting health of society and well-being from environmental crises and (d) disconnect economic growth from consumption of resource. These goals are among other ambitious targets implemented by the EU to different sectors in society. The EU allocates €1.8 trillion to finance the European green deal within a period of seven years. Preventing the generation of waste through demanding recycling/reusing and repair of products is one of the targets of the EU green deal (WECOOP 2023).

One of the most important issues missing in the international environmental law that is successfully managed by the EU is defining the legal meaning of waste as well as adding a new criterion called *End-of-waste*. According to the Waste Framework Directive (1998) article 6 (1) and (2) the waste is no longer regarded as a waste when it undergoes recycling/reusing or recovery as secondary material and when there is a demand in the market for this specific material. However, circulating the waste material back to the economy as a resource should be under a crucial condition of that recycling/reusing the waste material will not possess any threats to human health and the environment.

**Table 8.5** Current European directives and targets concerning waste management sector

| Directive | Year of adoption | Specified targets to member states that must be achieved within a planned time frame | Reference |
|---|---|---|---|
| Packaging and packaging Waste Directive 94/62/EC | 1994 | The main goal is to prevent the generation of packaging waste and promoting recycling of packages, currently 55% of recycling of glass, paper, metals, wood and plastic packages are achieved and the target is to reach to 65% in year 2025 and 70% in year 2030 of all kinds of packages | Packaging and packaging Waste Directive 94/62/EC (1994) |
| Waste Framework Directive 2008/98/EC | 1998 | The goal is to consider wastes as secondary resources through well-defined waste management programs by each member state by following the EU hierarchy and prevent the generation of wastes with clear binding targets. The current targets are recycling 55%, 60% and 65% of municipal solid wastes in years 2025, 2030 and 2035 respectively | Waste Framework Directive (2008) |
| End-of-Life Vehicles Directive 2000/53/EC | 2000 | This directive prohibit the use of hazardous compounds in the production of new vehicles like mercury, lead and cadmium and set recycling/reusing targets by member states. The goal is to recycle/reuse at least 85% of the weight of each end-of-life vehicle and 10% more should be recovered as energy based on the concept of Extended producer responsibility | End-of-Life Vehicles Directive (2000) |
| Batteries Directive 2006/66/EC | 2006 | By 2025 recycling:<br>– 75% by average weight of lead-acid batteries<br>– 65% by average weight of lithium-based batteries<br>– 80% by average weight of nickel–cadmium batteries<br>– Recycling of 50% by average weight of other waste batteries<br>By 2030 recycling:<br>– 80% by average weight of lead-acid batteries<br>– 70% by average weight of lithium-based batteries | |

(continued)

**Table 8.5** (continued)

| Directive | Year of adoption | Specified targets to member states that must be achieved within a planned time frame | Reference |
|---|---|---|---|
| Waste electrical and electronic equipment (WEEE) 2012/19/EU | 2012 | The aim of this Directive is to prevent the generation of WEEE and improve the collection and recycling/reusing. The Directive sets specific targets of recycling 65–85% of different electronic wastes by 2030 | Waste electrical and electronic equipment (2012) |
| Landfill Directive EU 2018/850 | 2018 | This directive aims to protect the environment from landfills impact by setting technical requirements to establish individual landfills based on the sort of waste (hazardous, non-hazardous and inert) as well as forcing member states to reduce landfilling of organic materials. In addition, member states must reduce landfilling to less than 10% by 2035 | Landfill Directive EU 2018/850 (2018) |

## 8.7 Integrated Waste Management System in the Context of Sustainable Development Goals (SDGs)

A crucial issue that human need to consider when discussing sustainability of natural resources is that we are living in a close system regarding natural materials because the main and the only source of these materials is the Earth. The Earth has finite natural resources which need to be protected from depletion due to unsustainable consumption. To guarantee the permanent function of the Earth to provide human with natural resources sustainability and circularity of resources is a vital strategy.

Sustainability of natural resources is one of the main challenges facing the management of municipal solid wastes. In 2015, the United Nations established the 2030 Agenda for Sustainable Development including 17 well-defined goals aiming to reach to sustainable development on the Earth to hinder the degradation of the environment, to protect human health and to guarantee prosperity of human (United Nations 2015). Waste management has connection with most of the 17 goals, however, a direct link can be to the following three goals:

Goal number 7: *Ensure access to affordable, reliable, sustainable and modern energy for all* (United Nations 2015). As it was already mentioned in Sect. 5.3, providing energy is connected to the generation of different forms of wastes as well as the use of different wastes as sources to provide energy such as incineration of solid wastes.

Goal number 9: *Build resilient infrastructure, promote inclusive and sustainable industrialization and foster innovation* (United Nations 2015). Industrial activities are recognized as the major emitter of solid wastes to the environment as well as the main consumer of natural resources including energy sources. The minimization, prevention, recycling and recovery of wastes by industries will contribute towards the sustainability of natural resources as well as protecting human and environmental health.

Goal number 12: *Ensure sustainable consumption and production patterns* (United Nations 2015). Minimization and prevention of waste streams through designing products with long lifetime, recyclability and easy to reuse is essential and is one of the targets of an integration solid waste management system.

The main aim of the integrated waste management system is to sustain the natural resources from depletion by keeping the resources in the economy loop as long time as possible. Such a system will, in firsthand, protect the environment from the contamination and degradation while, in the second hand, minimize the stream of wastes to landfills. However, such a system like the EU waste management hierarchy still proposes landfilling as the last step. As shown in Fig. 8.7, we are proposing an alternative system which can be called sustainable integrated waste management system that excluding the disposal of waste completely and implement mining landfills for secondary resources instead (Jani 2018). Such a sustainable system will import back all the lost materials dumped at landfills to the economy and provide alternative resources instead of harvesting new natural resources from the Earth. Landfill mining approach can also help to reduce the environmental impacts of landfills and open dumpsites (Jani et al. 2016). Nevertheless, mining landfills and open dumpsites is still facing technical and legal challenges that need to be addressed in the future.

## 8.8 Conclusions

The worlds municipal solid waste generation is challenging the United Nations sustainable development goals on different level especially when the focus is on protecting the environmental and human health or sustainability of resources. Generally, society is missing the impact of wastes when waste is generated in anthropogenic activities outside households. Awareness regarding invisible wastes generated with the production of daily usable items such as mobile phones or cloths needs to be highlighted on global level. Waste generation is not only responsible for contaminating land spaces due to landfilling in sanitary landfills or open dumpsites, but it is also contaminating surface water and groundwater and the atmosphere through the emissions of leachates and greenhouse gases respectively. The decomposition of organic wastes is connected to anerobic and aerobic reactions with gases such as carbon dioxide and methane while the movability of nano and micro sizes of organic and inorganic compounds can cause real environmental challenges when accumulated in considerable concentrations in different media such as water or soil. Landfilling, on the other hand, is responsible for losing the economic value of the disposed wastes as it will be complex to recall the buried wastes back as resources in the circular economy. International, regional and local regulations are tools to protect the environmental and

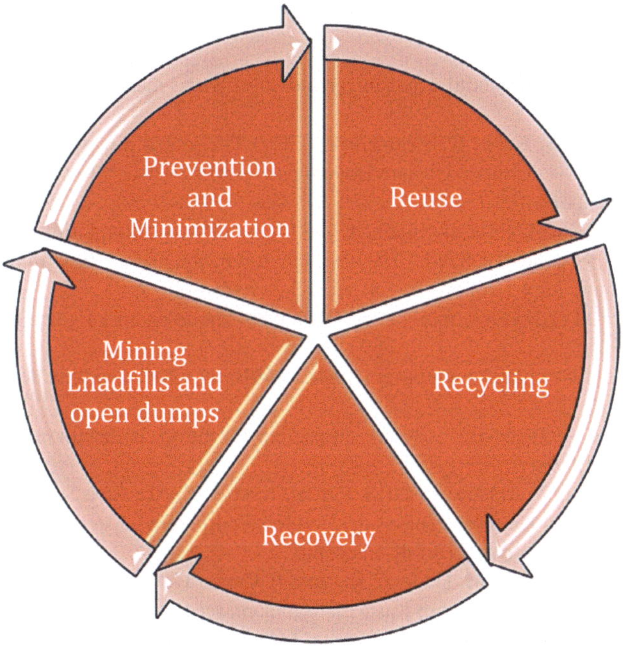


**Fig. 8.7** A sustainable integrated waste management system where landfilling and disposal of wastes are excluded, and wastes are valued as secondary resources through mining landfills

human health, however, the majority of the former and current international conventions and treaties are missing clear mechanism for reducing waste generation and minimizing of landfilling. In contrast, the European Union has implemented clear goals in this regard where member states need to reach to specific requirements and milestones as a crucial action to protect the environment in the future. Regardless, the efforts to reach to the sustainability of resources and reducing losing virgin materials as wastes, still a real development is needed on all basis (international, regional and countries) to implement an IWMS able to fit the need of each municipality. Such an integrated waste management system based on the European waste management hierarchy maybe able to reduce the impact of generated waste globally if steps will be carefully followed, as it was reviewed in this study.

**Acknowledgements** Authors would like to acknowledge the support given by the School of Business Society and Engineering, Division of Sustainable Environment and Construction at Mälardalen University.

**Conflict of Interest Statement** The authors have no conflicts of interest to declare.

## References


Bagchi A (2004) Design of landfills and integrated solid waste management. Wiley, New Jersey, USA

Balde P, Forti V, Gray V, Kuehr R, Stegmann P (2017) The global e-waste monitor 2017. https://www.itu.int/en/ITU-D/Climate-Change/Documents/GEM%202017/Global-E-waste%20Monitor%202017%20.pdf

Barles S (2014) History of waste management and the social and cultural representations of waste. In: Agnoletti M, Neri Serneri S (eds) The basic environmental history. Environmental history, vol 4. Springer, Cham. https://doi.org/10.1007/978-3-319-09180-8_7

Barsalou O, Picard M (2018) International environmental law in an era of globalized waste. Chin J Int Law 887–906

Basel Convention (1989) http://www.basel.int/Portals/4/Basel%20Convention/docs/text/BaselConventionText-e.pdf

Batteries Directive 2006/66/EC (2006) https://eur-lex.europa.eu/legal-content/EN/TXT/?uri=celex%3A32006L0066

BP Statistical Review of World Energy 2018. Report. June 2018. Accessed September 2023. https://www.bp.com/content/dam/bp/business-sites/en/global/corporate/pdfs/energy-economics/statistical-review/bp-stats-review-2018-full-report.pdf

Burlakovs J, Jani Y, Kriipsalu M, Gaile Z, Kaczala F, Celma G, Ozola R, Rozina L, Rudovica V, Hogland M et al (2018) On the zero waste management: recovery potential of elements including rare earth elements from fine fraction of waste. J Clean Prod 186:81–90

Carson R (1962) Silent spring. Houghton Mifflin, Boston, USA

CEWEP (Confederation of European Waste to Energy Plants), 2019, accessed on august 2023. Waste to energy climate roadmap. https://www.cewep.eu/wp-content/uploads/2022/06/CEWEP-WtE-Climate-Roadmap-2022.pdf.pdf

Chopra S, Khanna V (2014) Understanding resilience in industrial symbiosis networks: insights from network analysis. J Environ Manage 141(2014):86–94

Cudjoe D, Wang H (2022) Plasma gasification versus incineration of plastic waste: energy, economic and environmental analysis. Fuel Process Technol 237 (2022). https://doi.org/10.1016/j.fuproc.2022.107470

Directive (EU) 2019/904 of the European Parliament and of the Council of 5 June 2019 on the reduction of the impact of certain plastic products on the environment (Text with EEA relevance). Accessed September 2023

Domenech T, Davies M (2011) Structure and morphology of industrial symbiosis networks: the case of Kalundborg. Procedia Soc Behav Sci 10:79–89

Dupuy P, Viñuales J (2020) International environmental law, 2nd edn. Cambridge University Press, Cambridge. https://doi.org/10.1017/9781108399821

European Commission (2016) Sustainable Development: EU sets out its priorities. https://ec.europa.eu/commission/presscorner/detail/en/IP_16_3883. Accessed September 2023

European Parliament, European Parliament resolution of 10 February 2021 on the New Circular Economy Action Plan (2020/2077(INI)). https://www.europarl.europa.eu/doceo/document/TA-9-2021-0040_EN.html. Accessed September 2023

European Commission. Waste Framework Directive. https://environment.ec.europa.eu/topics/waste-and-recycling/waste-framework-directive_en. Accessed September 2023

Ehrenfeld J, Ni G (1997) Industrial ecology in practice, the evolution of interdependence at Kalundborg. J Ind Ecol 1:67–79

End-of-Life Vehicles Directive 2000/53/EC (2000) https://eur-lex.europa.eu/legal-content/EN/TXT/?uri=celex%3A32000L0053

EURELCO. https://eurelco.org/2018/09/30/data-launched-on-the-landfill-situation-in-the-eu-28/. Accessed September 2023

European parliament, Waste management in the EU: infographic with facts and figures. https://www.europarl.europa.eu/news/en/headlines/society/20180328STO00751/waste-management-in-the-eu-infographic-with-facts-and-figures. Accessed September 2023

Eurostat, Waste statistics. https://ec.europa.eu/eurostat/statistics-explained/index.php?title=Waste_statistics. Accessed September 2023

García López C, Ni A, Hernández Parrodi JC, Küppers B, Raulf K, Pretz T (2019) Characterization of landfill mining material after ballistic separation to evaluate material and energy recovery potential. Detritus 8(1):5–23. https://doi.org/10.31025/2611-4135/2019.13780

Gómez M, Grimes S, Qian Y, Feng Y, Fowler G (2023) Critical and strategic metals in mobile phones: a detailed characterisation of multigenerational waste mobile phones and the economic drivers for recovery of metal value. J Clean Prod 419:138099

Hernández Parrodi JC, Lucas H, Gigantino M, Sauve G, Esguerra JL, Einhäupl P, et al (2019) Integration of resource recovery into current waste management through (enhanced) landfill mining. Detritus 8(1):141–156. https://doi.org/10.31025/2611-4135/2019.13884

Hering R, Greeley A (1921) Collection and disposal of municipal refuse. McGraw-Hill, New York, USA

Hull RM, Krogmann U, Asce M et al (2005) Composition and characteristics of excavated materials from New Jersey landfill. J Environ Eng 3:478–490

ISWA (International Solid Waste Association) report (2016) A roadmap for closing waste dumpsites. http://ars.org.ar/news/ISWA_Roadmap_Report.pdf

Jani Y, Hogland W (2014) Waste glass in the production of cement and concrete-a review. J Environ Chem Eng 2:1767–1775

Jani Y, Kaczala F, Marchand C, Hogland M, Kriipsalu M, Hogland W, Kihl A (2016) Characterization of mined fine fraction and waste composition from a Swedish landfill. Waste Manage Res 34(12):1292–1299

Jani Y, Kriipsalu M, Pehme K, Burlakovs J, Hogland M, Denafas G, Hogland W (2017) Composition of waste at an early EU landfill of Torma in Estonia. IJEE 8(2):113–117

Jani Y (2018) Landfills and glass dumpsites as future bank accounts of resources: waste characterization and trace elements extraction. Linnaeus University Dissertations; 308. Växjö: Linnaeus University Press, ISBN: 9789188761163

Kaya M (2019) Electronic waste and printed circuit board recycling technologies. Springer, Cham. https://doi.org/10.1007/978-3-030-26593-9

Kayhanian M (1995) Biodegradability of the organic fraction of municipal solid waste in a high solids anaerobic digester. Waste Manage Res 13:123–136

Kaza S, Yao LC, Bhada-Tata P, Van Woerden F (2018) What a waste 2.0: a global snapshot of solid waste management to 2050. Urban Development. © Washington, DC: World Bank. http://hdl.handle.net/10986/30317 License: CC BY 3.0 IGO

Kaza S, Shrikanth S, Chaudhary S (2021) More growth, less garbage. Urban Development Series. © World Bank, Washington, DC. http://hdl.handle.net/10986/35998 License: CC BY 3.0 IGO

Krook J, Baas L (2013) Getting serious about mining the technosphere: a review of recent landfill mining and urban mining research. J Clean Prod 55:1–9

Kyoto Protocol (COP3) (1997) https://unfccc.int/sites/default/files/resource/docs/cop3/l07a01.pdf

Landfill Directive EU 2018/850 (2018) https://eur-lex.europa.eu/legal-content/EN/TXT/?uri=celex:32018L0850

Lipman Z (2015) Trade in hazardous waste. In Alam S, et al (eds) International environmental law and the global south, p 257
London Convention (1972) https://www.epa.gov/sites/default/files/2015-10/documents/lc1972.pdf
London Protocol (1996) https://www.epa.gov/sites/default/files/2015-10/documents/lpamended2006.pdf
McDougall F, White P, Franke M, Hindle P (2001) Integrated solid waste management: a life cycle inventory. Blackwell Science, Oxford, UK
Melosi M (1988) Technology diffusion and refuse disposal: the case of the British destructor. In: Tarr JA, Dupuy G (eds) Technology and the rise of the networked city in Europe and America. Temple University Press, Philadelphie, pp 207–226
Minamata Convention on Mercury (2013) https://minamataconvention.org/sites/default/files/2021-06/Minamata-Convention-booklet-Sep2019-EN.pdf
Mönkäre T, Palmroth M, Rintala A (2016) Characterization of fine fraction mined from two Finnish landfills. Waste Manage 47:34–39
Mumtaz H, Sobek S, Werle S, Sajdak M, Muzyka R (2023) Hydrothermal treatment of plastic waste within a circular economy perspective. Sustain Chem Pharm 32:100991
Packaging and packaging Waste Directive 94/62/EC (1994) https://eur-lex.europa.eu/legal-content/EN/TXT/?uri=celex%3A31994L0062
Pangallo D, Gelsomino A, Fazzino F, Pedullà A, Calabrò P (2023) Fate of biodegradable plastic during the anaerobic co-digestion of excess sludge and organic fraction of municipal solid waste. Waste Manage 168(2023):98–106
Pavoni J, Heer J, Hagerty D (1975) Handbook of solid waste disposal: materials and energy recovery. Krieger Pub Co; First Edition, USA
Quaghebeur M, Laenen B, Geysen D et al (2013) Characterization of landfilled materials: screening of the enhanced landfill mining potential. J Clean Prod 55:72–83
Ramasamy E, Abbasi A (2000) High-solids anaerobic digestion for the recovery of energy from municipal solid waste (MSW). Environ Technol 21(3):345–349
Rambo AB, Avfallshistoria (in Swedish) https://www.rambo.se/skola/renhallning-en-historia/. Accessed September 2023
Swedish Waste Management (2023) invisible waste- https://www.avfallsverige.se/in-english/invisible-waste/. Accessed September 2023
Swedish Waste Management (Avfall Sverige). Energiåtervinning, in Swedish, accessed September 2023. https://www.avfallsverige.se/fakta-statistik/avfallsbehandling/energiatervinning/
Swedish Waste Management (Avfall Sverige). Report- Swedish Waste Management 2022. https://www.avfallsverige.se/in-english/
de Sadeleer N (2006) The concept of waste. In Ormond T, Fuhr M, Barth R (eds) Environmental law and policy at the turn to the 21st century—liber Amicorum betty Gebers, vol 243, pp 249–250
Tan Q, Dong Q, Liu L, Song Q, Liang Y, Li J (2017) Potential recycling availability and capacity assessment on typical metals in waste mobile phones: a current research study in China. J Clean Prod 148:509–517
United Nations (2015) Global Sustainable Development Report 2015. https://sdgs.un.org/publications/global-sustainable-development-report-2015-advance-unedited-version-gsdr-2015-17874. Accessed September 2023
United Nations Framework Convention on Climate Change (UNFCCC) (1992) https://unfccc.int/files/essential_background/background_publications_htmlpdf/application/pdf/conveng.pdf
Valentine SV (2016a) Kalundborg symbiosis: fostering progressive innovation in environmental networks. J Clean Prod 118:65–77
Valentine S (2016b) Kalunborg symbiosis: fostering progressive innovation in environmental networks. J Clean Prod 118:65–77

Vermeşan H, Tiuc AE, Purcar M (2020a) Advanced recovery techniques for waste materials from IT and telecommunication equipment printed circuit boards. Sustainability 12(1):74. https://doi.org/10.3390/su120 10074

Vermeşan H, Tiuc AE, Purcar M (2020b) Advanced recovery techniques for waste materials from IT and telecommunication equipment printed circuit boards. Sustainability 12(1):74. https://doi.org/10.3390/su12010074

Waste electrical and electronic equipment (WEEE) 2012/19/EU (2012) https://eur-lex.europa.eu/legal-content/EN/TXT/?uri=CELEX:02012L0019-20180704

Waste Framework Directive 2008/98/EC. (1998) https://eur-lex.europa.eu/eli/dir/2008/98/oj/eng

WECOOP. https://wecoop.eu/regional-knowledge-centre/eu-policies-regulations/. Accessed September 2023

World Bank Group (2023) https://datatopics.worldbank.org/what-a-waste/trends_in_solid_waste_management.html

Xiao L, Zhang G, Zhu Y, Lin T (2017) Promoting public participation in household waste management: a survey based method and case study in Xiamen city, China. J Clean Prod 144:313–322

Zhao W, van der Voet E, Huppes G, et al (2209) Comparative life cycle assessments of incineration and non-incineration treatments for medical waste. Int J Life Cycle Assess 14:114–121. https://doi.org/10.1007/s11367-008-0049-1

# 9 Evaluation of the Water Quality of the Jucu/ES (Brazil) River, Through Physical–Chemical and Biological Parameters in Favor of the Sustainability

Taina Turial da Silva, Geisamanda Pedrini Brandão Athayde, Maria Tereza Weitzel Dias Carneiro, and Josimar Ribeiro

**Abstract**

The Jucu River is one of the main rivers of the state of Espírito Santo/Brazil. The conditions of the waters of the river were studied by physical–chemical parameters, such as temperature, electrical conductivity, dissolved oxygen, turbidity, total solids, dissolved and suspended solids, biochemical oxygen demand, chemical oxygen demand, total nitrogen, thermotolerant coliforms and *Escherichia coli*, trace elements, and phytotoxic effect using *Lactuca sativa* as a bioindicator. The results of the surface waters were compared with the resolution of the Environment National Council (CONAMA) n° 357/05, and the sediments with the resolution of CONAMA n° 420/09. The study revealed that the physicochemical parameters are in agreement with the resolutions. The application of the water quality index (WQI) showed that in the 2nd campaign, the waters were damaged due to rainfall, leaving the river with more contaminants. Besides, it was possible to observe that the stations JUC01, JUC05, and JUC06 are the ones that suffer the most from anthropogenic actions due to the high presence of *Escherichia coli*, which is related to fecal contamination. All the stations of the Jucu River presented contamination by dissolved iron, aluminum, and manganese in superficial waters, being above the allowed limit. The water samples showed a stimulating effect on phytotoxicity tests.

T. T. da Silva · G. P. B. Athayde · M. T. W. D. Carneiro · J. Ribeiro (✉)
Laboratório de Pesquisa e Desenvolvimento em Eletroquímica (LPDE), Department of Chemistry, Center of Exact Sciences, Federal University of Espírito Santo, Campus Goiabeiras, Av. Fernando Ferrari, Vitória, Espírito Santo 29075-910, Brazil
e-mail: josimar.ribeiro@ufes.br



M. Nasr and B. Ravindran (eds.), *Solid Waste Management in Delta Region for SDGs Fulfillment*, Deltas of the World, https://doi.org/10.1007/978-3-031-58253-0_9

## 9.1 Introduction

There has been an increase in pollution in these water bodies, where most of it is due to anthropogenic action, the dumping of liquid and solid waste in nature, causing serious problems related to the poor quality of water resources and also its scarcity (Buzelli and Cunha-Santino 2013).

The sources that lead to the degradation of water quality can be classified as punctual, which are those characterized by domestic and industrial effluents, and diffuse, which are residues from agriculture (fertilizers, herbicides, insecticides, fungicides, among others) and runoff (urban and rural) (Buzelli and Cunha-Santino 2013).

In Brazil, water potability standards for public supply are established by the National Environment Council (CONAMA), in the resolution of Conama n° 357/2005, which provides for the classification of water bodies and environmental guidelines for their framing.

The presence of trace elements in water can have significant impacts on human health and aquatic biota at high concentrations, occur naturally, and may come from anthropogenic sources, compromising water quality (Ferreira et al. 2023).

Biological analyses in water assess toxicity and genotoxicity, checking the reactions of living organisms to pollution and indicating the possible synergistic effect of various pollutants since physical and chemical analysis of water indicate only the presence and concentration of different pollutants (Christofoletti 2008).

The Jucu River is one of the main rivers that support the population of Espírito Santo, the source is localized in Serra do Castelo (ES/Brazil), having the stretch known as the north arm with an extension of 123 km and the southern arm with 80 km and the remaining 43 km represent the Jucu River stretch, from the meeting of the north and south arms to the mouth of '*Barra do Jucu*' (Terra and Prette-Santos 2009).

The Jucu River has changed since 2010, that is, due to the seasonality of the river, which is a natural variation of the environment, and also due to the drought that the state faced in 2015 and 2016. During this period there was a drop in flow in the Jucu River, due to lack of rain, increasing the concentration of river pollution, endangering the quality of life of aquatic animals, affecting humans, the river landscape, and destroying local biodiversity (Terra and Prette-Santos 2008; Turial da Silva et al. 2018).

In addition, on the Jucu River, there are sporting practices such as the traditional ecological descent of the Jucu River, which took place about 30 years ago in Vila Velha, which in 2016 was canceled due to the high level of sewage in the waters.

Considering the importance that the Jucu River has for the state of Espírito Santo, by supplying about 25% of the population, assessing the state of its waters through physical–chemical parameters, quantifying the presence of trace elements in surface waters and sediments, and the toxic effect of these waters are of great importance to the population.

Thus, the work aimed to evaluate water quality based on the limits established by current resolutions, using tools that could determine the physical–chemical parameters, in addition, of course, to make a brief analysis in favor of sustainability.

## 9.2 Methods

The study area covers six sampling stations, where surface water and margin sediment samples were collected (Fig. 9.1).

The collections were carried out in two campaigns, the 1st campaign was made on August 31, 2017, and the 2nd campaign was made on March 13, 2018.

The collections of the 1st campaign were made in August 2017, which is a winter period in the southeastern region of Brazil, considered a period of few rains. The collections of the 2nd campaign were held in March 2018, which is the summer period, considered rainy.

Stations 1, 2, 3, 5, and 6 are classified as waters of class 2 and station 4 as class 1 according to the State Agency of Water Resources (AGERH—*Agência Estadual de Recursos Hídricos*). Table 9.1 shows the location in terms of geographical coordinates (UTM—Urchin Tracking Module), and the description of the stations. The system used was the datum: Córrego Alegre. Area 24 K.

The collection stations of this paper are located in the mountain region of the state of Espírito Santo, except for the 1st station (JUC01) located in Viana, the metropolitan region of Vitória.

Accumulated rainfall data were provided by the National Institute of Meteorology (INMET—*Instituto Nacional de Meteorologia*). The months of the 1st campaign presented a cumulative rainfall of 30 mm and the 2nd campaign in March presented a cumulative rainfall of 175 mm.

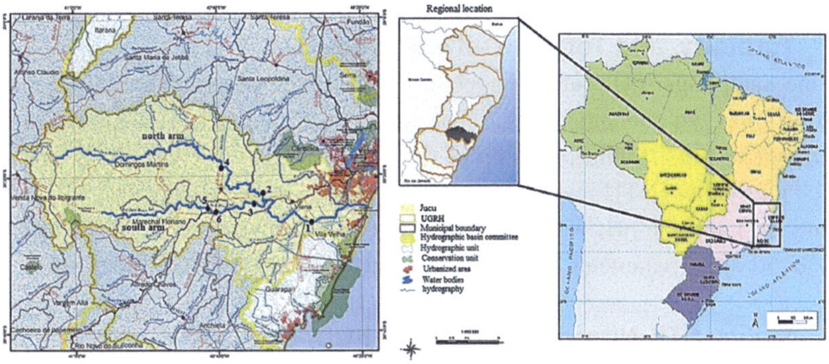


**Fig. 9.1** Location map of the collection stations in Jucu River

**Table 9.1** Location of monitored stations

| Stations | Coordinate (x) Longitude | Coordinate (y) Latitude | Description |
|---|---|---|---|
| JUC01 | −20.432013 | −40.456769 | Station after the junction of the north and south arms, near BR 101, Viana/ES |
| JUC02 | −20.376495 | −40.560023 | North arm of Jucu River, sampling station located in the city of Viana/ES. Rural area with native vegetation on the banks, station after receiving effluents from Domingos Martins/ES |
| JUC03 | −20.390485 | −40.577427 | South arm of Jucu River, located in Domingos Martins, after receiving effluents from Marechal Floriano/ES |
| JUC04 | −20.316488 | −40.665572 | North arm of Jucu River, located in the city of Domingos Martins, near ES-465 |
| JUC05 | −20.411641 | −40.681931 | South arm of Jucu River, located near BR 262, before receiving effluents from the city of Marechal Floriano/ES |
| JUC06 | −20.407381 | −40.671355 | South arm of Jucu River, sampling station located within the urban perimeter Marechal Floriano/ES, with little vegetation coverage |

In the collection period of the 1st campaign, the Jucu River flow was 7.398 L $s^{-1}$, and in the 2nd campaign, the flow rate was 34.454 L $s^{-1}$.

## 9.2.1 Physical–Chemical Parameters

### 9.2.1.1 Temperature, Dissolved Oxygen, Conductivity, Salinity, pH, Turbidity, and Total Dissolved Solids

During the samplings, the water temperature (°C | with an accuracy of ±0.3 °C) and the dissolved oxygen contents (mg $L^{-1}$ | accuracy of ±0.2 mg $L^{-1}$) were measured. The conductivity (μS $cm^{-1}$ | with precision of 0.1%), the salinity (ng $L^{-1}$ | with precision of ±3 ng $L^{-1}$), turbidity (NTU | with accuracy of ±5%), dissolved solids (mg $L^{-1}$ | accuracy = ±5 mg $L^{-1}$) and the pH, were determined from the subsurface Horiba Water Quality Checker U-50 multi-parameter (HORIBA Advanced Techno, Kyoto, Japan).

### 9.2.1.2 Solids, NTK Nitrogen, Coliforms, BOD, and COD

The total solids were determined by the gravimetric method in oven drying at a temperature of 105 °C until constant mass. Total nitrogen was analyzed according to the Kjeldahl procedure. Thermotolerant coliforms, biochemical oxygen demand (BOD), and chemistry oxygen demand (COD) were determined according to the methodology described in the Standard Methods for the Examination of Water and Wastewater (APHA 2005).

#### 9.2.1.3 Escherichia coli

For the *Escherichia coli* (*E. coli*) test, 10 mL of sample was added in an autoclave flask, 90 mL peptone water, and then added 2.3531 g of culture (generic medium for colorimetry). The solution was transferred to a counting plate which was sealed and incubated for 18 h at 37 °C. After this time, the plates were read, yielding the results of *E. coli* (APHA 2005).

#### 9.2.1.4 Organic Matter in Sediments

The organic matter was determined in the sediment samples by the muffle method established by Goldin (1987) with the following modifications: previous drying of the samples in the oven at 105 °C, for a period of 24 h (Carmo and Silva 2012).

#### 9.2.1.5 Trace Elements

Using the ICP-OES technique were evaluated Al, Ca, Cr, Fe, Mg, Mn, P, and Zn, and using the ICP-MS technique were evaluated Ni, Cd, Co, Cu, and Pb in water and sediment samples.

The decontamination procedure of the vials used for collection and analysis was according to the method of Rietzler et al. (2001).

The pseudo total trace elements in water were determined. The water samples were filtered with blue stripe filter paper and then submitted for analysis.

### 9.2.2 Procedure for the Determination of Trace Elements in Sediment by ICP-OES

The sediment samples were previously oven-dried at 60 °C until the mass remained constant, after which approximately 1.0 g of each 125 mL Erlenmeyer sediment sample was weighed. The samples were digested by adding 9 mL of distilled nitric acid and 3 mL of distilled hydrochloric acid. The digestion method used was modified EPA 3050B, where samples were placed on the plate for 15 min at 100 °C. After digestion, dilution to 50 mL was made, and when subjected to analysis, the sample was further diluted 3 times. This procedure was performed in three repetitions.

The trace elements Al, Ca, Fe, Mg, Mn, P, and Zn were quantified using inductively coupled plasma optical emission spectrometry (ICP-OES). The equipment used for the analyses of trace elements was the ICP-OES of Thermo Scientific Model iCAP 6000 (Thermo Fisher Scientific, Cambridge, England) coupled with dual vision configuration (axial and radial), equipped with an appropriate chamber and nebulizer for each array and an injection charge device (CID) as a detector. All are managed by iTEVA software (Thermo Fisher Scientific, Cambridge, England).

### 9.2.3 Procedure for the Determination of Trace Elements in Sediment by ICP-MS

The sediment samples were previously dried in an oven at 60 °C until the mass remained constant, after which it was sieved <0.063 mm. Approximately, 0.250 g of each pellet sample was weighed into decomposition tubes and 15 mL of nitric acid was added. Subsequently, it was digested using microwave heating and the digestion method used was EPA 3051A. After digestion, dilution to 35 mL was made, and when subjected to analysis the samples were diluted 100 times. This procedure was performed in three repetitions.

The stove used was Ethik Technology (Brazil), model 402-D, the analytical balance of Sartorius (Brazil), model ED224S with accuracy ±0.0001 g, and the microwave oven was Anton Paar (Austria), Multiwave GO model.

Cd, Co, Cu, Ni, and Pb metals were quantified using inductively coupled plasma mass spectrometry (ICP-MS), NexIon 300D equipment, and Perkin Elmer (USA).

### 9.2.4 Trials with *Lactuca sativa*

#### 9.2.4.1 Analysis of Germination Index (GI) and Root Growth Index (CR)

Seeds of *L. sativa* were exposed to water samples in Petri dishes at a controlled temperature of 24 °C and for a period of 48 h. After that, the time of germination index and the root growth index were obtained. Six samples of river water were analyzed. For the positive control the trifluralin 0.84 g $L^{-1}$ was used, which is a herbicide that inhibits the development of the plant, and for the negative control, the distilled water was used. In each Petri dish, 30 seeds of *L. sativa* were placed, and each sample was done in 5 replicates, giving a total of 1,200 seeds analyzed. Treatment of statistical data from the samples was performed by the ANOVA program followed by the Tukey test, with a 5% significance ($p > 0.05$) (Oliveira 2008).

## 9.3 Results and Discussion

The results of the physical and chemical parameters of the 1st and 2nd campaigns are shown in Tables 9.2 and 9.3.

**Table 9.2** Results of the analysis of the physical–chemical parameters of the 1st campaign

1ª campaign

| | JUC01 | JUC02 | JUC03 | JUC04 | JUC05 | JUC06 | CONAMA 357/05 c II |
|---|---|---|---|---|---|---|---|
| Time | 09:18 | 11:20 | 12:20 | 14:37 | 15:43 | 16:30 | – |
| Sample T (°C) | 21.30 ± 0.30 | 22.80 ± 0.30 | 23.00 ± 0.30 | 23.80 ± 0.30 | 21.00 ± 0.30 | 21.10 ± 0.30 | – |
| DO (mg $L^{-1}$) | 11.04 ± 0.20 | 12.96 ± 0.20 | 13.19 ± 0.20 | 11.66 ± 0.20 | 11.86 ± 0.20 | 13.52 ± 0.20 | >5 |
| EC (μS $cm^{-1}$) | 68.50 ± 0.68 | 48.50 ± 0.48 | 57.50 ± 0.57 | 45.50 ± 0.45 | 51.50 ± 0.51 | 62.00 ± 0.62 | – |
| pH | 6.97 ± 0.10 | 7.65 ± 0.10 | 7.69 ± 0.10 | 7.71 ± 0.10 | 7.64 ± 0.10 | 7.91 ± 0.10 | 6 a 9 |
| Turbidity (NTU) | 4.50 ± 0.23 | 3.6 ± 0.18 | 5.01 ± 0.26 | 6.30 ± 0.32 | 7.40 ± 0.37 | 16.7 ± 0.84 | <100 |
| TKN (mg $L^{-1}$) | 1.31 ± 0.29 | 1.51 ± 0.00 | 2.07 ± 0.07 | 0.71 ± 0.29 | 0.10 ± 0.00 | 1.56 ± 0.07 | 3.7 pH < 7.5 $\mid$ 2.0 pH 7.5–8 $\mid$ 1.0 pH 8.0–8.5 $\mid$ 0.5 pH > 8.5 |
| COD (mg $L^{-1}$) | 68.81 ± 26.79 | 155.15 ± 5.20 | 9.27 ± 2.52 | 83.7 ± 8.37 | 3.32 ± 0.50 | 9.27 ± 8.42 | – |
| BOD (mg $L^{-1}$) | Above | Above | <5 | Above | <5 | <5 | <5 |
| Coliforms NMP/100 mL | Present | Present | Absent | Present | Present | Present | – |
| Salinity (ng $L^{-1}$) | 0 | 0 | 0 | 0 | 0 | 0 | <5 ng $L^{-1}$ |
| TS (mg $L^{-1}$) | 119.88 ± 86.85 | 1173.32 ± 150.70 | 970.05 ± 23.57 | 1955.56 ± 156.40 | 103.35 ± 40.18 | 135.55 ± 109.61 | – |
| TDS (mg $L^{-1}$) | 38.50 ± 0.00 | 164.44 ± 0.00 | 26.00 ± 0.00 | 26.00 ± 0.00 | 28.50 ± 0.00 | 35.50 ± 0.00 | <500 |
| OM (%) | 10.31 ± 0.11 | 10.23 ± 0.96 | 13.28 ± 0.98 | 17.54 ± 0.32 | 9.37 ± 0.21 | 9.70 ± 0.13 | |

*T* temperature, *DO* dissolved oxygen, *EC* electrical conductivity, *TKN* total Kjeldahl nitrogen, *TS* total solids, *TDS* total dissolved solids, *BOD* biochemical oxygen demand, *COD* chemical oxygen demand, *OM* organic matter

**Table 9.3** Results of the analysis of the physical–chemical parameters of the 2nd campaign

| 2ª campaign | JUC01 | JUC02 | JUC03 | JUC04 | JUC05 | JUC06 | CONAMA 357/05 c II |
|---|---|---|---|---|---|---|---|
| Time | 08:20 | 09:30 | 10:00 | 11:50 | 12:45 | 13:20 | – |
| Sample T (°C) | 24.70 ± 0.30 | 26.0 ± 0.30 | 25.90 ± 0.30 | 26.60 ± 0.30 | 26.10 ± 0.30 | 27.10 ± 0.30 | – |
| DO (mg $L^{-1}$) | 11.26 ± 0.20 | 10.79 ± 0.20 | 12.2 ± 0.20 | 14.39 ± 0.20 | 10.58 ± 0.20 | 10.13 ± 0.20 | >5 |
| EC (μS $cm^{-1}$) | 64.00 ± 0.64 | 51.00 ± 0.51 | 46.00 ± 0.46 | 45.00 ± 0.45 | 45.00 ± 0.45 | 48.00 ± 0.48 | – |
| pH | 6.17 ± 0.10 | 7.00 ± 0.10 | 6.95 ± 0.10 | 7.18 ± 0.10 | 7.03 ± 0.10 | 6.80 ± 0.10 | 6 a 9 |
| Turbidity (NTU) | 154.00 ± 7.70 | 105.00 ± 5.25 | 120.00 ± 6.00 | 82.50 ± 4.12 | 86.60 ± 4.33 | 85.30 ± 4.27 | <100 |
| TKN (mg $L^{-1}$) | 0.50 ± 0.01 | 0.20 ± 0.10 | 0.08 ± 0.02 | 0.05 ± 0.00 | 0.05 ± 0.00 | 0.02 ± 0.00 | 3.7 pH < 7.5 \| 2.0 pH 7.5–8 \| 1.0 pH 8.0–8.5 \| 0.5 pH > 8.5 |
| COD (mg $L^{-1}$) | 52.88 ± 0.01 | 92.94 ± 0.02 | 30.12 ± 0.01 | 67.45 ± 0.03 | 91.12 ± 0.03 | 51.97 ± 0.01 | – |
| BOD (mg $L^{-1}$) | – | – | 16.00 | 9.00 | 30.00 | 48.00 | <5 |
| *E. coli* NMP/100 mL | $2.3 \times 10^3$ | $1.4 \times 10^2$ | $1.6 \times 10^2$ | $3.2 \times 10^2$ | $3.8 \times 10^2$ | $8.7 \times 10^4$ | – |
| Salinity (ng $L^{-1}$) | 0 | 0 | 0 | 0 | 0 | 0 | <5 ng $L^{-1}$ |
| TS (mg $L^{-1}$) | 153.33 ± 26.67 | 117.78 ± 15.39 | 133.33 ± 9.43 | 160 ± 63.60 | 148.89 ± 53.47 | 117.78 ± 40.73 | – |
| TDS (mg $L^{-1}$) | 32.00 ± 0.00 | 26.00 ± 0.00 | 23.00 ± 0.00 | 22.00 ± 00 | 23.00 ± 0.00 | 24.00 ± 0.00 | <500 |
| OM (%) | 9.98 ± 0.05 | 3.09 ± 0.04 | 12.13 ± 0.11 | 7.58 ± 0.17 | 9.17 ± 0.13 | 11.37 ± 0.23 | |
| *Escherichia coli* (*E. coli*/100 mL) | 2300 | 140 | 160 | 320 | 380 | 87,000 | <1000 (CONAMA 274/00) |

*T* temperature, *DO* dissolved oxygen, *EC* electrical conductivity, *TNK* total Kjeldahl nitrogen, *TS* total solids, *TDS* total dissolved solids, *BOD* biochemical oxygen demand, *COD* chemical oxygen demand, *OM* organic matter

### 9.3.1 Temperature

In Table 9.2, it is observed that the temperature varied from 21.0 to 24.0 °C in the 1st campaign. In the legislation, there are no maximum or minimum values stipulated for this variable, however, the results obtained are within the expected for the climatic condition of the region. According to INMET, the average temperature of the collection days was 24 °C.

The temperature variation in the 2nd campaign was from 24.7 to 27.1 °C (Table 9.3), and the minimum temperature measured by INMET for this day was 24.0 °C. Therefore, the results of the temperature of the 2nd campaign are in agreement with the average temperatures of the region.

The main factor that changes the temperature of the rivers is the release of industrial effluents. In the study area of the Jucu River, this parameter remained stable, within the standards established by the resolution of CONAMA nº 430/2011, that for the discharge of effluents into the water body, it should have a temperature below 40 °C. Thereafter, one can infer that, as some compounds are more toxic to aquatic life at higher temperatures, and there is a direct relationship between temperature and the presence of aquatic life.

### 9.3.2 Dissolved Oxygen

The values of dissolved oxygen (DO) should be higher than 5 mg $L^{-1}$ in the quality standard described in the resolution of CONAMA nº 357/05 for Class 2 waters and 6 mg $L^{-1}$ for Class 1 waters.

The dissolved oxygen is related to temperature that affects the solubility of $O_2$ in water, that is, with decreasing temperature the solubility of oxygen is higher, according to Henry's law, however, the temperature results in these locations being close together.

In the 1st campaign (Table 9.2), DO concentrations ranged from 11.04 to 13.52 mg $L^{-1}$ and the ideal variation for these waters should be from 8.40 to 9.08 mg $L^{-1}$ according to measured water temperature and with the saltiness of the sample. Therefore, the measured DO concentration was higher than the ideal, this increase may be related to the excessive growth of algae, or collection sites where there is agitation of the surface waters, promoting the aeration of the waters (Fiorucci and Filho 2005).

In the collection stations JUC02, JUC03, and JUC04, there is greater agitation of the waters, JUC01 station there is a greater presence of aquatic plants around it, which influences the increase of DO, JUC05, and JUC06 stations are in places of domestic pollution and industrial, but DO values for these stations are considered clean water values.

In the 2nd campaign (Table 9.3) there was a decrease in DO concentrations, this may be related to the rains of this period that promoted the increase of water turbidity. The turbidity influences the process of photosynthesis of aquatic plants since the result is the release of DO, so in high turbidity, this process of photosynthesis is impaired leading to a decrease in DO formation (Pereira 2004).

### 9.3.3 Electric Conductivity

The values of electrical conductivity (EC) in the 1st campaign varied from 45 to 77 μS $cm^{-1}$, and the 2nd campaign presented a variation from 45 to 64 μS $cm^{-1}$.

The EC represents the amount of dissolved salts capable of conducting electric current. The resolution of CONAMA n° 357/05 does not determine the specific value for this variable, but when the concentrations exceed 100 μS $cm^{-1}$, the environment may be impacted by anthropic actions, in addition, high values of EC may indicate corrosive characteristics of water (Junior and Simoura 2016).

According to Ferreira (2001), high conductivity is indicative of an abnormal pH and high salinity, and the quality of the treated water should be based on a conductivity between 50 and 400 μS $cm^{-1}$, having in this range water of excellent quality. Between 400 and 750 μS $cm^{-1}$, the water has a good quality, and between 750 and 1500 μS $cm^{-1}$ medium quality, although still usable. Waters with conductivity above 1500 μS $cm^{-1}$ will have excessive mineralization.

The results of EC of the 1st and 2nd campaigns (Tables 9.2 and 9.3) did not exceed 100 μS $cm^{-1}$, so it can be said that the waters of the Jucu River are in the water band of excellent quality.

The JUC01 station presented higher EC values in both campaigns. According to a study, Mcneil et al. (2005), evaluating the chemical characteristics of water and its spatial variation in Australia, have identified that sodium is the main cation present in the waters, being the geographic location the main aggravating factor, that is, water bodies closer to the coast can present values of dissolved salts due to salt intrusion by the wind. The JUC01 station is closer to the coast, its values of electrical conductivity may be higher than the other stations due to this factor.

### 9.3.4 pH

The pH of the 1st campaign (Table 9.2) samples ranged from 6.97 to 7.91. For the 2nd campaign, the pH variation was from 6.17 to 7.18. All results are within the limits of neutrality.

According to Lopes and Júnior (2010), the factors that can decrease the pH of a water body are releases of influences that release fulvic and humic acids resulting from the degradation of the organic matter present in the waters. And another factor according to Borges et al. (2003) is the type of soil where water travels.

The soil type of the regions where the Jucu River was collected is predominantly red-yellow Latosol, according to data from the Brazilian Soil Classification System (Embrapa 2006). The red-yellow Latosol differs from the yellow Latosol because it has a lower $Al_2O_3/Fe_2O_3$ ratio in the clay fraction and a higher percentage of free $Fe_2O_3$. Therefore, the red latosol has a higher $Al_2O_3/Fe_2O_3$ ratio in the clay fraction and a lower percentage of free $Fe_2O_3$ (Netto 2008). This type of soil is usually acidic (ranging in pH from 5 to 6), with low base, dystrophic, or aluminum saturation (Papa et al. 2011).

It is observed that for both campaigns, the JUC01 station had lower pH values than the other stations, which could be due to this station suffering more from erosion of the margins and surface runoff. In the 2nd campaign (Table 9.3), the pH values decreased slightly, and this may be related to the erosion of the margin and the silting of the river in that period due to the rains since the type of soil has an acid character.

### 9.3.5 Turbidity

According to the resolution of CONAMA n° 357/2005 turbidity values should be less than 40 NTU for class 1 waters and less than 100 NTU for Class 2 fresh waters. The turbidity was less than 20.00 NTU for a period of low rainfall, and it was relatively constant for all stations in the 1st campaign.

In the rainy campaign, the turbidity variation was 82.50 to 154.00 NTU. Buzelli and Cunha-Santino (2013) observed an increase in turbidity during the rainy campaign in the Barra Bonita reservoir, with a variation from 14.76 in the dry campaign to 41.26 in the rainy campaign, and in the same way, Abreu and Cunha (2017) observed in the Lower Jari River, Amapá, that in the dry campaign, turbidity varied from 5.65 to 7.01 NTU in September and between 4.53 and 11.53 NTU in December, while in the rainfall period ranged from 6.0 NTU to 17 NTU.

The increase in turbidity occurred due to the period of collection of the 2nd campaign in the rainy campaign, and this mainly causes sediment movement in shallow places such as the coastal zone and erosion of the margins.

According to the Jucu River Basin Committee, anthropogenic actions such as deforestation, sanitary sewage disposal, and industrial and agricultural effluents cause surface runoff to increase water turbidity, resulting in greater changes in the aquatic ecosystem, and promoting reduction in photosynthesis of aquatic plants.

### 9.3.6 Total Dissolved Solids

The values obtained for the total dissolved solids are well below the limit established by CONAMA, which is 500 mg $L^{-1}$ for class 2 freshwater. It presented a variation from

26.00 to 35.50 mg $L^{-1}$ in the 1st campaign (Table 9.2), and from 22.00 to 32.00 mg $L^{-1}$ in the 2nd campaign (Table 9.3).

With increase in places of greater pollution, can affect the aquatic community, altering the conditions of lightness of the water, making it difficult to perform photosynthesis, consequently damaging the other heterotrophic organisms dependent on the dissolved oxygen produced in the photosynthesis for respiration.

### 9.3.7 Total Nitrogen

Contamination of water by nutrients such as total nitrogen occurs mainly by domestic and industrial discharges and fertilizer releases.

All stations were below the limit, except for the JUC03 station, with a concentration of 2.07 mg $L^{-1}$. According to Von Sperling (2005), ammoniacal nitrogen can have anthropogenic origin, such as domestic, industrial, and animal droppings and fertilizers, justifying the direct relationship between increased chemical and biochemical oxygen demand, as well as thermotolerant coliforms.

The JUC05 and JUC06 stations are close to the center of the city of Marechal Floriano, where they receive direct discharges of domestic effluents that justify the high concentration of total nitrogen. Terra and Prette-Santos (2009) conducted a study of the southern arm of the Jucu River, and the results for nitrogen ranged from 0.7 to 0.9 mg $L^{-1}$ throughout July 2006.

The JUC03, JUC05, and JUC06 stations corresponded to the south arm and presented superior results, of 1.5, 7.0, and 2.0 mg $L^{-1}$ in July/August 2018, respectively. The results show that there was an increase in the concentration of nitrogen in these waters over the years in the dry season. Buzelli and Cunha-Santino (2013) in the Barra Bonita reservoir also observed this decrease in nitrogen in the rainy season.

In the 1st campaign, JUC03 and JUC06 stations presented nitrogen concentrations close to the limit established by the standard. The JUC03 station is very close to agricultural zones which may explain the nitrogen values, and the JUC06 station is within the city of Marechal Floriano (ES, Brazil) where it receives domestic and industrial evictions.

### 9.3.8 BOD and COD

The COD has higher values than the biochemical oxygen demand (BOD), that is, because the oxidizing agent used is very strong, oxidizing other compounds that are not oxidized in a microbiological way (Valente et al. 1997). The BOD degrades only biodegradable organic matter and can be considered an indicator of water quality when considering organic pollution (Buzelli and Cunha-Santino 2013).

For BOD, the limit value according to the resolution of CONAMA 357/05 is 5 mg $L^{-1}$ for class 1 and class 2 waters. And the legislation does not show limit values for COD. In the 1st campaign (Table 9.2), BOD and COD were high in JUC01, JUC02, and JUC04 stations; this COD value can be explained by the fact that these waters present higher amounts of organic matter, that is, typical of rural residential areas. For the 2nd campaign, the BOD values were all above the 5 mg $L^{-1}$ limit for all stations, and the COD results were higher, this may be due to the increase of predominantly organic loads, such as domestic effluents in nature, industry, and agriculture. Neves et al. (2021) observed that in the rainy season, the organic loads represent point pollution caused by wastewater, and diffuse pollution associated with runoff and carrying of fertilizers used in agricultural areas. Moreover, Zonta et al. (2008) also observed that in the rainy months, there was a deterioration in the water quality of the Alegre River, in the southern state of Espírito Santo, due to the increase of turbidity and BOD, being directly related to the use and occupation of the soil, often allowing the diagnosis of possible causes of environmental impacts on natural resources. According to studies by Turial da Silva et al. (2018), Alvarenga et al. (2016), and Terra and Prette-Santos (2009), in rainy seasons the BOD increased significantly.

### 9.3.9 *Escherichia coli*

Table 9.3 shows the results obtained for the 2nd campaign for *Escherichia coli*, which are microorganisms of the coliform group, indicating the possibility of pathogenic germs of intestinal origin (Junior and Simoura 2016).

Resolution of CONAMA nº 274/00 reports on water classifications considered appropriate for human bathing. If the concentration is greater than 2,500/100 mL of fecal coliforms or 2,000/100 mL of *Escherichia coli*, the water will be considered improper.

JUC06 and JUC01 stations presented high concentrations of *E. coli*, classifying this water improper for consumption.

The JUC05 and JUC06 stations receive greater discharges of domestic effluents since they are inside the city of Marechal Floriano, and Terra and Prette-Santos (2009), observed that at collection sites in the city of Marechal Floriano in 2006 presented coliform variation from 3,700 to 4,500 MPN/100 mL. Junior and Simoura (2016) found results for thermotolerant coliforms in the southern branch of the Jucu River in 2009 varying from 5,000 to 35,000 NMP/100 mL, revealing that these sites always showed concentrations well above the limit allowed by resolution of CONAMA nº 274/00.

Results achieved from *E. coli* and thermotolerant coliforms are in agreement with the work done by Ramos et al. (2016) who stated that during the rainy season, there is an increase in bacteria. The authors found that increased rainfall can interfere with sanitary conditions, regardless of season, or diffuse pollution.

**Table 9.4** Water quality index CETESB and the classification of waters of the Jucu River of the 1st and 2nd campaigns

| Stations | $1^a$ C. $WQI_{CETESB}$ | Classification | $2^a$ C. $WQI_{CETESB}$ | Classification |
|---|---|---|---|---|
| JUC01 | 90 | Great | 63 | Good |
| JUC02 | 69 | Good | 65 | Good |
| JUC03 | 81 | Great | 65 | Good |
| JUC04 | 74 | Good | 68 | Good |
| JUC05 | 63 | Good | 55 | Good |
| JUC06 | 88 | Great | 53 | Good |

### 9.3.10 Water Quality Index

The WQI analysis was done using QualiGraf_2018 software, Table 9.4 shows the $WQI_{CETESB}$ results of the samples analyzed in the 1st and 2nd campaigns.

In the 1st campaign, JUC05 station was the one that presented the lowest $WQI_{CETESB}$, with a value of 63, being classified as good waters. The JUC01 station was the one that presented better WQI with 90, with the classification of optimal water. Previous studies of the Jucu River, as reported by Junior and Simoura (2016) classified their waters as Good, with a WQI variation of 55–60.

The JUC04 station classification according to the classification is classified as class 1, which should have a WQI with an optimal water classification, but its waters were classified as good if compared to the other stations more contaminated by domestic effluents, since the WQI is sensitive to sewage-related pollution.

In the 2nd campaign, all the stations showed a decrease in their quality index, showing that in the rainy season, the waters receive more contamination from external factors.

Bonnet et al. (2008) correlated water quality and land use in a watershed in Goiás and also observed that during flood periods, WQI values were worse.

The parameters that most influenced the reduction of water quality in the 1st campaign for the JUC02 were thermotolerant coliforms and total phosphorus, for JUC04 it was thermotolerant coliforms and for JUC05 were thermotolerant coliforms and nitrogen. In the 2nd campaign the parameters, thermotolerant coliforms, total solids, and turbidity were the ones that influenced the drop in water quality at all stations of collection.

### 9.3.11 Trace Elements

Tables 9.5 and 9.6 shows the results of the dissolved trace elements of the surface water samples, respectively. In Table 9.7 the results of trace elements in sediment samples are

**Table 9.5** Results of trace elements in water samples in mg $L^{-1}$ from the 1st and 2nd campaigns and their respective RSD

| 1st campaign | | | | | | | 2nd campaign | | | | | | LD | LQ | CONAMA 357/05 |
|---|---|---|---|---|---|---|---|---|---|---|---|---|---|---|---|
| | JUC01 | JUC02 | JUC03 | JUC04 | JUC05 | JUC06 | JUC01 | JUC02 | JUC03 | JUC04 | JUC05 | JUC06 | | | |
| Al (mg $L^{-1}$) | 0.3092 | 0.2738 | 0.3061 | 0.3201 | 0.4245 | 0.3909 | 1.6335 | 1.2566 | 1.2324 | 1.1163 | 1.1028 | 1.1527 | 0.0010 | 0.0100 | 0.1000(D) |
| RSD (%) | 0.4397 | 2.2640 | 1.1093 | 0.5144 | 1.4950 | 0.9730 | 0.7137 | 0.6578 | 0.4263 | 0.9320 | 0.3543 | 0.8084 | | | |
| Ca (mg $L^{-1}$) | 25.8620 | 26.0890 | 26.3950 | 25.2820 | 29.3000 | 33.0180 | 10.2794 | 13.9183 | 11.4137 | 10.4843 | 11.9050 | 12.8180 | 0.0227 | 0.1000 | – |
| RSD (%) | 1.3809 | 0.4206 | 0.3835 | 0.7130 | 0.0432 | 0.7480 | 0.3095 | 0.2932 | 0.3518 | 0.7295 | 0.5267 | 0.2905 | | | |
| Fe (mg $L^{-1}$) | 1.4749 | 0.4673 | 0.6268 | 0.6051 | 0.8006 | 0.9116 | 1.7243 | 1.4523 | 1.4732 | 1.7307 | 1.5505 | 1.6132 | 0.0024 | 0.0100 | 0.3000(D) |
| RSD (%) | 0.6828 | 0.3652 | 0.9936 | 0.7694 | 0.7518 | 0.3991 | 0.3120 | 0.6293 | 0.4569 | 0.5042 | 0.4531 | 0.9789 | | | |
| Mg (mg $L^{-1}$) | 10.6803 | 10.4263 | 10.5667 | 10.0630 | 10.8960 | 12.4390 | 4.3424 | 5.7179 | 4.5811 | 4.2094 | 0.1180 | 0.1230 | 0.0700 | 0.1000 | – |
| RSD (%) | 0.4956 | 0.9691 | 0.1746 | 0.4438 | 0.5907 | 0.6567 | 2.1172 | 1.5131 | 4.6634 | 3.7271 | 0.0179 | 0.0259 | | | |
| Mn (mg $L^{-1}$) | 0.2474 | 0.1605 | 0.1821 | 0.1695 | 0.2294 | 0.2282 | 0.1257 | 0.1202 | 0.1181 | 0.1180 | 0.1230 | 0.1313 | 0.0025 | 0.0100 | 0.1000(T) |

(continued)

**Table 9.5** (continued)

| 1st campaign | | | | | | | 2nd campaign | | | | | | LD | LQ | CONAMA 357/05 |
|---|---|---|---|---|---|---|---|---|---|---|---|---|---|---|---|
| | JUC01 | JUC02 | JUC03 | JUC04 | JUC05 | JUC06 | JUC01 | JUC02 | JUC03 | JUC04 | JUC05 | JUC06 | | | |
| RSD (%) | 0.7734 | 0.5458 | 0.3315 | 0.0714 | 0.7556 | 0.4055 | 0.0289 | 0.0439 | 0.0203 | 0.0179 | 0.0259 | 0.0761 | | | |
| Zn (mg $L^{-1}$) | 0.0394 | 0.0369 | 0.0612 | 0.0598 | 0.0547 | 0.0746 | 0.0201 | 0.0206 | 0.0194 | 0.0572 | 0.0188 | 0.0185 | 0.0004 | 0.0100 | 0.1800(T) |
| RSD (%) | 1.5729 | 0.5169 | 0.9076 | 0.2824 | 0.2862 | 0.1450 | 0.0031 | 0.0169 | 0.0080 | 0.0154 | 0.0099 | 0.0088 | | | |
| Cr (mg $L^{-1}$) | <0.010 | <0.010 | <0.010 | <0.010 | <0.010 | <0.010 | <0.010 | <0.010 | <0.010 | <0.010 | <0.010 | <0.010 | 0.0006 | 0.01 | 0.05(T) |
| RSD (%) | 12.888 | 23.709 | 8.209 | 10.863 | 5.924 | 15.013 | 0.009 | 0.007 | 0.009 | 0.014 | 0.011 | 0.010 | | | |

**Table 9.6** Results of trace elements in water samples in μg $L^{-1}$ from the 1st and 2nd campaigns and their respective RSD

| 1st campaign | | | | | | | 2nd campaign | | | | | | LD | LQ | CONAMA 357/05 |
|---|---|---|---|---|---|---|---|---|---|---|---|---|---|---|---|
| | JUC01 | JUC02 | JUC03 | JUC04 | JUC05 | JUC06 | JUC01 | JUC02 | JUC03 | JUC04 | JUC05 | JUC06 | | | |
| Cd (μg $L^{-1}$) | 0.091 | 0.313 | 0.172 | 0.235 | 0.252 | 0.165 | 0.114 | 0.114 | 0.071 | 0.061 | 0.030 | 0.046 | $1.92 \times 10^{-5}$ | $6.39 \times 10^{-5}$ | 1(T) |
| RSD (%) | 9.400 | 2.000 | 5.700 | 1.900 | 3.400 | 2.000 | 4.500 | 0.700 | 1.800 | 2.900 | 8.200 | 3.900 | | | |
| Co (μg $L^{-1}$) | 1.410 | 1.961 | 1.051 | 1.758 | 1.144 | 1.040 | 0.620 | 0.139 | 0.267 | 0.245 | 0.215 | 0.469 | $9.40 \times 10^{-5}$ | $3.14 \times 10^{-4}$ | 50(T) |
| RSD (%) | 1.400 | 0.900 | 0.400 | 1.900 | 1.300 | 0.400 | 1.400 | 1.700 | 0.200 | 2.600 | 1.000 | 0.000 | | | |
| Cu (μg $L^{-1}$) | 1.709 | 2.082 | 2.069 | 2.086 | 2.446 | 2.222 | 58.115 | 42.388 | 51.217 | 49.542 | 59.159 | 78.913 | $1.94 \times 10^{-4}$ | $6.48 \times 10^{-4}$ | 9.00(D) |
| RSD (%) | 0.700 | 1.700 | 2.000 | 2.500 | 0.500 | 0.500 | 1.100 | 1.400 | 0.900 | 2.800 | 1.000 | 0.500 | | | |
| Pb (μg $L^{-1}$) | 0.464 | 0.669 | 0.416 | 0.448 | 1.085 | 0.566 | 3.176 | 3.239 | 3.549 | 2.851 | 3.598 | 3.640 | $4.40 \times 10^{-5}$ | $1.47 \times 10^{-4}$ | 10(T) |
| RSD (%) | 0.400 | 1.500 | 0.700 | 0.400 | 1.300 | 0.900 | 1.600 | 0.700 | 0.700 | 2.200 | 1.200 | 1.000 | | | |

(continued)

**Table 9.6** (continued)

| 1st campaign | | | | | | | 2nd campaign | | | | | | LD | LQ | CONAMA 357/05 |
|---|---|---|---|---|---|---|---|---|---|---|---|---|---|---|---|
| | JUC01 | JUC02 | JUC03 | JUC04 | JUC05 | JUC06 | JUC01 | JUC02 | JUC03 | JUC04 | JUC05 | JUC06 | | | |
| Ni (μg $L^{-1}$) | 1.829 | 2.281 | 2.002 | 2.199 | 2.152 | 2.115 | 1.337 | 0.543 | 0.721 | 0.642 | 0.520 | 0.725 | $8.80 \times 10^{-5}$ | $2.93 \times 10^{-4}$ | 25(T) |
| RSD (%) | 0.700 | 1.200 | 0.400 | 1.900 | 1.600 | 0.800 | 2.200 | 2.800 | 2.000 | 2.000 | 0.500 | 1.200 | | | |

*LQ* limit of quantification, *LD* limit of detection, *RSD* relative standard deviation, (*T*) totals, (*D*) dissolved

**Table 9.7** Results of trace elements in sediment samples from the 1st and 2nd seasons and their respective RSD

| 1st campaign | | | | | | | 2nd campaign | | | | | | LD | LQ | CONAMA 357/05 |
|---|---|---|---|---|---|---|---|---|---|---|---|---|---|---|---|
| | JUC01 | JUC02 | JUC03 | JUC04 | JUC05 | JUC06 | JUC01 | JUC02 | JUC03 | JUC04 | JUC05 | JUC06 | | | |
| Al (mg kg$^{-1}$) | 22,532.33 | 12,549.00 | 29,315.00 | 9637.10 | 21,746.00 | 24,183.00 | 22,195.00 | 8262.10 | 22,484.00 | 14,950.00 | 23,060.00 | 30,439.00 | 0.0010 | 0.5 | – |
| RSD (%) | 8.06 | 8.70 | 6.71 | 10.21 | 13.07 | 3.74 | 2.26 | 4.27 | 11.68 | 4.76 | 13.91 | 7.27 | | | |
| Ca (mg kg$^{-1}$) | 1632.37 | 3748.33 | 3454.20 | 914.86 | 1266.16 | 5280.17 | 1312.40 | 905.97 | 1996.63 | 952.38 | 2133.79 | 4413.73 | 0.0227 | 5 | – |
| RSD (%) | 6.28 | 16.52 | 6.44 | 10.11 | 11.98 | 2.80 | 10.82 | 1.82 | 9.64 | 3.24 | 8.43 | 6.72 | | | |
| Fe (mg kg$^{-1}$) | 29,865.00 | 33,301.50 | 50,513.50 | 25,812.50 | 38,649.50 | 40,513.50 | 38,389.00 | 18,118.50 | 35,726.50 | 28,240.50 | 34,573.50 | 44,821.00 | 0.0024 | 0.5 | – |
| RSD (%) | 0.85 | 13.23 | 13.58 | 32.28 | 7.56 | 2.77 | 2599.30 | 68.68 | 9.69 | 23.54 | 11.20 | 8.06 | | | |
| Mg (mg kg$^{-1}$) | 1972.10 | 3277.05 | 2045.15 | 805.47 | 1595.95 | 2366.35 | 626.49 | 1552.70 | 2401.65 | 1660.90 | 2012.75 | 2711.85 | 0.0700 | 5 | – |
| RSD (%) | 40.44 | 57.81 | 57.25 | 56.86 | 57.30 | 57.17 | 57.20 | 56.60 | 57.64 | 56.92 | 57.42 | 5.12 | | | |
| Mn (mg kg$^{-1}$) | 626.49 | 1376.45 | 872.14 | 592.49 | 330.47 | 351.80 | 401.62 | 101.37 | 244.23 | 186.15 | 351.80 | 500.65 | 0.0025 | 0.5 | – |
| RSD (%) | 55.70 | 57.17 | 56.39 | 56.06 | 53.48 | 55.95 | 56.95 | 60.01 | 74.39 | 50.17 | 54.18 | 55.08 | | | |

(continued)

**Table 9.7** (continued)

| 1st campaign | | | | | | | 2nd campaign | | | | | | LD | LQ | CONAMA 357/05 |
|---|---|---|---|---|---|---|---|---|---|---|---|---|---|---|---|
| | JUC01 | JUC02 | JUC03 | JUC04 | JUC05 | JUC06 | JUC01 | JUC02 | JUC03 | JUC04 | JUC05 | JUC06 | | | |
| Ni (mg $kg^{-1}$) | 13.13 | 8.06 | 11.42 | 5.10 | 8.53 | 9.80 | 12.48 | 6.53 | 10.70 | 7.59 | 9.80 | 11.34 | 0.0012 | 0.5 | 30.00 |
| RSD (%) | 9.76 | 13.01 | 11.16 | 13.30 | 15.30 | 5.70 | 3.22 | 4.33 | 14.72 | 7.99 | 13.27 | 10.22 | | | |
| Zn (mg $kg^{-1}$) | 74.29 | 34.90 | 73.01 | 24.81 | 48.29 | 80.15 | 56.25 | 23.55 | 54.06 | 32.26 | 66.50 | 90.12 | 0.0004 | 0.5 | 300.00 |
| RSD (%) | 9.74 | 12.78 | 10.03 | 15.07 | 15.63 | 5.80 | 2.78 | 2.61 | 13.70 | 9.05 | 13.73 | 10.42 | | | |
| Cd (mg $kg^{-1}$) | 1.16 | 0.64 | 1.24 | 0.51 | 0.96 | 1.09 | 0.93 | 0.39 | 0.84 | 0.63 | 0.89 | 1.22 | $1.4 \times 10^{-5}$ | 0.05 | 1.30 |
| RSD (%) | 9.33 | 15.37 | 12.93 | 23.14 | 15.39 | 5.14 | 8.38 | 11.64 | 24.85 | 7.53 | 19.98 | 9.89 | | | |
| Co (mg $kg^{-1}$) | 14.07 | 8.80 | 10.92 | 5.40 | 8.91 | 8.79 | 10.77 | 5.09 | 9.21 | 6.76 | 9.29 | 9.80 | 0.0003 | 0.5 | 25.00 |
| RSD (%) | 9.88 | 13.10 | 9.44 | 14.76 | 15.60 | 5.97 | 3.58 | 3.79 | 13.70 | 8.38 | 14.53 | 10.84 | | | |
| Cr (mg $kg^{-1}$) | 59.80 | 84.74 | 70.61 | 35.53 | 50.27 | 58.12 | 60.93 | 57.78 | 58.22 | 54.09 | 54.69 | 64.35 | 0.0006 | 0.5 | 75.00 |

(continued)

**Table 9.7** (continued)

| 1st campaign | | | | | | | 2nd campaign | | | | | | LD | LQ | CONAMA 357/05 |
|---|---|---|---|---|---|---|---|---|---|---|---|---|---|---|---|
| | JUC01 | JUC02 | JUC03 | JUC04 | JUC05 | JUC06 | JUC01 | JUC02 | JUC03 | JUC04 | JUC05 | JUC06 | | | |
| RSD (%) | 10.06 | 13.00 | 10.46 | 15.10 | 16.09 | 5.78 | 2.67 | 2.88 | 13.71 | 8.99 | 14.21 | 10.51 | | | |
| Cu (mg $kg^{-1}$) | 13.83 | 9.36 | 17.55 | 4.75 | 11.64 | 15.02 | 13.70 | 7.09 | 11.36 | 7.44 | 17.21 | 20.13 | 0.0002 | 0.05 | 60.00 |
| RSD (%) | 16.05 | 4.97 | 9.43 | 20.46 | 20.51 | 1.77 | 5.56 | 8.54 | 0.10 | 3.76 | 3.37 | 4.75 | | | |
| Pb (mg $kg^{-1}$) | 17.55 | 17.96 | 18.84 | 8.11 | 14.02 | 17.41 | 15.44 | 5.74 | 15.17 | 10.99 | 17.88 | 22.44 | 0.002 | 0.5 | 72.00 |
| RSD (%) | 9.75 | 14.32 | 10.49 | 14.85 | 16.82 | 5.01 | 3.68 | 3.72 | 16.12 | 7.69 | 13.44 | 9.23 | | | |
| OM (%) | 10.31 | 10.23 | 13.28 | 17.54 | 9.37 | 9.70 | 9.98 | 3.09 | 12.13 | 7.58 | 9.17 | 11.37 | | | |
| SD | 0.11 | 0.96 | 0.98 | 0.32 | 0.21 | 0.13 | 0.05 | 0.04 | 0.11 | 0.17 | 0.13 | 0.23 | | | |

*LQ* limit of quantification, *LD* limit of detection, *SD* standard deviation, *RSD* Relative standard deviation

shown. The water samples were compared with the resolution of CONAMA nº 357/05 and the sediment samples with the resolution of CONAMA nº 420/09.

Land use in the Jucu River Basin is mostly agricultural, although the various small industries located in the basin are located mainly in Viana, Cariacica, and Vila Velha, urbanized and anthropized regions, where the monitoring stations of this study are located.

In samples of water, the metals Al, Fe, and Mn were at all stations with concentrations above the permitted limit, and Ni in the 1st campaign was also above the limit allowed by the resolution of CONAMA nº 357/05 for waters of class 1 and 2.

Manganese ranged from 0.1118 to 0.2474 mg $L^{-1}$, where the maximum limit is 0.100 mg $L^{-1}$. Even though it is considered essential for aquatic animals and humans, too much can cause damage. Their availability is mainly due to soil erosion and industrial emissions. As its metal is widely used in diverse industries and fertilizers, the high concentration of Mn in the waters of the Jucu River may be due to nearby industries and also the use of the soil of the river environment by agriculture.

Ribeiro et al. (2012), also found Mn contamination dissolved in waters of the São Francisco/MG river, near the industries. Nienie et al. (2017) noted that water samples also showed Mn above the limit, but the origin of this metal would be from geochemical processes at the collection sites.

As the Jucu River basin soil is a red-yellow latosol, it can have considerable concentrations of iron and aluminum, JUC01 station is a station that suffers from margin erosion and surface runoff, which can increase the concentration of iron and aluminum in its bed.

There was an increase in the concentration of aluminum and iron in the 2nd season, which is probably associated with the rainy season Menezes et al. (2009), also observed the contamination of these elements in surface waters of the São Domingos River (RJ/ Brazil), where in the rainy season, this contamination was higher.

The concentrations of the other trace elements in sediment were below the soil prevention limit of resolution CONAMA 420/09.

In the sediment samples, the concentration of chromium was significant at all collection stations, being above the limit at station JUC02 in the 1st campaign, with 84.74 mg $kg^{-1}$. According to Rosolen et al. (2009) even though chromium is naturally present in the sediments, chromium reaches the sediment mainly through industrial effluents, such as leather treatment (tanning), paint and pigment manufacturing, wood preservatives and electroplating, fertilizers, mining activities, and urban and industrial waste. Near the JUC02 station was investigated, the presence of some industries that could have caused this high concentration of chromium in this period.

According to Ferreira (2001), trace elements have strong interactions with organic matter, humic and fulvic substances, which has a marked effect on the speciation and toxicity of these elements.

The organic matter content in the two campaigns was close (Tables 9.2 and 9.3), except for the JUC02 and JUC04 stations, where their contents were reduced in the 2nd campaign. Likewise, it was observed in the sediment samples that the elements showed similar trends to the organic matter content in all the sampling stations, except for the JUC04 stations, that for all elements this tendency was not observed, having in the 2nd campaign an increase of the concentration of all trace elements except Mn. This leads us to think that in the JUC04 stations, the elements do not have greater affinity with organic matter, but with another form of the sediment fraction, such as clay and silt for example.

### 9.3.12 Phytotoxicity

The results of the analyses of the germination index and the root growth index are shown in Fig. 9.2.

It can be observed that the water samples from the Jucu River did not present a phytotoxic effect on the *L. sativa* seeds, because the root growth was larger than the positive control.

It was evidenced by a stimulating effect in all stations of the river because the germination index was higher than the negative control. Rodrigues et al. (2013), Tigini et al. (2011), and Alvim et al. (2011) also observed this stimulating effect for plants exposed to domestic and industrial effluents. These authors argue that the organic load derived primarily from domestic effluent may have provided a greater supply of nutrients that contributed to increased root growth.

In the 1st season, the JUC05 station had the lowest root growth index compared to the other stations, which may be due to its low organic matter content, and in the WQI classification, its index was one of the lowest with 63. In the 2nd JUC04, JUC05, and JUC06 showed higher root growth, probably due to increased rainfall and the consequent availability of nutrients and organic matter from domestic and industrial.

### 9.3.13 Sustainable Development Goals (SDG)

The first studies relating to the level of water purity and pollution were carried out in Germany in 1848 (Ott 1978). These studies sought to synthesize water quality data, through the relationship between the level of water purity and pollution, with the occurrence of certain communities of aquatic organisms (Derísio 1992). Water contains a wide variety of constituents that can be measured in these quality monitoring programs, related to its physical, chemical, and biological aspects, and therefore, the assessment of water quality, the interpretation of data, as it involves multiple aspects, becomes complex, costly and presents difficulty in real-time application (Carvalho et al. 2016). To improve this assessment, the water quality index (WQI) has been developed and used throughout the world. It aims to summarize the variables analyzed in a number, which makes it possible to analyze the evolution of water quality and facilitates the interpretation of extensive lists of variables or indicators, which can help scientists and the rest of the planet with

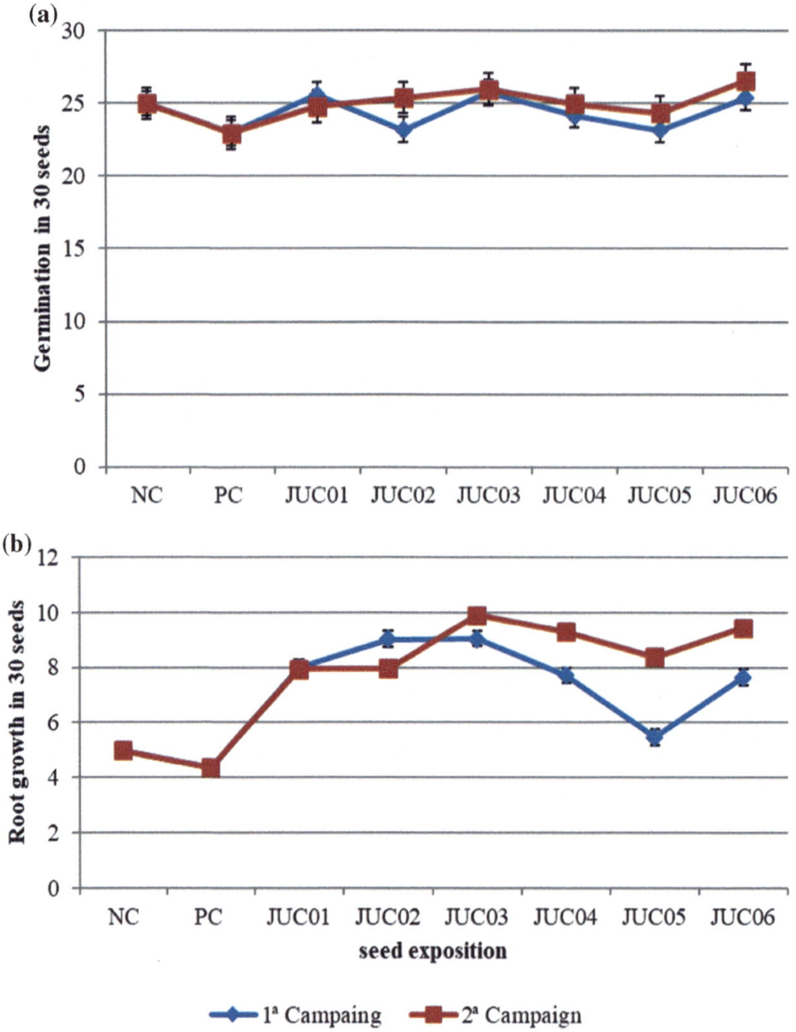


**Fig. 9.2** Results of the germination (**a**) and root growth (**b**) analysis in 30 *Lactuca sativa* seeds, exposed in samples from the 1st and 2nd seasons and in positive and negative control. Tukey's test ($p > 0.05$). CN: negative control and CP: positive control

issues related to water quality and Sustainable Development Goals (SDG) (Gastaldini and Souza 1994). Environmental performance indicators can be understood as parameters that provide information about an activity or a scenario, in relation to environmental factors, enabling analysis, conclusions, and strategic decision-making (Ferreira et al. 2016).

The WQI is a dimensionless number that expresses the quality of water for different purposes. This number is obtained from the aggregation of physicochemical, bacteriological, and chemical data using specific methodologies (ÍNDICE E INDICADORES DE QUALIDADE DE ÁGUA 2018). Due to their reductionist nature, in which several quality items are converted into a single grade or assessment, the indices are quite questionable, since they mask or attenuate the multiplicity of conditions that occur in a water body. On the other hand, the synthesis capacity provided by an index, as long as its intrinsic limitations are understood, is of great importance for communication with the public, facilitating its interrelationship with sustainability criteria (Lermontov 2009). With respect to indices that specifically deal with sustainability, and the quality of water in a river basin, the initial milestone occurred in 1965 when Horton developed, through a pioneering study, general indices, selecting and weighing analytical variables of quality of different waters (ÍNDICE E INDICADORES DE QUALIDADE DE ÁGUA 2018).

The WQI was defined by Horton in 1965, based on physical, chemical, and biological measurements, in which the related parameters were: dissolved oxygen, pH, fecal coliforms, electrical conductivity, alkalinity, and chlorides (Souza 2008). In 1970 Brown, McClelland, Deininger, and Tozer presented a water quality index very similar to the Horton index, and the study was funded by the National Sanitation Foundation called the Water Quality Index and/or National Sanitation Foundation Water Quality Index (NSFWQI). The index combined the opinions of 142 experts, based on the Rand Corporation's Delphi technique, via the responses to several questionnaires, tabulated and returned to each participant, to compare their response with that of the other participants to determine whether to reach a consensus. The experts and researchers evaluated and gave their opinions on the aggregation of water quality analytical variables to compose the NSFWQI. From thirty-five indicator variables, nine variables were selected to compose the index, dissolved oxygen, fecal coliforms, pH, biochemical oxygen demand, nitrate, total phosphorus, temperature, total dissolved substances, and turbidity (ÍNDICE E INDICADORES DE QUALIDADE DE ÁGUA 2018).

The WQI is particularly sensitive to sewage contamination, being a reference index normally associated with the quality of raw water collected for public supply after treatment (INSTITUTO MINEIRO DE GESTÃO DAS ÁGUAS 2014).

There are many studies that apply the water quality index together with other tools in their scientific work, as can be seen in the following studies. Zanini et al. (2017), used the water quality index and the average trophic state index to evaluate the quality of water from the Córrego Rico microbasin, which supplies the city of Jaboticabal (SP).

In Espírito Santo/Brazil, Bertossi et al. (2012) used WQI together with principal component analysis to evaluate the quality of surface and underground waters in the Horizonte stream basin, in Alegre-ES/Brazil. Effendia (2015) used the WQI together with the pollution index to evaluate the water quality of the Ciambulawung River, located in Indonesia. Pedroso and Colesanti (2017) also used the WQI to evaluate the water in the Ribeirão da Sand River basin (Goiás/Brazil); they also applied geoprocessing techniques aiming to spatialize the sampling points. Moreover, Şener et al. (2017) also used WQI to evaluate the quality of water from the Aksu River in Turkey for drinking, along with other physicochemical parameters. All these studies show without a shadow of a doubt that the WQI is a very powerful tool to assist decision-making with respect to sustainability criteria.

## 9.4 Conclusion

The Jucu River/ES is one of the main rivers used to supply the great Vitória, and its basin is very important for the entire population. Given this, it was possible to prove that the Jucu River is in regular condition, since during the rainy season the water consumption rate increased due to the large load of domestic effluents, industries and companies receiving the river, where the parameters are affected: turbidity, solids, BOD and COD and thermotolerant coliforms and *Escherichia coli.*

In the 2nd campaign, all stations had their waters reduced, the most affected stations were JUC01, JUC05, and JUC06.

The trace elements analyzed in the 1st campaign as Ca, Mg, Zn, Cr, Co, Cu, Ni, Cd, and Pb were all below the limit allowed for the resolution of CONAMA n° 357/05 for waters, and also for the resolution of CONAMA n° 420/09 for sediment. The surface waters of the Jucu River were contaminated by aluminum and iron, which is due to the soil characteristics around the river, with higher concentrations during the rainy season. The concentration of chromium was high in the sediment samples, passing the limit at the JUC02 station in the 1st campaign, being able to be related to the industries near the collection site.

Concerning the toxic effect on the waters of the Jucu River, the bioassay with *Lactuca sativa* revealed that there was no phytotoxic effect. The water samples caused a stimulating effect on root growth due to the concentration of nutrients from organic matter.

It is therefore recommended that the State Agency for Water Resources and other relevant bodies be vigilant and more aggressive in monitoring and enforcing effluent discharge laws to ensure minimum pollution in the Jucu River.

**Acknowledgements** To the Laboratory of Research and Development of Methodologies for Analysis of Petroleum of UFES (LabPetro), and to the Laboratory of Sanitation of the UFES for granting the laboratory and the reagents for the analysis. FAPES, CAPES, and CNPq for research support. The State Agency for Water Resources (AGERH) for the availability of the physical and chemical results.

**Conflict of Interest Statement** The authors declare that they have no known competing financial interests or personal relationships that could have appeared to influence the work reported in this paper.

**Declarations Funding** This research was funded by the Coordination for the Improvement of Higher Education Personnel—Brazil (CAPES); Espírito Santo Research and Innovation Foundation (FAPES); Brazilian National Council for Scientific and Technological Development (CNPq); and the Federal University of Espírito Santo (UFES).

## References

Abreu CHM, Cunha AC (2017) Qualidade da água e índice trófico em rio de ecossistema tropical sob impacto ambiental. Eng Sanit Ambient. Ambio. https://doi.org/10.1590/s1413-41522016144803

Alvarenga LA, de Mello CR, Colombo A, Cuartas LA, Chou SC (2016) Hydrological responses to climate changes in a headwater watershed. Ciência e Agrotecnologia 40(6):647–657. https://doi.org/10.1590/1413-70542016406027716

Alvim LB, Kummrow F, Beijo LA, Lima CAA, Barbosa S (2011) Avaliação da citogenotoxicidade de efluentes têxteis utilizando Allium cepa L. Revista Ambiente & Água - Interdiscip J Appl Sci. https://doi.org/10.4136/ambi-agua.198

APHA, Awwa, WEF (2005) Standard methods for examination of water and wastewater. American Public Health Association, Washington

Bertossi APA, Cecílio RA, Neves MA, Garcia GO (2012) Qualidade da água em microbacias hidrográficas com diferentes coberturas do solo no sul do Espírito Santo. Revista Árvore 37:107–117

Bonnet BRP, Ferreira LG, Lobo FC (2008) Relações entre qualidade da água e uso do solo em Goiás: uma análise à escala da Bacia Hidrográfica. Revista Árvore. https://doi.org/10.1590/S0100-67622008000200014

Borges MJ, Galbiatti JA, Ferraudo AS (2003) Monitoramento da qualidade hídrica e eficiência de interceptores de esgotos em cursos d'água urbanos da Bacia Hidrográfica do Córrego Jaboticabal. Revista Brasileira de Recursos Hídricos. https://doi.org/10.21168/rbrh.v8n2.p161-171

Buzelli G, Cunha-Santino M (2013) Análise e diagnóstico da qualidade da água e estado trófico do reservatório de Barra Bonita, SP. Revista Ambiente & Água. https://doi.org/10.4136/ambi-agua.930

Carmo DL, Silva CA (2012) Métodos de quantificação de carbono e matéria orgânica em resíduos orgânicos. R Bras Ci Solo. https://doi.org/10.1590/S0100-06832012000400015

Carvalho A, Balduino A, Maciel G, Picanço A (2016) Avaliação da poluição em rios utilizando índices de qualidade da água: um estudo de caso no Ribeirão São João em Porto Nacional –TO. São Paulo, UNESP, Geociências 35:472–484

Christofoletti C (2008) Avaliação dos potenciais citotóxico, genotóxico e mutagênico das águas de um ambiente lêntico, por meio dos sistemas-teste de allium cepa e oreochromis niloticus. Dissertação, Universidade Estadual Paulista "Júlio De Mesquita Filho" Instituto de Biociências

Derísio JC (1992) Introdução ao controle da poluição ambiental. Editora da CETESB, São Paulo

Effendia H (2015) Water quality status of Ciambulawung River, Banten Province, based on pollution index and NSF-WQI. Procedia Environ Sci 24:228–237. https://doi.org/10.1016/j.proenv.2015.03.030

EMBRAPA - Empresa Brasileira de Pesquisa Agropecuária (2006) Sistema brasileiro de classificação de solos. EMBRAPA-SPI, Rio de Janeiro

Ferreira RJS (2001) Determinação de elementos traço em sedimentos de rios: caso da bacia do Baixo Itajaí-Açu. Dissertação de mestrado. Universidade Federal de Santa Catarina, Florianópolis

Ferreira MA, Santos EC, Lima JE (2016) Índice de avaliação ambiental: Uma análise a partir de modelos estatísticos multivariados. Revista Iberoamericana de Economía Ecológica 26:177–190

Ferreira, Rafael de Queiroz, et al. Química A 3rd ed, Sead (2023) https://acervo.sead.ufes.br/materiais/quimica/quimica-ambiental/.

Fiorucci AR, Filho EB (2005) Oxigênio dissolvido em sistemas aquáticos. Química nova na escola 22:10–16

Gastaldini MCC, Souza MDS (1994) "Diagnóstico do Reservatório do Vacacaí- Mirim através de Índices de Qualidade de Água", 1° Seminário sobre Qualidade de Águas Continentais no Mercosul, Porto Alegre

Goldin A (1987) Reassessing the use of loss-on-ignition for estimating organic matter content in noncalcareous soils. Commun Soil Sci Plant Anal. 18:1111–1116.

ÍNDICE E INDICADORES DE QUALIDADE DE ÁGUA: Revisão da Literatura, 2006. Disponível em: http://www.cprh.pe.gov.br/downloads/indice-agua-volume1.pdf. Acessado em: 8 de agosto de 2018

INSTITUTO MINEIRO DE GESTÃO DAS ÁGUAS (2014) Monitoramento da qualidade das águas superficiais de Minas Gerais em 2013: resumo executivo/Instituto Mineiro de Gestão das Águas. Instituto Mineiro de Gestão das Águas, Belo Horizonte, 17 pp

Junior BSG, Simoura LT (2016) Cobertura florestal e qualidade da água na bacia hidrográfica do rio Jucu, estado do Espírito Santo, Brasil. Scientia Plena. https://doi.org/10.14808/sci.plena.2016.010201

Lermontov A (2009) Novo Índice de Qualidade das Águas com uso da Lógica e Inferência Nebulosa. Tese de Doutorado. Universidade do Federal do Rio de Janeiro. Programa de Pós-graduação em Tecnologia de Processos Químicos e Bioquímicos da Escola de Química. Rio de Janeiro/RJ, Brasil. EQ/UFRJ

Lopes FW, Júnior AP (2010) Influência das condições naturais de pH sobre o índice de qualidade das águas (IQA) na bacia do Ribeirão de Carrancas. Geografias 6(2):134–147

Mcneil VH, Cox ME, Pred M (2005) Assessment of chemical water types and their spatial variation using multi-stage cluster analysis, Queensland, Austrália. J Hydrol. https://doi.org/10.1016/j.jhydrol.2004.12.014

Menezes JM, Prado RB, Júnior GCS, Mansur KL, Oliveira ES (2009) Qualidade da água e sua relação espacial com as fontes de contaminação antrópicas e naturais: bacia hidrográfica do rio São Domingos – RJ. Eng Agríc. https://doi.org/10.1590/S0100-69162009000400019

Netto ITP (2008) Qualidade física e química de um latossolo vermelho-amarelo sob pastagens com direfentes períodos de uso. Dissertação de mestrado. Universidade de Brasília-BR

Neves GL, Guimarães TT, Anjinho PS, Barbosa MAGA, Santos ARD, Virgens Filho JS, Mauad FF (2021) Spatial and seasonal assessment of water quality in the Lobo Stream River Basin, Brazil using multivariate statistical techniques. Anais Da Academia Brasileira de Ciências 93(suppl 4). https://doi.org/10.1590/0001-3765202120210072

Nienie AB, Sivalingam P, Laffite A, Nglinkoto P, Otamonga J, Matand A, Mulaji CK, Mubedi JI, Mpiana PT, Poté J (2017) Seasonal variability of water quality by physicochemical indexes and traceable metals in suburban area in Kikwit, Democratic Republic of the Congo. Int Soil Water Conserv Res. https://doi.org/10.1016/j.iswcr.2017.04.004

Oliveira AFG (2008) Testes estatísticos para comparação de médias. Revista Eletrônica Nutritime 6:777

Ott WR (1978) Environmental indices: theory and practice. Ann Arbor Science, Ann, Arbor, Michigan, p 371

Papa RA, Lacerda MPC, Campos PM, Goedert WJ, Ramos MLG, Kato E (2011) Qualidade de latossolos vermelhos e vermelho-amarelos sob vegetação nativa de cerrado. Pesq Agropec Trop. https://doi.org/10.5216/pat.v41i4.13234

Pedroso LB, Colesanti MTM (2017) Determinação do índice de qualidade de água da bacia hidrográfica do ribeirão da areia – goiás, em período de estiagem. Caminhos de Geografia 18:219–230

Pereira RDS (2004) Identificação e Caracterização das Fontes de Poluição em Sistemas Hídricos. Revista Eletrônica de Recursos Hídricos (RERH) 1:20–36

Ribeiro E, Junior A, Horn A, Trindade W (2012) Metais pesados e qualidade da água do rio São Francisco no segmento entre três marias e Pirapora - MG: Índice de contaminação. Geonomos. https://doi.org/10.18285/geonomos.v20i1.27

Rietzler AC, Fonseca AL, Lopes GP (2001) Heavy metals in tributaries of Pampulha Reservoir, Minas Gerais. Bras J Biol. https://doi.org/10.1590/S1519-69842001000300004

Rodrigues LCDA, Barbosa S, Pazin M, Maselli BDS, Beijo LA, Kummrow F (2013) Fitotoxicidade e citogenotoxicidade da água e sedimento de córrego urbano em bioensaio com Lactuca sativa. Revista Brasileira de Engenharia Agrícola e Ambiental. https://doi.org/10.1590/S1415-436620 13001000012

Rosolen V, Herpin U, Coelho NMM, Coelho LM, Brito JLS, Silva LA, Lima SC (2009) Qualidade dos sedimentos no rio Uberabinha (Uberlândia, MG) e implicações ambientais. Revista Brasileira de Geociências 39(1):151–159

Şener Ş, Şener E, Davraz A (2017) Evaluation of water quality using water quality index (WQI) method and GIS in Aksu River (SW-Turkey). Sci Total Environ 584585:131–144. https://doi.org/ 10.1016/j.scitotenv.2017.01.102

Souza META (2008) Proposição de um Índice de Qualidade da Água Bruta Afluente a Estações Convencionais de Tratamento de Água. Dissertação de Mestrado. Universidade Federal de Minas Gerais. Programa de Pós-graduação em Saneamento, Meio Ambiente e Recursos Hídricos. Belo Horizonte/MG, Brasil. UFMG

Terra V, Prette-Santos R (2008) Perspectivas da avaliação da qualidade da água em rios por intermédio de parâmetros físicos, químicos e biológicos. Natureza On line 6(2):63–65

Terra V, Prette-Santos R (2009) Monitoramento do rio Jucu Braço Sul: Caracterização e avaliação dos parâmetros físico-químicos e microbiológicos. Natureza On line 7:5–11

Tigini V, Giansanti P, Mangiavillano A, Pannocchia A, Varese GC (2011) Evaluation of toxicity, genotoxicity and environmental risk of simulated textile and tannery wastewaters with a battery of biotests. Ecotoxicol Environ Saf 74:866–873. https://doi.org/10.1016/j.ecoenv.2010.12.001

Turial da Silva T, Soares do Espírito Santo KT, Matsumoto S, Ribeiro J (2018) Monitoring of the water quality of the foz of Santa Maria da Vitória river and the Vitória bay-ES, by systematic studies of physical-chemical and biological parameters. Rev Virtual de Quim 10(5):1373–1392. https://doi.org/10.21577/1984-6835.20180094

Valente JPS, Padilha PM, Silva AMM (1997) Oxigênio dissolvido (OD), demanda bioquímica de oxigênio (DBO) e demanda química de oxigênio (DQO) como parâmetros de poluição no ribeirão Lavapés/Botucatu – SP. Eclet Quím. https://doi.org/10.1590/S0100-467019970001 00005

Von Sperling M (2005) Princípios do tratamento biológico de águas residuárias: introdução à qualidade das águas e ao tratamento de esgotos. 3rd edn, vol 1. UFMG, Belo Horizonte, 452 pp

Zanini HLT, Amaral LA, Zanini JR, Tavares LHS (2017) Caracterização da água da microbacia do córrego rico avaliada pelo índice de qualidade de água e de estado trófico. Engenharia Agrícola 30:732–741

Zonta JH, Zonta JB, Rodrigues JI, Reis EF (2008) Qualidade das águas do Rio Alegre, Espírito Santo. Rev Ciên Agron 39(1):155–161

# Solid Waste Management and Sustainability in Deltas: Concluding Remarks and Outlook

10

Mahmoud Nasr

**Abstract**

While Deltas represent about 5% of the global land surface, they occupy over 550 million people (i.e., a high population density). The Delta area, which is characterized by flat topography, fertile alluvial soils, and abundant natural resources, is vital for human habitation. However, this area is subjected to diverse natural and anthropogenic threats, owing to climate change vulnerability. The delicate equilibrium of delta ecosystems is threatened as anthropogenic activities escalate (e.g., illegal solid-waste dumping), necessitating innovative solutions. This chapter includes the main concluding remarks on the evolution of waste management practices within Deltas, emphasizing the significance of sustainable approaches to balance the ecological and socio-economic aspects. As Deltas continue to confront dynamic and seasonal challenges, this book tends to provide a foundation for promoting resilient and sustainable practices in solid waste management (SWM) within these vital ecosystems.

M. Nasr (✉)
Environmental Engineering Department, Egypt-Japan University of Science and Technology (E-JUST), Alexandria, Egypt
e-mail: mahmoud-nasr@alexu.edu.eg; mahmoud.nasr@ejust.edu.eg

Sanitary Engineering Department, Faculty of Engineering, Alexandria University, Alexandria, Egypt

 
M. Nasr and B. Ravindran (eds.), *Solid Waste Management in Delta Region for SDGs Fulfillment*, Deltas of the World, https://doi.org/10.1007/978-3-031-58253-0_10

## 10.1 Introduction

Deltas are landforms that form at the mouth of a river, typically where the river flows into an ocean, sea, or lake, and they are influenced by the deposition of sediment (Elsadek et al. 2024). Deltas are often fertile and are vital for agriculture, as they receive significant amounts of sediment from the river, creating rich soil. The shape and size of a delta can vary depending on factors such as the volume of water and sediment carried by the river, tidal influences, and coastal processes (Youssef et al. 2024). These Deltas are ecologically significant and play important roles in various aspects of the regions they are located in. Valuable ecosystems, such as wetlands, coastal forests, and marshes, are distributed in the Deltas. However, the pollution of water, soil, and air increases in the delta regions, owing to the various anthropogenic activities responsible for degrading the environmental component. Deltas are vulnerable to sediment compaction through heavy structures and underground resource extraction, further influencing and impacting water and sediment supplies (Deng and Liu 2024). For instance, hydropower dams lead to changes in flood pulse and sediment transport.

The Deltas condition can be illustrated through three research domains (i) advancing biophysical classifications of deltas, (ii) understanding Deltas as coupled social–ecological systems, and (iii) analyzing and informing social and environmental vulnerabilities in delta regions. The progress towards the achievement of the recently endorsed sustainable development goals (SDGs) and accompanying targets can be effectively assessed using integrated and collaborative approaches that account for interactions across social, ecological, and physical systems. These delta systems are highly influenced by dumping solid wastes in unauthorized areas. For instance, the considerable increase in urbanization and population growth in most deltas globally has been recently linked with an exponential rise in the amount of solid waste produced (Sonu et al. 2023). This elevated trend in solid waste generation imposes substantial challenges to the SWM systems and infrastructure worldwide (Cucchiella et al. 2014). Implementing and assessing the SDGs in deltaic systems is an essential strategy, offering reliable solid waste management (SWM) and facilities to clients (Saqr et al. 2024). Clients should receive excellent services from specialists in the field of waste recycling, promoting efficient and sustainable waste management methods (Arteaga et al. 2023). The service includes eliminating or reducing waste at the source, reusing waste, increasing recovery and recycling rates, reducing hazardous content in waste, as well as reducing the business cost of managing and disposing of waste (Awino and Apitz 2023).

This book contains a comprehensive survey of the various categories of solid residues collected from domestic households, agricultural/agro-industrial sources, and manufacturing industries. The book also includes sufficient information about the broadly employed waste management techniques, such as physical, chemical, physico-chemical, thermal, thermo-chemical, biological, and bio-chemical methods. It also offers knowledge about SWM and sustainable development in Deltas was correlated with the 17 United Nations

SDGs. The main barriers, limitations, challenges, and opportunities associated with the fulfillment of the sustainability agenda in waste management were reported.

## 10.2 Clustering of Most Common Keywords Associated with Solid Waste Management and Sustainability in Deltas

A bibliometric study was used to analyze data on the main topics of the published articles about "Waste", "Sustainability", and "Deltas" searched within The SCOPUS search engine. VOSviewer was used to demonstrate a broad understanding of SWM to meet sustainability in Deltas. The main topics addressing the waste management strategy can be described as follows (Fig. 10.1):

- integrated assessment for sustainable development goals of solid waste gasification and pyrolysis systems (Chu et al. 2023)
- use of machine learning and artificial intelligence for optimizing anaerobic fermentation of organic solid waste
- bibliometric studies on the importance of microalgae and municipal waste in bioenergy products
- occurrence and monitoring technologies for exergy and sustainability analysis of various typical solar energy applications

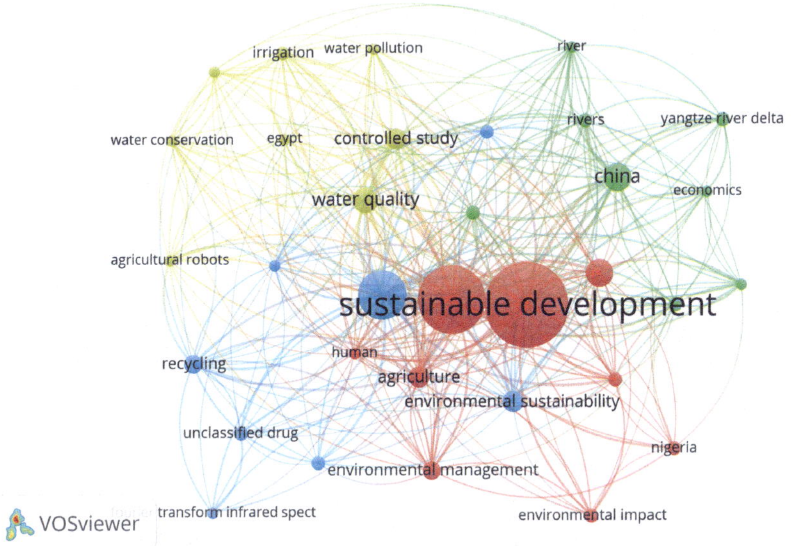


**Fig. 10.1** Clustering of keywords obtained from bibliometric analysis data using query string: TITLE-ABS-KEY (waste AND sustainability AND deltas)

- sustainable valorization of tires, plastics, and food processing wastes for biochar production (Ezz et al. 2023)
- promotion of net-zero emissions from decarbonization in the waste recycling industry (Dadebo et al. 2023a)
- developing green and sustainable concrete: assessment of strength and durability of concrete containing different agricultural waste as agro-cement
- resource recovery from SWM in a water-energy-food nexus perspective (Hasan et al. 2023)
- circular economy approach towards sustainability of plastic waste management (Dadebo et al. 2023b)
- vermicomposting in SWM: a sustainable, cost-efficient, and environment-friendly approach (Chowdhury and Sarkar 2023)
- anaerobic digestion (AD), composting, and fermentation of biowaste (Eraky et al. 2023)
- sustainable management of electronic waste (e.g., desktop monitors, laptops, and tablets) in developing countries
- Industry 4.0-based sustainable circular economy approach for smart waste management systems (Karthik et al. 2023)
- biofuels from agricultural wastes and byproducts, maintaining environmental sustainability (e.g., agricultural waste is one of the widely used feedstock (as a substrate) for biogas production under AD (Torres-Lozada et al. 2023)).

## 10.3 Solid Waste Composition

Solid wastes are generated in large quantities in most delta systems worldwide because of the exponential increase of urbanization and population growth. The main composition of municipal solid waste (MSW) is summarized in Table 10.1. Food, vegetables, paper, and plastic wastes have lower calorific value than oil, gas, and coal.

## 10.4 Negative Impacts of Illegal Solid Waste Disposal on the Environment

Unplanned and nonsegregated solid waste disposal sites, receiving considerable quantities of solid waste, are considered sources of pollution to several plants and animals (Peng et al. 2023). The growth and dominance of most microorganisms, including bacteria, fungi, cyanobacteria, and algae (Ansari et al. 2023), are oppositely impacted by indiscriminate waste disposal and inadequate waste collection (Santulli et al. 2023). Various environmental and socio-economic benefits can be achieved through the proper management of solid wastes (Wang et al. 2023a). Greenhouse gas (GHG) emissions are of

**Table 10.1** Typical sources of municipal solid waste (MSW)

| Category | Common source |
|---|---|
| Kitchen waste | Food waste including fruit and vegetable residues |
| Agro-food residues | Solid food processing wastes |
| Yard waste | Leaves, grass, tree trimmings, and burning barbeque wood |
| Paper | Scrap paper, newspaper, magazines, books, paper bags, wet wipes, tissue paper and napkins, parchment paper, wrapping paper, cardboard, and packaging boxes |
| Plastic | Bottles, container lids and caps, boxes, clear wraps, Zip lock bags, polyethylene bags, polystyrene, cable wire, pipes, CDs, DVDs, and vinyl records |
| Glass | Bottles, broken glassware, utensils, mirrors, light bulbs, decorative pieces, and window panels |
| Metal | Cans and tins, aluminum foils, utensils, cookware, container lids, appliances, railings, bicycles |
| Electronic waste | Thrashed computers, monitors, tablets, phones, watches, and batteries |
| Inert materials | Construction, demolition and renovation wastes, drywalls and ash |
| Miscellaneous | Broken appliances, textiles, leather, rubber, pet litter, personal hygiene products, health care products, pharmaceuticals, cosmetics, discarded furniture (Ngulube et al. 2024) |

great concern as they contribute to changes in climatic conditions and are regarded as a major environmental issue (Aboughaly and Fattah 2023). GHG emissions, comprising $CO_2$, methane ($CH_4$), and nitrous oxide ($N_2O$) from unregulated or unlicensed waste incinerators or pyrolysis systems, are significant contributors to global anthropogenic air pollution.

## 10.5 Solid Waste Conversion Options

The recent elevation in population growth and urbanization has been accompanied by increasing the amounts of solid wastes dumped into the environmental matrix, especially in most developing countries (Bhatia et al. 2023). The composition of household solid waste includes approximately 44, 13, 11, 10, 9, 6, and 7% of organics, plastic, paper, glass, rubber, Hazardous material, and others (Mannaa et al. 2023), requiring various waste conversion techniques (Table 10.2).

Anaerobic digesters receive organic-based substrates from different sources (Shi et al. 2021), including (i) animal waste (urine, manure, cow dung, poultry litter, and chicken droppings), (ii) plant waste (crop residue, peels, forest/landscape biomass, roadside vegetation, and exhausted aquatic plants), (iii) industrial and municipal wastes (city garbage,

**Table 10.2** Commonly employed treatment methods for solid organic waste management

| Method | Description | Drawbacks |
|---|---|---|
| Physical methods | Cutting, chopping, grinding, pulverizing, ultrasonication, irradiation, and steam explosion | Needs higher energy inputs to maintain certain temperature and pressure |
| Chemical methods | Using certain chemicals such as acids, bases, and oxidizers, reducers | Some buffering reagents are unavailable or expensive, and they may transfer through the environmental matrix |
| Biological methods | Microbial and enzymatic activity to achieve the degradation of organic fraction of solid waste | Bacterial acclimatization and adaptation entail extended retention time<br>Operational factors (pH and temperature) should be optimized to enhance microbial growth and activity because microorganisms are sensitive to variations in environmental conditions |

biogas slurry, sewage, sugar industry, paper mill waste), and (v) household wastes (e.g., food and kitchen wastes) (Tawfik et al. 2024). According to the content of total solids (TS) in the feedstock and inoculum, AD can be operated under either dry or wet conditions. For instance, a solids input of TS over 20% can be used to describe the dry digestion process (Chiumenti et al. 2018). Various strategies, such as feedstock pre-treatment, co-digestion, conductive material supplementation, and percolate recirculation (improving mass transfer), can be used to enhance the methanogenesis pathway of dry AD. The pretreatment strategies can be mechanical treatment (grinding and chopping), chemical treatment (alkali, and oxidant), biological treatment (fungal, enzyme, and microbial consortia), and thermal treatment (high- and low-temperature levels). In parallel, wet digestion is used to describe a liquid phase medium (slurry, manure, and sewage sludge) with lower TS contents (Emmanouilidou et al. 2023). Dry AD consumes lower amounts of water and generates fewer slurry digestate quantities compared to wet-based ones. As such, dry AD can be efficiently used in areas that suffer from insufficient water resources and unavailability of land for nutrient recovery from digestate. For instance, bio-stabilization/composting, drying, and pyrolysis are employed for nutrient recovery from the solid fraction of digestates. Other methods, such as evaporation, membrane separation, precipitation, and $NH_3$ stripping, are applied for nutrient recovery from the liquid digestate. Nutrient recovery can also be established using microalgae, fungal, hydroponic culture, and black soldier fly larva cultivation in digestates (Wang et al. 2023b). Composting of dry digestate undergoes multiple biochemical reactions performed by the microbial communities under an aerobic condition, giving a nutrient-rich material (Syarifinnur et al. 2023). The composite should contain sufficient fractions of carbon, nutrients, and minerals, providing a suitable organic conditioner to the soil matrix (Midden et al. 2023). The efficiency of compositing of solid digestate is influenced by several parameters, including moisture content, C/N ratio, bulk density, porosity, medium temperature, enzymatic activity, and salinity index (Palansooriya et al. 2023).

## 10.6 Recovery Material from Solid Waste Management (SWM)

Waste management system plays a significant role in sustainable development and the circular economy transition (Zidan et al. 2023). Waste management can be addressed with the minimization of waste, the biological transformation of waste into value-added products, and the detoxification of toxic pollutants present in solid waste. Life cycle assessment (LCA) is a technique used for assessing environmental aspects and potential impacts associated with a product or service. As such, LCA is a tool employed to assess potential environmental impacts throughout a product's life cycle (i.e., from natural resource acquisition) via the management of solid wastes (including disposal and recycling) (Banik et al. 2023).

Waste valorization is a thermal process that uses waste as raw material to generate electricity and heat through combustion. Gases generated are then filtered and eliminated, complying with the strictest environmental regulations for emissions. The commonly used valorization techniques are summarized in Table 10.3. The pyrolysis of solid waste is recognized as one of the popular routes for delivering carbon-rich material, known as biochar (Blanchard et al. 2023). This biochar can be employed as an adsorbent phase to capture undesirable species from industrial effluents, a precursor for manufacturing highly efficient catalytic material, and as a soil conditioner (Nanda et al. 2023). Biochar can also be utilized as a cost-effective and supportive material for enhancing microorganism metabolism and DIET mechanism in AD systems (Wei et al. 2024). Solid characterization is an essential step for selecting the most appropriate way for conversion into valuable products. The characterization techniques comprise the determination of surface morphology, calorific values, thermogravimetric kinetics, elemental analysis (C, H, O, N, and S), and proximate composition (e.g., moisture and ash contents) (Czekała et al. 2023).

## 10.7 Advantages and Limitations of Solid Waste Management (SWM) Techniques

Biochemical conversion involves the use of microorganisms or their enzymes to breakdown biomass (Table 10.4). AD and fermentation are the widely used biochemical technologies for waste conversion. During AD, microbes convert the organic fraction of wastes into biogas, i.e., 40–65% $CH_4$, 35–55% $CO_2$, and other gases present in trace amounts like $H_2$ and $H_2S$. Biogas production is a multistep process that describes the progressive transformation of complex organic compounds (liquid or solid wastes) into

**Table 10.3** Commonly employed techniques for converting solid wastes to valuable products

| Method | Process |
| --- | --- |
| Pyrolysis | It is the chemical decomposition of organic compounds in the absence of oxygen. It represents the first stage in the gasification phase for combustion, mainly generating biochar (Chang et al. 2023) |
| Controlled pyrolysis | It is the process by which organic matter is partly oxidized, thereby producing syngas |
| Incineration | Incineration or direct combustion is another technology available for waste treatment. Direct combustion or incineration generates energy in the form of heat. This energy is reusable and can be used to produce electricity, steam, or hot water (Chen et al. 2023) |
| Waste-to-fuel | It is a treatment process in which certain types of fuel are produced from waste, usually plastic waste |
| Waste-to-energy (WtE) | It involves processing waste to obtain energy that can be used within the same plant for internal processes or else to be sold to external parties |

low-molecular-weight products by various microbial communities. The fermentation of waste involves the conversion of sucrose into fructose and glucose via hydrolysis enzymes followed by their transformation into ethanol. Thermal conversion of organic waste material, such as plastics, or biomass under the influence of oxygen, into crude synthesis gas yields a hydrogen product (Thi et al. 2024).

## 10.8 Sustainable Development Goals (SDGs) and Circular Economy in Relation to Solid Waste Management (SWM)

Waste management should fulfill the sustainable development goals (SDG) including:

- the end of poverty and hunger
- better standards of education and healthcare, particularly as it pertains to water quality and better sanitation
- achievement of gender equality
- sustainable economic growth while promoting jobs and stronger economies
- sustainability to include the health of the land, air, and sea (Heiba et al. 2023a)

**Table 10.4** Notable advantages and disadvantages of municipal solid waste (MSW) conversion technologies

| Process | Advantages | Disadvantages |
|---|---|---|
| Conventional incineration | • Widely used technology for MSW conversion<br>• Mature technology with established industrial infrastructures<br>• Converts MSW to combined heat and power, electricity and steam<br>• Commercially viable<br>• Low operating costs<br>• Faster process than biological conversion | • Generates massive amounts of GHGs and pollutants<br>• Heavy metals from MSW are retained in fly ash, bottom ash, and flue gases, which require expensive sequestration techniques<br>• Non-combustible ash may generate clinkers and corrode incinerator walls, thus increasing the maintenance cost |
| Hydrothermal incineration | • Clean technology due to the involvement of supercritical water for the hydrothermal flame generation<br>• Aids in the remediation of highly recalcitrant, complex, and hazardous pollutants in MSW<br>• Results in a partial conversion of pollutants to energy<br>• Applications can be extended to extraterrestrial organic waste remediation<br>• Faster process than biological conversion | • Requires high temperatures and high pressures to maintain supercritical conditions<br>• Corrosion of reactor from salts and metals present in MSW<br>• Issues in the recovery of waste heat<br>• Lack of literature due to the new technology and requires more research |

(continued)

**Table 10.4** (continued)

| Process | Advantages | Disadvantages |
|---|---|---|
| Pyrolysis | • Converts MSW to bio-oil, biochar, and producer gas (Fang et al. 2023)<br>• Biochar can be used for soil application, activation, pollutant adsorption, carbon sequestration, and several other environmental applications<br>• Bio-oil can be catalytically upgraded to synthetic transportation fuels<br>• Bio-oil can be used as a precursor for the recovery of biochemicals<br>• Moderate operating cost<br>• Faster process than biological conversion | • Metals present in MSW can either catalyze or retard the pyrolysis process, further affecting product yields<br>• Product yield depends on operating conditions (e.g. temperature, heating rate, vapor residence time, reaction time, reactor type, and MSW properties)<br>• MSW properties greatly influence the chemistry and fuel properties of bio-oil<br>• MSW requires drying to enhance the energy content of bio-oil. This drying step can add to the overall process expenditures |
| Thermochemical and hydrothermal liquefaction | • Converts MSW to bio-crude oil (Carregosa et al. 2023)<br>• Adding acid and base catalysts can enhance product yields<br>• Liquefaction-derived bio-crude oil has superior fuel properties and energy density compared to pyrolysis-derived bio-oil<br>• Bio-crude oil serves as a precursor for biochemicals<br>• Hydrothermal liquefaction operates in subcritical water. Hence, the drying of MSW is optional | • Metals existing in MSW can either catalyze or delay the liquefaction process, influencing product yields<br>• Product yield depends on operating conditions (e.g. temperature, heating rate, feed concentration, reaction time, catalyst type and loading, reactor type and MSW properties)<br>• MSW properties greatly influence the chemistry and fuel properties of bio-oil |

(continued)

**Table 10.4** (continued)

| Process | Advantages | Disadvantages |
|---|---|---|
| Thermochemical and hydrothermal gasification | • Converts MSW to $H_2$-rich syngas and char products (Ahmad et al. 2023)<br>• The presence of combustible gases (i.e., $H_2$, $CH_4$, and CO) in the syngas can enhance its heating value<br>• Operation can occur in subcritical and supercritical water. As such, the drying of MSW is not required<br>• Results in the enhanced dissolution of organics, paper, and plastics in MSW<br>• $H_2$ present in MSW-derived syngas can be used as an energy carrier or energy vector and has applications in fuel cells, bio-oil upgrading, and production of hydrocarbons and chemicals | • Requires high temperatures and high pressures to maintain supercritical conditions<br>• Corrosion of reactor from salts and metals present in MSW<br>• Issues in the recovery of waste heat (Halba et al. 2023)<br>• Tar and char may cause the plugging of the reactor<br>• Operating conditions (e.g. temperature, feed concentration, heating rate, reaction time, catalyst type and loading, reactor type and MSW properties) should be optimized |
| Anaerobic digestion | • Clean technology requiring methanogenic bacteria<br>• The presence of undesired microorganisms is restricted due to the anaerobic environment<br>• Low operating costs and less maintenance<br>• Mature and traditional technology<br>• Potential to boost the rural economy with self-sustained and integrated bioenergy and MSW-management infrastructure<br>• Bio-sludge can be used for agricultural applications to enhance soil fertility and carbon content<br>• Bio-sludge can be diverted to thermochemical conversion for energy generation | • A time-intensive process requiring days to weeks due to slower microbial metabolism<br>• Requires strictly anaerobic conditions<br>• Biogas ($CH_4$) is considered as a potent GHG<br>• Create foul and obnoxious odor in case of leakage of the bioreactor<br>• The existence of inhibitory compounds in MSW can inhibit microbial growth and impact methanogenesis |

The three pillars of sustainability:

- environmental sustainability refers to humans living within and as part of a finite set of natural resources.
- practices for economic sustainability contribute to the economic development of a country or an organization within a country while protecting the environmental and social angles.
- social sustainability considers the impacts of business actions on people (Heiba et al. 2023b).

A circular economy represents an economic system that is structured with the objectives of reduction in waste and consumption of resources. The circular economy can be achieved through various proper SWM strategies, generating the following valuable items:

- biofuel is defined as liquid, solid, or gaseous fuel produced by the conversion of biomass such as bioethanol from sugar cane or corn, charcoal or woodchips, and biogas from anaerobic decomposition of wastes. Biofuels comprise fuels derived from organic materials such as plants and animals. These plant- and animal-based fuels include gaseous fuels, such as $H_2$ and $CH_4$, and liquid fuels such as methanol ($CH_3OH$), ethanol ($C_2H_5OH$), biodiesel, and acetone ($C_3H_6O$).
- agricultural wastes can be utilized as a feedstock resource (green biomass) for biogas production through an AD process (Mengqi et al. 2023). This renewable resource is characterized for surface morphology and elemental composition and used to determine various value-added materials.
- biomass feedstock resources (e.g., forestry residues) can be converted to adsorbent materials, such as biochar, and used to remediate contaminated sites (Mengqi et al. 2023).

## 10.9 Challenges and Barriers in Employing of Solid Waste Management (SWM)

The unrestricted disposal of solid wastes tends to pose serious environmental risks to soils and food chains (Table 10.5). Groundwater pollution from solid waste dumping in landfills without proper liners or leachate collection systems (Akhtar et al. 2023). Greenhouse gas emissions represent a major and complex concern for the future use of fossil energy in general and coal in particular. They include water vapor ($H_2O$), $CO_2$, nitrous oxide ($N_2O$), $CH_4$, hydrofluorocarbons (HFCs), perfluorocarbons (PFCs), and sulfur hexafluoride ($SF_6$).

**Table 10.5** Challenges and barriers facing the implementation of solid waste management (SWM)

| Challenge | Justification/description |
|---|---|
| Financial and economic | • Budget and initial investment for constructing the facility and infrastructure and purchasing the equipment (e.g., pump, sensors, heater, and mixer)<br>• Running expenses for electricity bill payments/feed-in tariff, chemicals/consumables, and water utilities<br>• Payback period required to recover the initial investment, which should be shorter than the project's lifetime<br>• Carbon emission trading, footprint, and market<br>• Shadow prices of pollution, also known as marginal abatement cost, allow economists and policymakers to estimate the economic value of environmental resources that do not have explicit market prices<br>• Subsidies and incentives<br>• Taxes, internal rate of return, and net present value |
| Technical | • Insufficient advanced and modern technologies, especially in less-developed countries<br>• Accessible and availability of waste feedstock<br>• Highly skilled and well-trained workers<br>• Operational personnel's knowledge and awareness<br>• Optimization of operational factors using computer-based models<br>• Feasibility, simplicity, and reliability of thermal treatment systems<br>• Infrastructure network (biomethane injection into the main gas grid or use as a road fuel)<br>• Decentralization of decision-making<br>• Scarcity of integrated information on waste<br>• Capacity factor and household per capita waste generation rate<br>• Research and development efforts<br>• Safety and legislation in biogas production and storage |

(continued)

**Table 10.5** (continued)

| Challenge | Justification/description |
|---|---|
| Environmental | • Large amounts of GHG emissions generated from waste incineration (He et al. 2023)<br>• Waste accumulation in open fields, drains, and roadside areas<br>• Land availability for cultivating biofuel crops<br>• Waste dumping into unauthorized locations (e.g., water bodies)<br>• Disruption of ecological processes and system<br>• Noise pollution status, bad smell of waste, effectiveness of pollution control<br>• Nitrate leachate from solid waste landfilling, reaching groundwater reservoirs (Dar et al. 2023)<br>• Aquatic species growth is influenced by dumping solid wastes into surface water (Demarema et al. 2024) |
| Social | • Awareness and information about bioenergy and the green economy<br>• Cultural and behavior of illegal waste disposal and garbage collection (e.g., using different types of dustbins for relevant waste, for instance yellow, blue, green, and red for plastic, paper, glass, and metal)<br>• Public and community acceptance of solid waste recycling in household activities<br>• Capacity building (community participation and strengthening of managerial systems with proper policy and legal frameworks) to implement SWM at the local level<br>• Women's participation in waste-to-energy projects |

(continued)

**Table 10.5** (continued)

| Challenge | Justification/description |
|---|---|
| Policies and regulations | • Enabling policies and regulations on waste recycling<br>• Promulgate and enforce regulations for solid waste collection and provide stable political regimes<br>• Provide financial assistance and maintain open international economic systems<br>• Green building technologies adoption, codes, and regulations<br>• Incentive support from government programs to avoid dumping of wastes into a waterway<br>• Financial aid in research development grants to develop guidelines governing waste-to-energy facilities<br>• Effective legislation and governance systems to keep toxins out of the environment (e.g., providing penalties for improper disposal of garbage) |
| Administrative and institutional | • Institutional/organizational support and capacity<br>• Developing national economy, culture, and currency<br>• Involvement of stakeholders, decisionmakers, and policy-makers in SWM<br>• Bureaucratic licensing and review procedures for reusing materials that would otherwise be discarded |

## 10.10 Conclusions

Deltaic systems are among the most dynamic and productive environments on Earth, home to a human population density many-fold that of the world average. Deltas play a central role in food and water security but are increasingly facing hazards such as submergence, riverine and coastal flooding, and coastal erosion. The advantages offered by deltaic environments, such as freshwater and seawater access, fertile soils, and rich biodiversity, underscore the urgency of addressing waste management issues. This chapter offers concluding remarks on the synergy of solid waste management, environmental sustainability, and the well-being of delta communities. It succeeded in introducing a comprehensive perspective on the challenges faced by densely populated deltas and proposes updated strategies to enhance waste management efficiency while fostering sustainability.

**Acknowledgements** The author acknowledges Nasr Academy for Sustainable Environment (NASE).

**Contributions** MN: conceptualization, and writing—original draft.

**Ethics Declarations**

**Conflict of Interest** The author has no relevant financial or non-financial interests to disclose.

**Ethical Approval** Not applicable.

**Consent to Participate** Not applicable.

**Consent for Publication** The author consents for publication.

## References

Aboughaly M, Fattah I (2023) Environmental analysis, monitoring, and process control strategy for reduction of greenhouse gaseous emissions in thermochemical reactions. Atmosphere 14(4):655. https://doi.org/10.3390/atmos14040655

Ahmad A, Ahmad M, Md Ali U, Ken K (2023) Gasification char residues management: assessing the characteristics for adsorption application. Arab J Chem 16(9):104993. https://doi.org/10.1016/j.arabjc.2023.104993

Akhtar S, Hollaender H, Yuan Q (2023) Impact of heat and contaminants transfer from landfills to permafrost subgrade in arctic climate: a review. Cold Reg Sci Technol 206:103737. https://doi.org/10.1016/j.coldregions.2022.103737

Ansari F, Nasr M, Rawat I, Bux F (2023) Meeting sustainable development goals (SDGs) through progression of pilot-scale algal system to commercial raceway pond (300,000 L). Biomass Convers Biorefin 13(9):7663–7676. https://doi.org/10.1007/s13399-021-01661-0

Arteaga C, Silva J, Yarasca-Aybar C (2023) Solid waste management and urban environmental quality of public space in Chiclayo, Peru. City Environ Interact 20:100112. https://doi.org/10.1016/j.cacint.2023.100112

Awino F, Apitz S (2023) Solid waste management in the context of the waste hierarchy and circular economy frameworks: an international critical review. Integr Environ Assess Manag. https://doi.org/10.1002/ieam.4774

Banik J, Chakraborty D, Rizwan M, Shaik A, Chandan M (2023) Review on disposal, recycling and management of waste polyurethane foams: a way ahead. Waste Manag Res 41(6):1063–1080. https://doi.org/10.1177/0734242X221146082

Bhatia L, Jha H, Sarkar T, Sarangi P (2023) Food waste utilization for reducing carbon footprints towards sustainable and cleaner environment: a review. Int J Environ Res Public Health 20(3):2318. https://doi.org/10.3390/ijerph20032318

Blanchard C, Harris P, Pocock C, McCabe B (2023) Food and garden organic waste management in Australia: co-benefits for regional communities and local government. Sustainability (Switzerland) 15(13):9901. https://doi.org/10.3390/su15139901

Carregosa I, Carregosa J, Silva W, Santos T, Wisniewski A Jr (2023) Thermochemical conversion of aquatic weed biomass in a rotary kiln reactor for production of bio-based derivatives. J Anal Appl Pyrol 173:106048. https://doi.org/10.1016/j.jaap.2023.106048

Chang Y-J, Chang J-S, Lee D-J (2023) Gasification of biomass for syngas production: research update and stoichiometry diagram presentation. Biores Technol 387:129535. https://doi.org/10.1016/j.biortech.2023.129535

Chen B, Perumal P, Illikainen M, Ye G (2023) A review on the utilization of municipal solid waste incineration (MSWI) bottom ash as a mineral resource for construction materials. J Build Eng 71:106386. https://doi.org/10.1016/j.jobe.2023.106386

Chiumenti A, da Borso F, Limina S (2018) Dry anaerobic digestion of cow manure and agricultural products in a full-scale plant: efficiency and comparison with wet fermentation. Waste Manag 71:704–710. https://doi.org/10.1016/j.wasman.2017.03.046

Chowdhury A, Sarkar A (2023) Vermicomposting—the sustainable solid waste management. Waste Manag Resour Recycl Dev World 701–719.https://doi.org/10.1016/B978-0-323-90463-6.00013-0

Chu Z, Li Y, Zhang C, Fang Y, Zhao J (2023) A review on resource utilization of oil sludge based on pyrolysis and gasification. J Environ Chem Eng 11(3):109692. https://doi.org/10.1016/j.jece.2023.109692

Cucchiella F, D'Adamo I, Gastaldi M (2014) Sustainable management of waste-to-energy facilities. Renew Sustain Energy Rev 33:719–728. https://doi.org/10.1016/j.rser.2014.02.015

Czekała W, Nowak M, Piechota G (2023) Sustainable management and recycling of anaerobic digestate solid fraction by composting: a review. Biores Technol 375:128813. https://doi.org/10.1016/j.biortech.2023.128813

Dadebo D, Atukunda A, Ibrahim M, Nasr M (2023a) Integrating chemical coagulation with fixed-bed column adsorption using rice husk-derived biochar for shipboard bilgewater treatment: scale-up design and cost estimation. Chem Eng J Adv 16:100520. https://doi.org/10.1016/j.ceja.2023.100520

Dadebo D, Ibrahim M, Fujii M, Nasr M (2023b) Sequential treatment of surfactant-laden wastewater using low-cost rice husk ash coagulant and activated carbon: modeling, optimization, characterization, and techno-economic analysis. Bioresour Technol Rep 22:101464. https://doi.org/10.1016/j.biteb.2023.101464

Dar S, Sharma P, Shukla B (2023) Enhanced landfill mining waste valorization techniques: a methodological review. Lect Notes Civ Eng 281:109–125. https://doi.org/10.1007/978-981-19-4731-5_10

Demarema S, Nasr M, Ookawara S, Abdelhaleem A (2024) Enhanced synergistic system for the persulfate activation under visible light using novel N-ZnO photocatalyst supported on Lantana camara-based biochar. Chemosphere 349:140840. https://doi.org/10.1016/j.chemosphere.2023.140840

Deng M, Liu H (2024) Driving factors and potential offshore pollution of plastic mulch residue in farmland in the Yellow River Delta, China. Agr Ecosyst Environ 364:108902. https://doi.org/10.1016/j.agee.2024.108902

Elsadek E, Zhang K, Hamoud Y, Mousa A, Awad A, Abdallah M, Shaghaleh H, Hamad A, Jamil M, Elbeltagi A (2024) Impacts of climate change on rice yields in the Nile River Delta of Egypt: a large-scale projection analysis based on CMIP6. Agric Water Manag 292:108673. https://doi.org/10.1016/j.agwat.2024.108673

Emmanouilidou E, Mitkidou S, Agapiou A, Kokkinos N (2023) Solid waste biomass as a potential feedstock for producing sustainable aviation fuel: a systematic review. Renew Energy 206:897–907. https://doi.org/10.1016/j.renene.2023.02.113

Eraky M, Nasr M, Elsayed M, Ai P, Tawfik A (2023) Synergistic interaction of tween 20 and magnesium@ functionalized graphene oxide nano-composite for dual productivity of biohydrogen and biochar from onion peel waste. Renew Energy 216:119082. https://doi.org/10.1016/j.renene.2023.119082

Ezz H, Ibrahim M, Fujii M, Nasr M (2023) Dual biogas and biochar production from rice straw biomass: a techno-economic and sustainable development approach. Biomass Convers Biorefin 13:10807–10821. https://doi.org/10.1007/s13399-021-01879-y

Fang L, Huang T, Lu H, Wu X-L, Chen Z, Yang H, Wang S, Tang Z, Li Z, Hu B, Wang X (2023) Biochar-based materials in environmental pollutant elimination, $H_2$ production and $CO_2$ capture applications. Biochar 5(1):42. https://doi.org/10.1007/s42773-023-00237-7

Halba A, Thengane S, Arora P (2023) A critical outlook on lignocellulosic biomass and plastics co-gasification: a mini-review. Energy Fuels 37(1):19–35. https://doi.org/10.1021/acs.energyfuels.2c02907

Hasan M, Anzar N, Sharma P, Malode S, Shetti N, Narang J, Kakarla R (2023) Converting biowaste into sustainable bioenergy through various processes. Bioresour Technol Rep 23:101542. https://doi.org/10.1016/j.biteb.2023.101542

He D, Hu H, Jiao F, Zuo W, Liu C, Xie H, Dong L, Wang X (2023) Thermal separation of heavy metals from municipal solid waste incineration fly ash: a review. Chem Eng J 467:143344. https://doi.org/10.1016/j.cej.2023.143344

Heiba Y, Nasr M, Fujii M, Mohamed A, Ibrahim M (2023a) Improving irrigation schemes using sustainable development goals (SDGs)-related indicators: a case study of tomato production in pot-scale experimentation. Environ Dev Sustain. https://doi.org/10.1007/s10668-023-03357-z

Heiba Y, Ibrahim M, Mohamed A, Fujii M, Nasr M (2023b) Developing smart sustainable irrigation matrix (SIM)-based model for selection of best irrigation techniques: a framework to achieve SDGs. J Clean Prod 420:138404. https://doi.org/10.1016/j.jclepro.2023.138404

Karthik M, Sreevidya L, Nithya Devi R, Hemalatha G, Yamini R (2023) An efficient waste management technique with IoT based smart garbage system. Mater Today: Proc 80:3140–3143. https://doi.org/10.1016/j.matpr.2021.07.179

Mannaa M, Mansour A, Park I, Lee D-W, Seo Y-S (2023) Insect-based agri-food waste valorization: agricultural applications and roles of insect gut microbiota. Environ Sci Ecotechnol 17:100287. https://doi.org/10.1016/j.ese.2023.100287

Mengqi Z, Shi A, Ajmal M, Ye L, Awais M (2023) Comprehensive review on agricultural waste utilization and high-temperature fermentation and composting. Biomass Convers Biorefin 13(7):5445–5468. https://doi.org/10.1007/s13399-021-01438-5

Nanda S, Sarker T, Kang K, Li D, Dalai A (2023) Perspectives on thermochemical recycling of end-of-life plastic wastes to alternative fuels. Materials 16(13):4563

Ngulube K, Abdelhaleem A, Fujii M, Nasr M (2024) Synergism of artificial intelligence and techno-economic for sustainable treatment of methylene blue dye-containing wastewater by photocatalysis. Sustainability 16:529. https://doi.org/10.3390/su16020529

Palansooriya K, Dissanayake P, Igalavithana A, Tang R, Cai Y, Chang S (2023) Converting food waste into soil amendments for improving soil sustainability and crop productivity: a review. Sci Total Environ 881:163311. https://doi.org/10.1016/j.scitotenv.2023.163311

Peng L, Li Y, Li Q, Liang C, Nasr M, Xu Y, Liu Y, Zhou Y (2023) The effect of free ammonia on ammonium removal and $N_2O$ production in a consortium of microalgae and partial nitritation cultures. Chem Eng J 474:145572. https://doi.org/10.1016/j.cej.2023.145572

Santulli C, Fragassa C, Pavlovic A, Nikolic D (2023) Use of sea waste to enhance sustainability in composite materials: a review. J Marine Sci Eng 11(4):855. https://doi.org/10.3390/jmse11040855

Saqr A, Nasr M, Fujii M, Yoshimura C, Ibrahim M (2024) Delineating suitable zones for solar-based groundwater exploitation using multi-criteria analysis: a techno-economic assessment for meeting sustainable development goals (SDGs). Groundw Sustain Dev 25:101087. https://doi.org/10.1016/j.gsd.2024.101087

Shi Z, Zhang L, Yuan H, Li X, Chang Y, Zuo X (2021) Oyster shells improve anaerobic dark fermentation performances of food waste: hydrogen production, acidification performances, and microbial community characteristics. Biores Technol 335:125268. https://doi.org/10.1016/j.biortech.2021.125268

Sonu G, Rani DP, Abhimanyu R, Umapathi S, Rustagi Y, Huh V, Gupta A, Kaushik VC (2023) Agro-waste to sustainable energy: a green strategy of converting agricultural waste to nano-enabled energy applications. Sci Total Environ 875:2023. https://doi.org/10.1016/j.scitotenv.2023.162667

Syarifinnur S, Nuraini Y, Prasetya B, Handayanto E (2023) Comparing compost and vermicompost produced from market organic waste. Int J Recycl Organ Waste Agric 12(3):279–289. https://doi.org/10.30486/ijrowa.2022.1944251.1368

Tawfik A, Nasr M, Ni S-Q (2024) Paper mill sludge rich enzymes and microbial community promotes the hydrogenesis of black liquor containing furfural. Int J Hydrog Energy 52(Part D):58–73. https://doi.org/10.1016/j.ijhydene.2023.01.242

Thi T, Bui T, Truong L (2024) Effects of syngas from various biomass gasification on combustion of spark ignition engine. GMSARN Int J 18(1):123–129

Torres-Lozada P, Manyoma-Velásquez P, Gaviria-Cuevas J (2023) Prioritization of waste-to-energy technologies associated with the utilization of food waste. Sustainability (Switzerland) 15(7):5857. https://doi.org/10.3390/su15075857

van Midden C, Harris J, Shaw L, Sizmur T, Pawlett M (2023) The impact of anaerobic digestate on soil life: a review. Appl Soil Ecol 191:105066. https://doi.org/10.1016/j.apsoil.2023.105066

Wang W, Chang J-S, Lee D-J (2023a) Digestate-derived carbonized char and activated carbon: application perspective. Biores Technol 381:129135. https://doi.org/10.1016/j.biortech.2023.129135

Wang W, Chang J-S, Lee D-J (2023b) Anaerobic digestate valorization beyond agricultural application: current status and prospects. Biores Technol 373:128742. https://doi.org/10.1016/j.biortech.2023.128742

Wei Z, Wei Y, Liu Y, Niu S, Xu Y, Park J-H, Wang J (2024) Biochar-based materials as remediation strategy in petroleum hydrocarbon-contaminated soil and water: performances, mechanisms, and environmental impact. J Environ Sci (China) 138:350–372. https://doi.org/10.1016/j.jes.2023.04.008

Youssef Y, Gemail K, Atia H, Mahdy M (2024) Insight into land cover dynamics and water challenges under anthropogenic and climatic changes in the eastern Nile Delta: inference from remote sensing and GIS data. Sci Total Environ 913:169690. https://doi.org/10.1016/j.scitotenv.2023.169690

Zidan A, Nasr M, Fujii M, Ibrahim M (2023) Environmental and economic evaluation of downflow hanging sponge reactors for treating high-strength organic wastewater. Sustainability (Switzerland) 15(7):6038. https://doi.org/10.3390/su15076038

# GPSR Compliance

*The European Union's (EU) General Product Safety Regulation (GPSR) is a set of rules that requires consumer products to be safe and our obligations to ensure this.*

*If you have any concerns about our products, you can contact us on ProductSafety@springernature.com*

In case Publisher is established outside the EU, the EU authorized representative is:

Springer Nature Customer Service Center GmbH
Europaplatz 3
69115 Heidelberg, Germany

The manufacturer's authorised representative in the EU is Springer Nature Customer Service Centre GmbH, Europaplatz 3, 69115 Heidelberg, Germany. If you have any concerns regarding our products, please contact ProductSafety@springernature.com

Printed and bound by CPI Group (UK) Ltd, Croydon, CR0 4YY

01/10/2026

02225899-0019